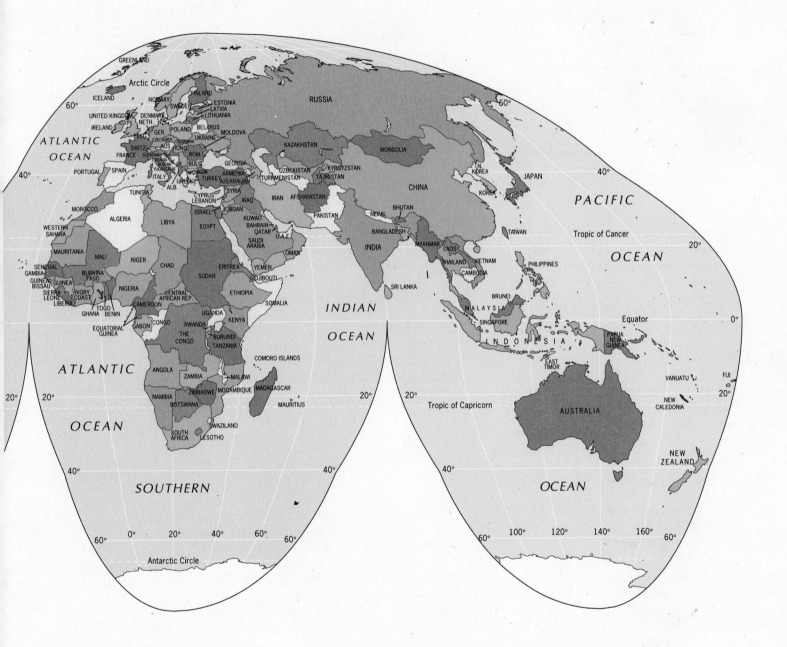

Geography links the study of human societies and their natural-environmental settings through a revealing, spatial approach, aided in this book by a unique series of thematic maps. From glaciation to globalization, desertification to devolution, Mexico to Madagascar, *Concepts and Regions in Geography, 1e* puts processes and places in spatial context. The focus shifts seamlessly from the global to the regional to the local, yielding insights into the forces that shape the human mosaic of our planet.

Inside these covers lies an information highway to geographic literacy. The textbook, CD, and Web site will help you to understand geographic concepts, prepare for examinations, and ultimately improve your grade. To create an individualized learning strategy that combines use of text and media according to your specific strengths, complete the Learning Styles Survey on the Concepts and Regions Web site found at www.wiley.com/college/deblij.

As you work through the text, special icons direct you to learning resources available on the GeoDiscoveries CD-ROM. Each icon will lead you to additional content, activities and quizzing.

GEOGRAPHER'S TOOLBOX

The **Photo Gallery** provides numerous photographs, most taken by H.J. de Blij during his continuing geographic fieldwork. The senior author's field notes are provided to help you view each photograph through the eyes of a professional geographer. Locator maps add further spatial context.

More to Explore essays fall into two categories: **Systematic Essays** and **Issues in Geography**. Regional geography allows us to integrate concepts and principles from the disciplines major subfields to create an overall image of our divided world. These constitute topical or systematic geography, and are described in the Systematic Essays that supplement each chapter. Issues in Geography essays explore interesting and current topics that are specific to each of the world's realms.

The enclosed CD-ROM has 36 interactive **Map Quizzes** to help you master the locations of countries, provinces, states, cities and physical features.

The **Expanded Regional Coverage** of every world region is available on the enclosed CD-ROM. Although *Concepts and Regions in Geography, 1e* is the most compact world regional geography textbook available, the enclosed CD-ROM has all of the regional coverage available in H.J. de Blij and Peter Muller's best-selling textbook, *Geography: Realms, Regions and Concepts, 10e*.

LEARNING ACTIVITIES

1 As you read the textbook, you will notice that each chapter begins with a list of key **Concepts, Ideas, and Terms**. These are noted by numbers in the margins that correspond to the introduction of each item in the text. The CD-ROM provides simulated flashcards and multiple-choice quizzes to help you study and memorize their definitions.

Each **GeoDiscoveries** module uses videos, animations, quizzing, and critical thinking exercises to help you learn core content and improve your grade. These modules challenge you to form spatial questions and apply the tools of geography to find answers.

In the back of this book you will find four pages which map, in detail, the chapter-specific resources available on both the enclosed CD-ROM and the *Concepts and Regions in Geography, 1e* Web site, found at **http://www.wiley.com/college/deblij**.

Please let us know if you have any comments or questions by sending an e-mail to: **ConceptsAndRegions@wiley.com**.

Concepts and Regions in
GEOGRAPHY

First Edition

H. J. de Blij

Michigan State University

Peter O. Muller

University of Miami

With Media Integration by
Eugene J. Palka

United States Military Academy

The first map figure in each chapter that appears in this text
comes from *Goode's World Atlas*, Rand McNally,
R.L. 93-S-115, and is used with permission.

JOHN WILEY & SONS, INC.

ACQUISITIONS EDITOR: Ryan Flahive
MARKETING MANAGER: Kevin Molloy
EDITORIAL ASSISTANT: Denise Powell
SENIOR PRODUCTION EDITOR: Kelly Tavares
SENIOR DESIGNER: Karin Kincheloe
PHOTO EDITOR: Jennifer MacMillan
ILLUSTRATION EDITOR: Sigmund Malinowski
TEXT DESIGN: Lee Goldstein
FRONT/BACK COVER PHOTO: Keith Macgregor/FPG International
ELECTRONIC ILLUSTRATIONS: Mapping Specialists

This book was set in Times Roman by UG / GGS Information
Services, Inc. and printed and bound by Von Hoffman Press.

ISBN 0-471-09303-3

Printed in the United States of America

10 9 8 7 6 5 4 3 2 1

Preface

Concepts and Regions in Geography is drawn from our full-length text, *Geography: Realms, Regions, and Concepts*, which for more than three decades has reported (and sometimes anticipated) trends in the discipline of Geography and developments in the world at large. Through ten preceding editions, *Regions* has explained the modern world's great geographic realms and their physical and human contents, and has introduced geography itself, the discipline that links the study of human societies and natural environments through a fascinating, spatial approach. From old ideas to new, from environmental determinism to expansion diffusion from decolonization to devolution, *Regions* has provided geographic perspective on our transforming world.

The book before you is an extension of *Regions*, updated to accommodate the newest twenty-first century educational technologies. In fact, you are actually holding two books in one: a condensed printed text, and a complete *learning system* that bolsters the text with CD-ROM, Web, and other advanced media resources.

As with *Regions*, this integrated learning system is an information highway to geographic literacy. The first edition of the textbook appeared in 1971, at a time when school geography in the United States (though not in Canada) was a subject in decline. It was a precursor of a dangerous isolationism in America, and geographers foresaw the looming cost of geographic illiteracy. Sure enough, the media during the 1980s began to report that polls, public surveys, tests, and other instruments were recording a lack of geographic knowledge at a time when our world was changing ever faster and becoming more competitive by the day. Various institutions, including the National Geographic Society, banks, airline companies, and a consortium of scholarly organizations mobilized to confront an educational dilemma that had resulted substantially from a neglect of the very topics this book is about.

Before we can usefully discuss such commonplace topics as our "shrinking world," our "global village," and our "distant linkages," we should know what the parts are, the components that do the shrinking and linking. This is not just an academic exercise. You will find that much of what you encounter in this book is of immediate, practical value to you—as a citizen, a consumer, a traveller, a voter, a jobseeker. North America is a geographic realm with intensifying global interests and involvements. Those interests and involvements require countless, often instantaneous decisions. Such decisions must be based on the best possible knowledge of the world beyond our continent. That knowledge can be gained by studying the layout of our world, its environments, societies, resources, policies, traditions, and other properties—in short, its regional geography.

Realms and Concepts

This book is organized into thirteen chapters. The Introduction discusses the world as a whole, outlining the physical stage on which the human drama is being played out, providing environmental information, demographic data, political background, and economic geographical context. Each of the remaining twelve chapters focuses on one of the world's major geographic realms.

Geographic concepts and ideas are placed in their regional settings in all 13 chapters. Most of these approximately 150 concepts are primarily geographical, but others are ideas about which, we believe, students of geography should have some knowledge. Although such concepts are listed on the opening page of every chapter, we have not, of course, enumerated every geographic notion used in that chapter. Many colleagues, we suspect, will want to make their own realm-concept associations, and as readers will readily perceive, the book's organization is quite flexible. It is possible, for example, to focus almost exclusively on substantive regional material, or, alternatively, to concentrate mainly on conceptual issues.

Pedagogy

We continue to devise ways to help students learn important geographic concepts and ideas, and to make sense of our complex and rapidly changing world. Continuing special features from *Regions* include the following:

Atlas Maps. As in previous editions, a comprehensive map of the region opens each chapter. The maps are reproduced from the 20th revised edition (2000) of *Goode's World Atlas* (the maps for Chapters 6 and 12 have been specially created in the Atlas style). Each of these maps is assigned the first figure number in each chapter, which better facilitates the integration of this cartographic material into the text.

Concepts, Ideas, and Terms. Each chapter begins with a boxed sequential listing of the key geographic concepts, ideas, and terms that appear in the pages that follow. These are noted by numbers in the margins (e.g., 1) that correspond to the introduction of each item in the text.

Two-Part Chapter Organization. To help the reader to logically organize the material within chapters, we have broken the regional chapters into two distinct parts: first, "Defining the Realm" includes the general physiographic, historical, and human-geographic background common to the realm, and the second section, "Regions of the Realm," presents each of the distinctive regions within the realm (denoted by the symbol ▶).

List of Regions. Also on the chapter-opening page, a list of the regions within the particular realm provides a preview and helps to organize the chapter. For ease of identification, the triangular symbol (shown at the end of the previous paragraph) that denotes the regions list here also appears beside each region heading in the chapter.

Major Geographic Qualities. Near the beginning of each realm chapter, we list, in boxed format, the major geographic qualities that best summarize that portion of the Earth's surface.

Appendices and Glossary. At the end of the book, the reader will find three sections that enrich and/or supplement the main text: (1) *Appendix A*, a guide to Using the Maps; (2) *Appendix B*, an overview of Career Opportunities in Geography; and (3) an extensive *Glossary*. The general index follows. A geographical index or *gazetteer* of the place names contained in our maps now appears in the book's Web site.

GeoDiscoveries CD-ROM. This robust media tool contains an Interactive Globe that allows students to explore and understand the world by changing the face of this three-dimensional globe using 5 distinct textures. This CD also contains several quizzes per chapter that test student understanding of map features. The regional material contains *Presentations* that use videos, animations, and other resources to focus on key concepts from the chapter; *Interactivities* that engage students in concept-based exercises; and *Assessment* self-tests that allow students to measure their comprehension of the concept being explored. There is extensive expanded coverage from each realm as well, including text, illustrations, and maps.

Web site. Additional resources for students include annotated web links, web quizzes with feedback, links to webcams and live radio from around the globe, blank outline maps of each region, and a learning-style survey that provides students' feedback on their preferred method of learning and how the book and media pedagogy teach to these styles. Additional resources for instructors include the Test Bank, Guide to Virtual Field Trips, Virtual Field Trips, Lesson Outlines, Concepts-Ideas-Terms, and Using Geographic Qualities.

Ancillaries

A broad spectrum of print and electronic ancillaries are available to accompany *Concepts and Regions in Geography*. Additional information, including prices and ISBNs for ordering, can be obtained by contacting John Wiley & Sons.

Data Sources

For all matters geographical, of course, we consult *The Annals of the Association of American Geographers, The Professional Geographer, The Geographical Review, The Journal of Geography*, and many other academic journals published regularly in North America—plus an array of similar periodicals published in English-speaking countries from Scotland to New Zealand.

As with every new edition of this book, all quantitative information was updated to the year of publication and checked rigorously. Hundreds of other modifications were made, many in response to readers' and reviewers' comments. The stream of new spellings of geographic names continues, and we pride ourselves in being a reliable source for current and correct usage.

The statistical data that constitute Table I-1 (pp. 23–28) are derived from numerous sources. As users of such data are aware, considerable inconsistency marks the reportage by various agencies, and it is often necessary to make informed decisions on contradictory information. For example, some sources still do not reflect the rapidly declining rates of population increase or life expectancies in AIDS-stricken African countries. Others list demographic averages without accounting for differences between males and females in this regard.

In formulating Table I-1 we have used among our sources the United Nations, the Population Reference Bureau, the World Bank, the Encyclopaedia Britannica *Books of the Year*, the *Economist* Intelligence Unit, the *Statesman's Year-Book*, and the *The New York Times Almanac*.

The urban population figures—which also entail major problems of reliability and comparability—are mainly drawn from the most recent database published by the United Nations' Population Division. For cities of less than 750,000, we developed our own estimates from a variety of other sources. At any rate, the urban population figures used here are estimates for 2002 and they represent *metropolitan-area totals* unless otherwise specified.

Cartography

This newest version of the text continues the innovation begun in the Seventh Edition, when atlas-style maps from the most recently available edition of *Goode's World Atlas* (currently the 20th, published in 2000) were first used as opening maps for each chapter. In the Eighth Edition, two maps were specifically drawn in the Rand McNally style to serve as matching openers: those of North Africa/Southwest Asia and the Pacific Realm. The South Asia map was substantially expanded from its *Goode's* base.

Users of this book should note that the spelling of some names on these thematic maps does not always match that on the *Goode's World Atlas* maps. This is not unusual; you will even find inconsistencies among various atlases. Almost invariably, we have followed the very latest standards set by the United States Board of Geographic Names.

FOR SALE TO THE STUDENT

Student Study Guide. Text co-author Peter O. Muller and his geographer daughter, Elizabeth Muller Hames, have written a popular Study Guide to accompany the book that is packed with useful study and review tools. For each chapter in the textbook, the Study Guide gives students and faculty access to chapter objectives, content questions-and-answers, outline maps of each realm, sample tests, and more.

Goode's Atlas from Rand McNally.   We are delighted to be able to continue offering the *Goode's Atlas* at a deeply-discounted price when shrink-wrapped with the text. Economies of scale allow us to provide this at a net price that is close to our cost. Our partnership with Rand McNally and the widely-popular *Goode's Atlas* is an arrangement that is exclusive to John Wiley & Sons.

Microsoft Encarta Interactive Atlas CD-ROM. This award-winning atlas CD-ROM will captivate the imaginations of students and engage them in a spatial adventure, all the while exposing them to an abundance of resources appropriate for university-level geography. Our arrangement with Microsoft enables us to offer the *Encarta Interactive Atlas* at a cost that is less than one-third the suggested retail price when shrink-wrapped with this text.

Annenberg/CPB.   Power of Place: World Regional Geography **Study Guide, Third Edition**. The Third Edition of the *Power of Place: World Regional Geography Study Guide* updates content and references so that the Annenberg/Power of Place Telecourse and Video Series connects to this book. It was written by Gil Latz, Portland State University.

For Instructors

PowerPoint Slides.   Available for this edition, these electronic files outline the main concepts of each chapter in *Regions* in a highly visual manner. These presentations are available on the Instructor's Web Site and the Resource CD-ROM, and can be uploaded to presentation programs such as PowerPoint, or to any popular word processing program.

Instructor's Manual.   Distributed on-line to instructors via a secure, password-protected Instructor's Web Site, the *Instructor's Manual* by Wendy Shaw, Southern Illinois University, Edwardsville, provides outlines, descriptions, and key terms to help professors organize the concepts in the book for classroom use.

Test Bank.   Prepared by long-term Test Bank author Ira Sheskin, University of Miami, the *Test Bank* contains over 3000 test items including multiple-choice, fill-in, matching, and essay questions. It is distributed via the secure Instructor's Web Site as electronic files, which can be saved into all major word processing programs.

Computerized Test Bank.   An easy to use program that can be used to create and customize exams.

Student Web Site.   This comprehensive on-line resource will contain chapter-based self-quizzes and extensive links to Web material providing real-world examples and additional research tools.

Course Management.   On-line course management assets are available to accompany the Tenth Edition of *Regions*.

OTHER RESOURCES FOR THE CLASSROOM

Overhead Transparencies and Slides. The book's maps and diagrams are available in their entirety for either transparency or slide projection in beautifully rendered, 4-color format.

Concepts and Regions in Geography Resource CD-ROM. This rich resource contains animations, videos, PowerPoint presentations, the Test Bank, and the Instructor's Manual. Organized by chapter, the Resource CD-ROM has a tested, intuitive interface that allows for easy file management and presentation building. If the instructor prefers to use programs such as PowerPoint in the classroom, the text, map, and photo files can be uploaded easily from the Resource CD-ROM into your presentation program.

ACKNOWLEDGMENTS

Over the more than three decades since the publication of the First Edition of *Geography: Realms, Regions, and Concepts*, we have been fortunate to receive advice and assistance from literally hundreds of people. One of the rewards associated with the publication of a book of this kind is the steady stream of correspondence and other feedback it generates. Geographers, economists, political scientists, education specialists, and others have written us, often with fascinating enclosures. We make it a point to respond personally to every such letter, and our editors have communicated with many of our correspondents as well. Moreover, we have considered every suggestion made—and many who wrote or transmitted their reactions through other channels will see their recommendations in print in this edition.

STUDENT RESPONSE

A good part of the correspondence we receive comes from student readers. On this occasion, we would like to extend our deep appreciation to the several million students around the world who have studied from the first ten editions of our text. In particular, we thank the students from more than 100 different colleges and universities across the United States who took the time to send us their opinions.

Generally, students have told us that they found the pedagogical devices quite useful. We have kept the study aids the students cited as effective: a boxed list of each chapter's key concepts, ideas, and terms (now numbered for quick reference in both the box and text margins); a box summarizing each realm's major geographic qualities; and an extensive Glossary.

FACULTY FEEDBACK

Faculty members from a large number of North American colleges and universities continue to supply us with vital feedback and much-appreciated advice. Our publishers commissioned a number of reviews, and we are most grateful to the following professors for showing us where the written text could be strengthened and made more precise:

RANDY BERTOLAS, Wayne State College
JONATHAN C. COMER, Oklahoma State University
MICHAEL CORNEBISE, University of Tennessee

FIONA M. DAVIDSON, University of Arkansas

MEL DROUBAY, University of West Florida

DAVID J. KEELING, Western Kentucky University

MOHAMEDEN OULD-MEY, Indiana State University

THOMAS W. PARADIS, Northern Arizona University

JAMES W. PENN, JR., University of Florida

JOHN D. REILLY, University of Florida

THOMAS C. SCHAFER, Fort Hays State University

In addition, several faculty colleagues from around the world assisted us with earlier editions, and their contributions continue to grace the pages of this book. Among them are:

JAMES P. ALLEN, California State University, Northridge

STEPHEN S. BIRDSALL, University of North Carolina

J. DOUGLAS EYRE, University of North Carolina

FANG YONG-MING, Shanghai, China

EDWARD J. FERNALD, Florida State University

RAY HENKEL, Arizona State University

RICHARD C. JONES, University of Texas at San Antonio

GIL LATZ, Portland State University (Oregon)

IAN MACLACHLAN, University of Lethbridge (Alberta)

MELINDA S. MEADE, University of North Carolina

HENRY N. MICHAEL, Temple University (Pennsylvania)

CLIFTON W. PANNELL, University of Georgia

J. R. VICTOR PRESCOTT, University of Melbourne (Victoria)

JOHN D. STEPHENS, University of Washington

CANUTE VANDER MEER, University of Vermont

We also received input from a much wider circle of academic geographers. The list that follows is merely representative of a group of colleagues across North America to whom we are grateful for taking the time to share their thoughts and opinions with us:

MEL AAMODT, California State University—Stanislaus

R. GABRYS ALEXSON, University of Wisconsin—Superior

NIGEL ALLAN, University of California—Davis

JAMES P. ALLEN, California State University, Northridge

JOHN L. ALLEN, University of Connecticut

JERRY R. ASCHERMANN, Missouri Western State College

JOSEPH M. ASHLEY, Montana State University

THEODORE P. AUFDEMBERGE, Concordia College (Michigan)

EDWARD BABIN, University of South Carolina—Spartanburg

MARVIN W. BAKER, University of Oklahoma

THOMAS F. BAUCOM, Jacksonville State University (Alabama)

GOURI BANERJEE, Boston University (Massachusetts)

J. HENRY BARTON, Thiel College (Pennsylvania)

STEVEN BASS, Paradise Valley Community College (Arizona)

KLAUS J. BAYR, University of New Hampshire—Manchester

JAMES BELL, Linn Benton Community College (Oregon)

WILLIAM H. BERENTSEN, University of Connecticut

ROYAL BERGLEE, Indiana State University

RIVA BERLEANT-SCHILLER, University of Connecticut

THOMAS BITNER, University of Wisconsin

WARREN BLAND, California State University—Northridge

DAVIS BLEVINS, Huntington College (Alabama)

S. BO JUNG, Bellevue College (Nebraska)

MARTHA BONTE, Clinton Community College (Idaho)

GEORGE R. BOTJER, University of Tampa (Florida)

R. LYNN BRADLEY, Belleville Area College (Illinois)

KEN BREHOB, Elmhurst, Illinois

JAMES A. BREY, University of Wisconsin—Fox Valley

ROBERT BRINSON, Santa Fe Community College (Florida)

REUBEN H. BROOKS, Tennessee State University

LARRY BROWN, Ohio State University

LAWRENCE A. BROWN, Troy State—Dothan (Alabama)

ROBERT N. BROWN, Delta State University (Mississippi)

STANLEY D. BRUNN, University of Kentucky

RANDALL L. BUCHMAN, Defiance College (Ohio)

DIANN CASTEEL, Tusculum College (Tennessee)

JOHN E. COFFMAN, University of Houston (Texas)

DAWYNE COLE, Grand Rapids Baptist College (Michigan)

JONATHAN C. COMER, Oklahoma State University

BARBARA CONNELLY, Westchester Community College (New York)

WILLIS M. CONOVER, University of Scranton (Pennsylvania)

OMAR CONRAD, Maple Woods Community College (Missouri)

BARBARA CRAGG, Aquinas College (Michigan)

GEORGES G. CRAVINS, University of Wisconsin

ELLEN K. CROMLEY, University of Connecticut

JOHN A. CROSS, University of Wisconsin—Oshkosh

WILLIAM CURRAN, South Suburban (Illinois)

ARMANDO DA SILVA, Towson State University (Maryland)

DAVID D. DANIELS, Central Missouri State University

RUDOLPH L. DANIELS, Morningside College (Iowa)

SATISH K. DAVGUN, Bemidji State University (Minnesota)

JAMES DAVIS, Illinois College

JAMES L. DAVIS, Western Kentucky University

KEITH DEBBAGE, University of North Carolina—Greensboro

MOLLY DEBYSINGH, California State University, Long Beach

DENNIS K. DEDRICK, Georgetown College (Kentucky)

STANFORD DEMARS, Rhode Island College

THOMAS DIMICELLI, William Paterson College (New Jersey)

D.F. DOEPPERS, University of Wisconsin—Madison

ANN DOOLEN, Lincoln College (Illinois)

STEVEN DRIEVER, University of Missouri—Kansas City

WILLIAM ROBERT DRUEN, Western Kentucky University

ALASDAIR DRYSDALE, University of New Hampshire

KEITH A. DUCOTE, Cabrillo Community College (California)

WALTER N. DUFFET, University of Arizona

CHRISTINA DUNPHY, Champlain College (Vermont)

ANTHONY DZIK, Shawnee State University (Kansas)

DENNIS EDGELL, Firelands BGSU (Ohio)

JAMES H. EDMONSON, Union University (Tennessee)

M.H. EDNEY, State University of New York—Binghamton

HAROLD M. ELLIOTT, Weber State University (Utah)

JAMES ELSNES, Western State College

DINO FIABANE, Community College of Philadelphia (Pennsylvania)

G.A. FINCHUM, Milligan College (Tennessee)

IRA FOGEL, Foothill College (California)

ROBERT G. FOOTE, Wayne State College (Nebraska)

G.S. FREEDOM, McNeese State University (Louisiana)

RONALD FORESTA, University of Tennessee

OWEN FURUSETH, University of North Carolina—Charlotte

RICHARD FUSCH, Ohio Wesleyan University

GARY GAILE, University of Colorado—Boulder

EVELYN GALLEGOS, Eastern Michigan University & Schoolcraft College

JERRY GERLACH, Winona State University (Minnesota)

LORNE E. GLAIM, Pacific Union College (California)

SHARLEEN GONZALEZ, Baker College (Michigan)

DANIEL B. GOOD, Georgia Southern University

GARY C. GOODWIN, Suffolk Community College (New York)

S. GOPAL, Boston University (Massachusetts)

ROBERT GOULD, Morehead State University (Kentucky)

GORDON GRANT, Texas A&M University

DONALD GREEN, Baylor University (Texas)

GARY M. GREEN, University of North Alabama

MARK GREER, Laramie County Community College (Wyoming)

STANLEY C. GREEN, Laredo State University (Texas)

W. GREGORY HAGER, Northwestern Connecticut Community College

RUTH F. HALE, University of Wisconsin—River Falls

JOHN W. HALL, Louisiana State University—Shreveport

PETER L. HALVORSON, University of Connecticut

MERVIN HANSON, Willmar Community College (Minnesota)

ROBERT J. HARTIG, Fort Valley State College (Georgia)

JAMES G. HEIDT, University of Wisconsin Center—Sheboygan

CATHERINE HELGELAND, University of Wisconsin—Manitowoc

NORMA HENDRIX, East Arkansas Community College

JAMES HERTZLER, Goshen College (Indiana)

JOHN HICKEY, Inver Hills Community College (Minnesota)

THOMAS HIGGINS, San Jacinto College (Texas)

EUGENE HILL, Westminster College (Missouri)

LOUISE HILL, University of South Carolina—Spartanburg

MIRIAM HELEN HILL, Indiana University Southeast

SUZY HILL, University of South Carolina—Spartanburg

ROBERT HILT, Pittsburg State University (Kansas)

SOPHIA HINSHALWOOD, Montclair State University (New Jersey)

PRISCILLA HOLLAND, University of North Alabama

ROBERT K. HOLZ, University of Texas—Austin

R. HOSTETLER, Fresno City College (California)

LLOYD E. HUDMAN, Brigham Young University (Utah)

JANIS W. HUMBLE, University of Kentucky

WILLIAM IMPERATORE, Appalachian State University (North Carolina)

RICHARD JACKSON, Brigham Young University (Utah)

MARY JACOB, Mount Holyoke College (Massachusetts)

GREGORY JEANE, Samford University (Alabama)

SCOTT JEFFREY, Catonsville Community College (Maryland)

JERZY JEMIOLO, Ball State University (Indiana)

SHARON JOHNSON, Marymount College (New York)

SARA MAYFIELD, San Jacinto College, Central (California)

DAVID JOHNSON, University of Southwestern Louisiana

JEFFREY JONES, University of Kentucky

MARCUS E. JONES, Claflin College (South Carolina)

MOHAMMAD S. KAMIAR, Florida Community College, Jacksonville

MATTI E. KAUPS, University of Minnesota—Duluth

COLLEEN KEEN, Gustavus Adolphus College (Minnesota)

GORDON F. KELLS, Mott Community College

SUSANNE KIBLER-HACKER, Unity College (Maine)

JAMES W. KING, University of Utah

JOHN C. KINWORTHY, Concordia College (Nebraska)

ALBERT KITCHEN, Paine College

TED KLIMASEWSKI, Jacksonville State University (Alabama)

ROBERT D. KLINGENSMITH, Ohio State University—Newark

LAWRENCE M. KNOPP, JR., University of Minnesota—Duluth

TERRILL J. KRAMER, University of Nevada

ARTHUR J. KRIM, Cambridge, Massachusetts

ELROY LANG, El Camino Community College (California)

CHRISTOPHER LANT, Southern Illinois University—Carbondale

A.J. LARSON, University of Illinois—Chicago

LARRY LEAGUE, Dickinson State University (North Dakota)

DAVID R. LEE, Florida Atlantic University

JOE LEEPER, Humboldt State University (California)

YECHIEL M. LEHAVY, Atlantic Community College (New Jersey)

JOHN C. LEWIS, Northeast Louisiana University

CAEDMON S. LIBURD, University of Alaska—Anchorage

T. LIGIBEL, Eastern Michigan University

Z.L. LIPCHINSKY, Berea College (Kentucky)

ALLAN L. LIPPERT, Manatee Community College (Florida)

JOHN H. LITCHER, Wake Forest University (North Carolina)

LI LIU, Stephen F. Austin State University (Texas)

WILLIAM R. LIVINGSTON, Baker College (Michigan)

CYNTHIA LONGSTREET, Ohio State University

TOM LOVE, Linfield College (Oregon)

K.J. LOWREY, Miami University (Ohio)

ROBIN R. LYONS, University of Hawai'i—Leeward Community College

SUSAN M. MACEY, Southwest Texas State University

CHRISTIANE MAINZER, Oxnard College (California)

HARLEY I. MANNER, University of Guam

JAMES T. MARKLEY, Lord Fairfax Community College (Virginia)

SISTER MAY LENORE MARTIN, Saint Mary College (Kansas)

GARY MANSON, Michigan State University

KENT MATHEWSON, Louisiana State University

DICK MAYER, Maui Community College (Hawai'i)

DEAN R. MAYHEW, Maine Maritime Academy

J.P. MCFADDEN, Orange Coast College (California)

BERNARD MCGONIGLE, Community College of Philadelphia (Pennsylvania)

PAUL D. MEARTZ, Mayville State University (North Dakota)

DALTON W. MILLER, JR., Mississippi State University

RAOUL MILLER, University of Minnesota, Duluth

INES MIYARES, Hunter College, CUNY (New York)

BOB MONAHAN, Western Carolina University

KEITH MONTGOMERY, University of Wisconsin—Marathon

JOHN MORTON, Benedict College (South Carolina)

ANNE MOSHER, Syracuse University (New York)

BARRY MOWELL, Broward Community College (Florida)

ROBERT R. MYERS, West Georgia College

YASER M. NAJJAR, Framingham State College (Massachusetts)

JEFFREY W. NEFF, Western Carolina University

DAVID NEMETH, University of Toledo (Ohio)

RAYMOND O'BRIEN, Bucks County Community College (Pennsylvania)

JOHN ODLAND, Indiana University

JOSEPH R. OPPONG, University of North Texas

RICHARD OUTWATER, California State University, Long Beach

PATRICK O'SULLIVAN, Florida State University

BIMAL K. PAUL, Kansas State University

JAMES PENN, Southeastern Louisiana University

PAUL PHILLIPS, Fort Hays State University (Kansas)

MICHAEL PHOENIX, ESRI (California)

JERRY PITZL, Macalester College (Minnesota)

BILLIE E. POOL, Holmes Community College (Mississippi)

VINTON M. PRINCE, Wilmington College (North Carolina)

RHONDA REAGAN, Blinn College (Texas)

DANNY I. REAMS, Southeast Community College (Nebraska)

JIM RECK, Golden West College (California)

ROGER REEDE, Southwest State University (Minnesota)

JOHN RESSLER, Central Washington University

JOHN B. RICHARDS, Southern Oregon State College

DAVID C. RICHARDSON, Evangel College (Missouri)

SUSAN ROBERTS, University of Kentucky

WOLF RODER, University of Cincinnati

JAMES ROGERS, University of Central Oklahoma

PAUL A. ROLLINSON, AICP, Southwest Missouri State University

JAMES C. ROSE, Tompkins/Cortland Community College (New York)

THOMAS E. ROSS, Pembroke State University (North Carolina)

THOMAS A. RUMNEY, State University of New York—Plattsburgh

GEORGE H. RUSSELL, University of Connecticut

RAJAGOPAL RYALI, Auburn University at Montgomery (Alabama)

PERRY RYAN, Mott Community College

ADENA SCHUTZBERG, Middlesex Community College (Massachusetts)

SIDNEY R. SHERTER, Long Island University (New York)

NANDA SHRESTHA, Florida A&M University

WILLIAM R. SIDDALL, Kansas State University

DAVID SILVA, Bee County College (Texas)

DEBRA STRAUSSFOGEL, University of New Hampshire

MORRIS SIMON, Stillman College (Alabama)

KENN E. SINCLAIR, Holyoke Community College (Massachusetts)

ROBERT SINCLAIR, Wayne State University (Michigan)

EVERETT G. SMITH, Jr., University of Oregon

RICHARD V. SMITH, Miami University (Ohio)

CAROLYN D. SPATTA, California State University—Hayward

M.R. SPONBERG, Laredo Junior College (Texas)

DONALD L. STAHL, Towson State University (Maryland)

ELAINE STEINBERG, Central Florida Community College

D.J. STEPHENSON, Ohio University Eastern

HERSCHEL STERN, Mira Costa College (California)

REED F. STEWART, Bridgewater State College (Massachusetts)

NOEL L. STIRRAT, College of Lake County (Illinois)

GEORGE STOOPS, Mankato State University (Minnesota)

JOSEPH P. STOLTMAN, Western Michigan University

PHILIP SUCKLING, University of Northern Iowa

CHRISTOPHER SUTTON, Western Illinois University

T. L. TARLOS, Orange Coast College (California)

MICHAEL THEDE, North Iowa Area Community College

DERRICK J. THOM, Utah State University

CURTIS THOMSON, University of Idaho

S. TOOPS, Miami University (Ohio)

ROGER T. TRINDELL, Mansfield University of Pennsylvania

DAN TURBEVILLE, East Oregon State College

NORMAN TYLER, Eastern Michigan University

GEORGE VAN OTTEN, Northern Arizona University

C.S. VERMA, Weber State College (Utah)

GRAHAM T. WALKER, Metropolitan State College of Denver

DEBORAH WALLIN, Skagit Valley College (Washington)

MIKE WALTERS, Henderson Community College (Kentucky)

J.L. WATKINS, Midwestern State University (Texas)

P. GARY WHITE, Western Carolina University (North Carolina)

W.R. WHITE, Western Oregon University

GARY WHITTON, Fairbanks, Alaska

GENE C. WILKEN, Colorado State University

STEPHEN A. WILLIAMS, Methodist College

P. WILLIAMS, Baldwin-Wallace College

MORTON D. WINSBERG, Florida State University

ROGER WINSOR, Appalachian State University (North Carolina)

WILLIAM A. WITHINGTON, University of Kentucky

A. WOLF, Appalachian State University, N.C.)

JOSEPH WOOD, University of Southern Maine

RICHARD WOOD, Seminole Junior College (Florida)

GEORGE I. WOODALL, Winthrop College (North Carolina)

STEPHEN E. WRIGHT, James Madison University (Virginia)

LEON YACHER, Southern Connecticut State University

DONALD J. ZEIGLER, Old Dominion University (Virginia)

In assembling this newest edition, we are indebted to the following people for advising us on a number of matters:

THOMAS L. BELL, University of Tennessee

KATHLEEN BRADEN, Seattle Pacific University

JESUS CAÑAS, Research Deparment, Federal Reserve Bank of Dallas, El Paso Branch

STUART E. CORBRIDGE, University of Miami/London School of Economics

WILLIAM V. DAVIDSON, Louisiana State University

JAMES D. FITZSIMMONS, U.S. Bureau of the Census (D.C.)

GARY A. FULLER, University of Hawai'i

RICHARD J. GRANT, University of Miami

MARGARET M. GRIPSHOVER, University of Tennessee

TRUMAN A. HARTSHORN, Georgia State University

PHILIP L. KEATING, Indiana University

DAVID LEY, University of British Columbia

RICHARD LISICHENKO, Fort Hays State University (Kansas)

GLEN M. MACDONALD, University of California, Los Angeles

IAN MACLACHLAN, University of Lethbridge (Alberta)

DALTON MILLER, Mississippi State University

ANNE MOSHER, Syracuse University

VALIANT C. NORMAN, Lexington Community College (Kentucky)

PAI YUNG-FENG, New York City

EUGENE J. PALKA, U.S. Military Academy (New York)

JOSEPH L. SCARPACI, JR., Virginia Tech

ROLF STERNBERG, Montclair State University (New Jersey)

COLLEEN J. WATKINS, Linfield College (Oregon)

BARBARA A. WEIGHTMAN, California State University, Fullerton

KRISTOPHER D. WHITE, University of Connecticut

For assistance with the map of North American indigenous peoples (p. 104), we are greatly indebted to Jack Weatherford, Professor of Anthropology at Macalester College (Minnesota); Henry T. Wright, Professor and Curator of Anthropology at the University of Michigan; and George E. Stuart, President of the Center for Maya Research (North Carolina). The map of Russia's federal regions (p. 85) could not have been compiled without the invaluable help of David B. Miller, Senior Edit Cartographer at the National Geographic Society, and Leo Dillon of the Russia Desk of the U.S. Department of State. And special thanks, too, go to Charles Pirtle, Professor of Geography at Georgetown University's School of Foreign Service for his advice on Chapter 4, and to Charles Fahrer of the Department of Geography at the University of South Carolina for his suggestions on Chapter 6.

We also record our appreciation to those geographers who ensured the quality of this book's ancillary products: Ira M. Sheskin (University of Miami) prepared the *Test Bank* and manipulated a large body of demographic data to derive the tabular display in Table I-1; Eugene J. Palka (U.S. Military Academy) prepared the *Instructor's Manual* and supervised the media integration; and Elizabeth Muller Hames (M.A. in Geography, University of Miami) co-authored the *Study Guide* and prepared the Geographical Index found on the Web site. At the University of Miami's Department of Geography, Peter Muller is most grateful for the advice and support he continues to receive from all his faculty colleagues: Tom Boswell, Stuart Corbridge, Richard Grant, Jennifer Mandel, Jan Nijman, Jennifer Papp, Ira Sheskin, and Mark White, plus GIS Lab Manager Chris Hanson. Moreover, his departmental office staff tirelessly performs an array of critical supporting tasks; in addition to Assistant to the Chair/Office Manager Hannibal Burton, he wishes to thank Melissa Blankson, Scarleth Padilla, and Brian Jones.

PERSONAL APPRECIATION

We are privileged to work with a team of professionals at John Wiley & Sons that is unsurpassed in the college textbook publishing industry. As authors we are acutely aware of these talents on a daily basis during the crucial production stage, especially the outstanding coordination and leadership skills of Senior Production Editor Kelly Tavares, Illustration Editor Sigmund Malinowski, Photo Coordinator Jennifer MacMillan, and Production Assistants Rebecca Rothaug and Carmen Hernandez. Others who played a leading role in this process were Senior Designer Karin Kincheloe, Copy Editor Betty Pessagno, Photo Director Marge Graham, Photo Manager Hilary Newman, Photo Researchers Alexandra Truitt and Jerry Marshall, and Don Larson of Mapping Specialists, Ltd. in Madison, Wisconsin. We appreciated the leadership of Geography Editor Ryan Flahive, who was the prime mover in launching and supervising this exciting project, and was ably assisted throughout by Tom Kulesa, Martin Batey, Mark Gerber, and especially Denise Powell. We also thank our marketing managers, Clay Stone and Kevin Molloy, for their efforts on behalf of this project. Beyond this immediate circle, we acknowledge the support and encouragement we have received over the years from others at Wiley including Vice-President and Publisher Anne Smith, Vice-President of Production Ann Berlin, and Executive Vice-President and Publisher Kaye Pace.

Finally, and most of all, we thank our wives, Bonnie and Nancy, for yet again seeing us through the challenging schedule of our seventh collaboration on this volume in the past 18 years.

H. J. de Blij
Boca Grande,
Florida

Peter O. Muller
Coral Gables,
Florida

April 9, 2002

Brief Contents

Contents

I / World Regional Geography

St. Michael the Archangel

Sitka, Alaska: vacated by ice, settled by Tlingit, invaded by Russians, purchased by Americans, now sustained by tourism.

by the arms manufacturers in the wealthy core countries. During the Cold War, dictatorial regimes in the periphery became close allies of both superpowers, and proxy wars were fought in Asia, Africa, and the Americas. By the time the Cold War ended and representative government began to return to some (though not all) of these countries, they were so deeply in debt to the rich core countries that they had no prospect of recovery.

External debt is not confined to the countries in the periphery, and even the richer countries (including the United States) have national debts. What matters is a country's ability to service its debt and still have the capacity to pay for its other needs. South Korea, for example, currently has the highest per-capita debt in the world, but its per-capita *gross national product* (*GNP*, the total value of all goods and services produced by the citizens of a country, within or outside of its boundaries, during a calendar year) is more than three times as high. Nicaragua, a Middle American Cold War victim, is much worse off: its per-capita debt is nearly *four times* its per-capita GNP. When a destructive hurricane struck Nicaragua in 1998, the country was left totally dependent on outside help. It had no reserves nor, given its debts, did it have the ability to borrow.

Several Middle American countries suffer from heavy debt burdens, but the realm most severely afflicted in this respect in Subsaharan Africa. As we note in Chapter 7, postcolonial tropical Africa suffers from a combination of conditions and circumstances that mires most of its countries in debt-ridden poverty.

Globalization

In August 1999 a Frenchman named José Bové, leader of the so-called Peasant Federation, became a national hero. He achieved his stardom because a group of his sympathizers destroyed a McDonald's restaurant being built in the town of Millau; when Mr. Bové came to trial ten months later, some 40,000 supporters rallied to his cause. To Mr. Bové, McDonald's **48** is a symbol of **globalization**, a process he views as the Americanization of France's traditions, and trashing it was a matter of cultural self-defense.

Why does an apparently beneficial process arouse such heated passions? Globalization breaks down barriers to international trade, stimulates commerce, brings jobs to remote places, and promotes social, cultural, political, and other kinds of exchanges. High-tech workers in India are employed by computer firms based in California. Japanese cars are assembled in Thailand. American shoes are made in China. Thousands of McDonald's restaurants serve their familiar (and standard) menus to customers from Tokyo to Tel Aviv.

Opponents of globalization argue that the negatives far outweigh the positives. Those high-tech workers in India are paid a fraction of what their Californian counterparts earn, so that many of them will want to leave India for better pay abroad at a time when the country has great need of their skills. The wages and working conditions for the Chinese shoe-factory workers are far below those acceptable to American labor. And as for those McDonald's restaurants, they promote the consumption of junk food over better traditional fare. Opposition to globalization in the United States often centers on the loss of American jobs when corporations move their factories to foreign countries where labor is cheaper.

In a way, the current globalization process is a revolution—but it is not the first of its kind. The first "globalization revolution" occurred during the nineteenth and early twentieth centuries, when Europe's colonial expansion spread ideas, inventions, products, and habits around the world. Colonialism transformed the world as the European powers built cities, transport networks, dams, irrigation systems, power plants, and other facilities, often with devastating impact on local traditions, cultures, and economies. From goods to games (soap to soccer) people in much of the world started doing similar things. The largest of all colonial empires, that of Britain, made English a worldwide language, a key element in the current, second globalization process.

The present globalization is even more revolutionary than the colonial phase because it is driven by more modern, higher-speed communications. When the British colonists planned the construction of their ornate Victorian government and public buildings in (then) Bombay, the architectural drawings had to be

prepared in London and sent by boat to India. When the Chinese government in the 1980s decided to create a Manhattan-like commercial district on the riverfront in Shanghai, the plans were drawn in the United States, Japan, and Western Europe and transmitted to Shanghai via the Internet. One container ship carrying products from China to the American market hauls more cargo than a hundred colonial-era boats.

And, as the pages that follow will show frequently, the world's national-political boundaries are becoming increasingly porous. Economic alliances enable manufacturers to send raw materials and finished products across borders that once inhibited such exchanges. Groups of countries forge unions whose acronyms (NAFTA, Mercosur) stand for freer trade. The ultimate goal of the World Trade Organization is to lower the remaining trade barriers the world over, boosting not just regional commerce but also global trade.

As with all revolutions, the overall consequences of the present globalization process are uncertain. Critics underscore that one of its outcomes is a growing gap between rich and poor, a polarization of wealth that will destabilize the world. Core-periphery contrasts are intensified, not lessened, by globalization as the poor in peripheral societies are exploited by core-based corporations. Proponents argue that, as with the Industrial Revolution, it will take time for the benefits to spread—but that globalization's ultimate effects will be advantageous to all.

Indeed, the world is functionally shrinking, and we will find evidence for this throughout the book. But the "global village" still retains its distinctive neighborhoods, and two revolutionary globalizations have failed to erase their particular properties. In the chapters that follow we use the vehicle of geography to visit and investigate them.

THE REGIONAL FRAMEWORK

At the beginning of this Introduction, we outlined a map of the great geographic realms of the world

(Fig. I-2). We then addressed the task of dividing these realms into regions, and we used criteria ranging from physical geography to economic geography. The result is Figure I-11. Before we begin our survey, here is a summary of the 12 geographic realms and their regional components.

Europe (1)

Territorially small and politically fragmented, Europe remains disproportionately influential in global affairs. A core geographic realm, Europe has five regions: Western Europe, the British Isles, Northern (Nordic) Europe, Mediterranean Europe, and Eastern Europe.

Russia (2)

Territorially enormous and politically unified, Russia was the dominant force in the former Soviet Union that disbanded in 1991. Undergoing a difficult transition from dictatorship to democracy and from communism to capitalism, Russia is geographically complex and changing. We define four regions: the Russian Core, the Eastern Frontier, Siberia, and the Far East.

North America (3)

Another realm in the global core, North America consists of the United States and Canada. We identify nine regions: the North American Core, the Maritime Northeast, French Canada, the Continental Interior, the South, the Southwest, the Western Frontier, the Northern Frontier, and the Pacific Hinge.

Middle America (4)

Nowhere in the world is the contrast between core and periphery as sharply demarcated as it is between

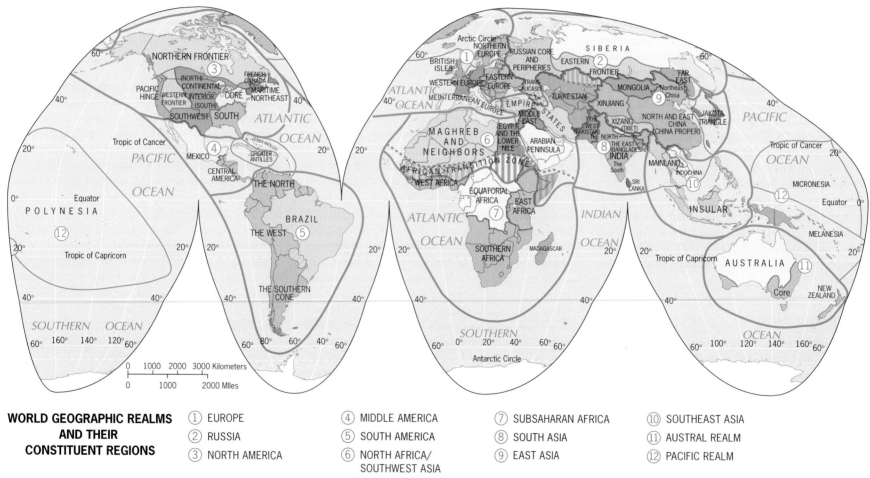

WORLD GEOGRAPHIC REALMS AND THEIR CONSTITUENT REGIONS

① EUROPE
② RUSSIA
③ NORTH AMERICA
④ MIDDLE AMERICA
⑤ SOUTH AMERICA
⑥ NORTH AFRICA/ SOUTHWEST ASIA
⑦ SUBSAHARAN AFRICA
⑧ SOUTH ASIA
⑨ EAST ASIA
⑩ SOUTHEAST ASIA
⑪ AUSTRAL REALM
⑫ PACIFIC REALM

FIGURE I-11

North and Middle America. This small, fragmented realm clearly divides into four regions: Mexico, Central America, and the Greater Antilles and Lesser Antilles of the Caribbean.

South America (5)

The continent of South America also defines a geographic realm in which Iberian (Spanish and Portuguese) influences dominate the cultural geography but Amerindian imprints survive. We recognize four regions: Brazil, the realm's giant; the North, composed of Caribbean-facing states; the Andean West, with its strong Amerindian influences; and the Southern Cone.

North Africa/Southwest Asia (6)

This vast geographic realm has several names, extending as it does from North Africa into Southwest and, indeed, Central Asia. Some geographers call it *Naswasia* or *Afrasia*. There are seven regions: Egypt and the Lower Nile, the Maghreb, and the African Transition Zone in North Africa; the Middle East, the Arabian Peninsula, the Empire States, and 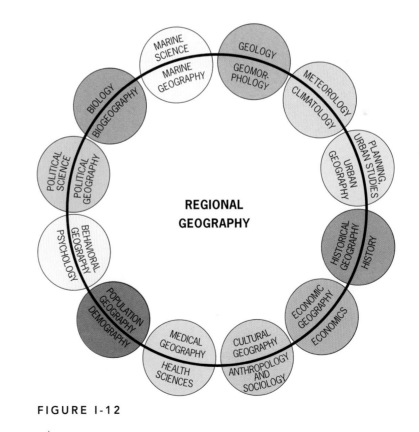 Turkestan in Southwest Asia.

Subsaharan Africa (7)

Between the African Transition Zone and the southernmost Cape of South Africa lies Subsaharan Africa. The realm consists of four regions: West Africa, East Africa, Equatorial Africa, and Southern Africa.

South Asia (8)

Physiographically one of the most clearly defined geographic realms, South Asia has a complex cultural geography. It consists of five regions: India at the center; Pakistan to the west; Bangladesh to the east; the mountainous North; and the peninsular South, which includes island Sri Lanka.

East Asia (9)

The vast East Asian geographic realm extends from the deserts of Central Asia to the tropical coasts of the South China Sea and from Japan to Xizang (Tibet). We identify five regions: China Proper, including North Korea; Xizang (Tibet) in the southwest; desert Xinjiang in the west; Mongolia in the north; and the Jakota Triangle (Japan, South Korea, and Taiwan) in the east.

Southeast Asia (10)

Southeast Asia is a varied mosaic of natural landscapes, cultures, and economies. Influenced by India, China, Europe, and the United States, it includes dozens of religions and hundreds of languages plus economies representing both core and periphery. Physically, Southeast Asia consists of a peninsular mainland and an arc consisting of thousands of islands. Its two regions (Mainland and Insular) are 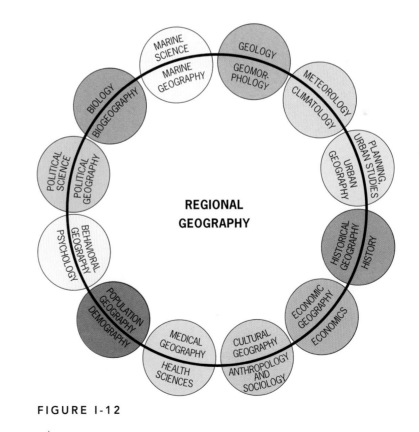 based on this distinction.

Austral Realm (11)

Australia and its neighbor New Zealand form the Austral geographic realm by virtue of continental dimensions, insular separation, and dominantly Western cultural heritage. The four regions are defined by physical as well as cultural geography: a highly urbanized, two-part core and a vast, desert-dominated interior in Australia; and two main islands in New Zealand that exhibit considerable geographic contrast.

**THE RELATIONSHIP BETWEEN
REGIONAL AND SYSTEMATIC GEOGRAPHY**

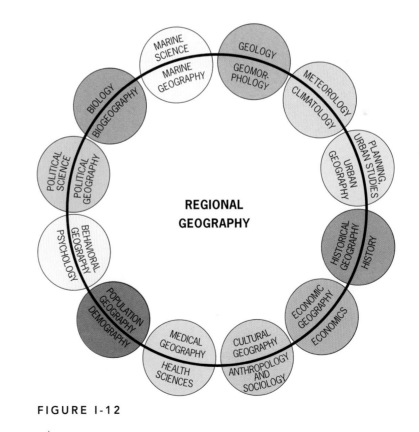

FIGURE I-12

Pacific Realm (12)

The vast Pacific Ocean, larger than all the landmasses combined, contains tens of thousands of islands large and small. Dominant cultural criteria warrant three regions: Melanesia, Micronesia, and Polynesia.

THE PERSPECTIVE OF GEOGRAPHY

As this introductory chapter demonstrates, our world regional survey is no mere description of places and areas. We have combined the study of realms and regions with a look at geography's ideas and concepts—the notions, generalizations, and basic theories that make the discipline what it is. We continue this method in the chapters ahead so that we will become better acquainted with the world and with geography. By now you are aware that geography is a wide-ranging, multifaceted discipline. It is often described as a social science, but that is only half the story: in fact, geography straddles the divide between the social and the physical (natural) sciences. Many of the ideas and concepts you will encounter have to do with the multiple interactions between human societies and natural environments.

49 **Regional geography** allows us to view the world in an all-encompassing way. As we have seen, regional geography borrows information from many sources to create an overall image of our divided world. Those sources are not random. They form topical or **50** **systematic geography**. Research in the systematic fields of geography makes our world-scale generalizations possible. As Figure I-12 shows, these systematic fields relate closely to those of other disciplines. Cultural geography, for example, is allied with anthropology; it is the spatial perspective that distinguishes cultural geography. Economic geography focuses on the spatial dimensions of economic activity; political geography concentrates on the spatial imprints of political behavior. Other systematic fields include historical, medical, behavioral, environmental, agricultural, and coastal geography. We will also draw on information from biogeography, marine geography, population geography, and climatology (as we did earlier in this chapter).

✳ These systematic fields of geography are so named because their approach is global, not regional. Take the geographic study of cities, urban geography. Urbanization is a worldwide process, and urban geographers can identify certain human activities that all cities in the world exhibit in one form or another. But cities also display regional properties. The model Japanese city is quite distinct from, say, the African city. Regional geography, therefore, borrows from the systematic field of urban geography, but it injects this regional perspective.

✳ In the following chapters we call upon these systematic fields to give us a better understanding of the world's realms and regions. As a result, you will gain insights into the discipline of geography as well as the regions we investigate. This will prove that geography is a relevant and practical discipline when it comes to comprehending, and coping with, our fast-changing world.

TABLE I-1   23

Table I-1
AREA AND DEMOGRAPHIC DATA FOR THE WORLD'S STATES

	Land Area 1000 (sq mi)	Population 2002 (Millions)	Population 2010 (Millions)	Population Density Arithmetic	Population Density Physiologic	Birth Rate	Death Rate	Natural Increase	Doubling Time (years)	Infant Mortality per 1,000 (births)	Life Expectancy Males (years)	Life Expectancy Females (years)	Percent Urban Pop	Per Capita GNP ($US)
WORLD	**51510.8**	**6238.1**	**6866.8**	**117.0**		**22**	**9**	**1.4%**	**51**	**57**	**64**	**68**	**45**	**$4,890**
Europe	**2197.2**	**582.6**	**580.4**	**265.2**		**10**	**11**	**0.0%**		**7**	**72**	**79**	**73**	**$16,518**
Albania	10.6	3.5	3.9	329.1	1,567.4	18	5	1.3%	55	22	69	74	46	$810
Austria	31.9	8.1	8.1	253.9	1,493.6	10	10	0.0%		5	75	81	65	$26,830
Belarus	80.1	9.9	9.7	123.6	426.2	9	14	−0.5%		11	63	74	70	$2,180
Belgium	11.8	10.2	10.3	866.1	3,608.9	11	10	0.1%	770	6	75	81	97	$25,380
Bosnia	19.7	3.8	4.0	194.8	1,391.6	13	8	0.5%	141	12	71	76	40	
Bulgaria	42.7	8.1	7.5	189.7	512.8	8	14	−0.6%		14	67	74	68	$1,220
Croatia	21.6	4.6	4.5	212.5	1,012.1	11	12	−0.1%		8	69	76	54	$4,620
Cyprus	3.6	0.9	0.9	253.0	2,108.4	14	8	0.6%	124	8	74	79	64	$11,920
Czech Rep.	29.8	10.3	10.2	344.3	839.6	9	11	−0.2%		5	71	78	77	$5,150
Denmark	16.4	5.3	5.5	323.8	539.7	12	11	0.1%	472	5	74	79	85	$33,040
Estonia	16.3	1.4	1.4	85.0	386.5	8	13	−0.5%		9	64	75	69	$3,360
Finland	117.6	5.2	5.2	44.3	553.8	11	10	0.1%	433	4	74	81	60	$24,280
France	212.4	59.9	61.6	281.9	854.3	13	9	0.4%	204	5	75	82	74	$24,210
Germany	134.9	81.9	81.2	607.4	1,840.6	9	10	−0.1%		5	74	80	86	$26,570
Greece	49.8	10.6	10.5	212.9	1,120.3	10	10	0.0%		7	75	81	59	$11,740
Hungary	35.7	9.9	9.6	277.3	543.8	9	14	−0.5%		9	66	75	64	$4,510
Iceland	38.7	0.3	0.3	7.9		15	7	0.8%	81	3	77	82	92	$27,830
Ireland	26.6	3.8	4.1	144.6	1,112.1	15	9	0.6%	116	6	73	79	58	$27,135
Italy	113.5	57.7	55.6	508.2	1,639.5	9	10	−0.1%		6	75	81	90	$20,090
Latvia	24.0	2.4	2.3	98.8	365.9	8	14	−0.6%		11	64	76	69	$2,420
Liechtenstein	0.1	0.1	0.1	507.0		14	7	0.7%	105	18	67	78		
Lithuania	25.0	3.7	3.6	147.7	301.4	10	11	−0.1%		9	67	77	68	$2,540
Luxembourg	1.0	0.4	0.5	403.2	1,680.0	13	9	0.4%	198	5	74	80	88	$45,100
Macedonia	9.8	2.0	2.1	206.9	4,139.0	15	8	0.7%	112	16	70	75	59	$1,290
Malta	0.1	0.4	0.4	3,360.1		12	8	0.4%	182	5	74	80	89	$10,100
Moldova	12.7	4.3	4.4	338.6	638.8	11	11	0.0%		18	63	70	46	$380
Netherlands	13.1	16.0	16.5	1,223.5	4,531.4	13	9	0.4%	193	5	75	81	61	$24,780
Norway	118.5	4.5	4.7	38.2	1,273.4	13	10	0.3%	217	4	76	81	74	$34,310
Poland	117.5	38.6	38.6	328.5	699.0	10	10	0.0%		9	69	78	62	$3,910
Portugal	35.5	10.0	9.7	281.7	1,083.4	11	11	0.0%		5	72	79	48	$10,670
Romania	88.9	22.4	21.7	251.5	613.3	11	12	−0.1%		21	66	73	55	$1,360
Slovakia	18.6	5.4	5.4	290.9	938.4	11	10	0.1%	866	9	69	77	57	$3,700
Slovenia	7.8	2.0	2.0	255.9	2,132.5	9	10	−0.1%		5	71	79	50	$9,780

	Land Area 1000 (sq mi)	Population 2002 (Millions)	Population 2010 (Millions)	Population Density Arithmetic	Population Density Physiologic	Birth Rate	Death Rate	Natural Increase	Doubling Time (years)	Infant Mortality per 1,000 (births)	Males (years)	Females (years)	Percent Urban Pop	Per Capita GNP ($US)
Spain	192.8	39.5	38.4	204.9	682.9	9	9	0.0%		6	74	82	64	$14,100
Sweden	158.9	8.9	9.0	55.9	798.5	10	11	−0.1%		4	77	82	84	$25,580
Switzerland	15.3	7.1	7.3	465.9	4,659.1	11	9	0.2%	315	5	77	83	68	$39,980
Ukraine	223.7	48.9	47.4	218.6	377.0	8	14	−0.6%		13	63	74	68	$980
United Kingdom	93.3	59.9	61.6	642.2	2,568.9	12	11	0.1%	546	6	74	80	89	$21,410
Yugoslavia	26.9	10.7	10.7	397.8	1,325.9	11	11	0.0%		10	70	75	52	
Russia	**6550.7**	**143.2**	**135.7**	**21.9**	**273.2**	**8**	**15**	**−0.7%**		**17**	**61**	**73**	**73**	**$2,260**
Armenia	10.9	3.8	3.9	351.4	2,067.2	10	6	0.4%	161	15	71	78	67	$460
Azerbaijan	33.4	7.8	8.6	234.7	1,303.9	15	6	0.9%	77	17	68	75	52	$480
Georgia	26.9	5.5	5.2	204.9	2,276.3	9	8	0.1%	462	15	69	76	56	$970
North America	**7567.5**	**316.4**	**342.6**	**41.8**		**14**	**9**	**0.5%**	**124**	**7**	**74**	**80**	**75**	**$28,230**
Canada	3849.7	31.0	33.0	8.1	174.4	11	7	0.4%	178	6	76	81	78	$19,170
United States	3717.8	285.4	309.6	76.8	415.1	15	9	0.6%	120	7	74	79	75	$29,240
Middle America	**1021.9**	**179.4**	**202.5**	**175.6**		**25.2**	**6**	**2.0%**	**35**	**37**	**72**	**73**	**66**	
Antigua and Barbuda	0.2	0.1	0.1	607.2	3,373.4	22	6	1.6%	45	17	69	74	37	$8,450
Bahamas	3.9	0.3	0.3	79.4	7,940.4	21	5	1.6%	45	18	70	77	84	
Barbados	0.2	0.3	0.3	1,782.4	4,817.3	14	9	0.5%	130	14	72	77	38	
Belize	8.8	0.3	0.3	36.0	1,797.8	32	5	2.7%	26	34	70	74	50	$2,660
Costa Rica	19.7	3.7	4.6	189.4	3,156.3	22	4	1.8%	39	13	75	79	45	$2,770
Cuba	42.8	11.3	11.4	265.5	1,106.1	14	7	0.7%	103	7	73	78	75	
Dominica	0.3	0.1	0.1	350.4	3,893.0	16	8	0.8%	83	15	75	80		$3,150
Dominican Rep.	18.7	8.8	10.1	469.2	2,234.2	28	6	2.2%	32	47	67	71	62	$1,770
El Salvador	8.0	6.6	7.9	824.1	3,052.4	30	7	2.3%	29	35	67	73	58	$1,850
Grenada	0.1	0.1	0.1	805.0	5,366.8	29	6	2.3%	30	14	68	73	34	$3,250
Guadeloupe	0.7	0.4	0.4	629.0	4,492.8	17	6	1.1%	61	10	73	80	48	
Guatemala	41.9	13.5	17.0	321.6	2,679.7	37	7	2.9%	24	45	61	67	39	$1,640
Haiti	10.6	6.6	7.8	624.5	3,122.4	33	16	1.7%	40	103	47	51	34	$410
Honduras	43.2	6.4	7.3	148.9	992.9	33	6	2.7%	25	42	66	71	45	$740
Jamaica	4.2	2.7	2.9	637.8	4,555.4	22	7	1.5%	45	24	70	73	50	$1,740
Martinique	0.4	0.4	0.4	993.2	12,415.6	15	6	0.9%	81	9	75	82	81	
Mexico	737.0	103.6	115.2	140.6	1,171.7	24	4	2.0%	36	32	69	75	74	$3,840
Netherlands Antilles	0.3	0.2	0.2	659.4	6,594.3	17	6	1.1%	62	14	72	78		
Nicaragua	46.9	5.4	6.7	115.4	1,281.8	36	6	3.0%	23	40	66	71	63	$370
Panama	28.7	3.0	3.3	104.5	1,493.0	22	5	1.7%	41	21	72	77	56	$2,990
Puerto Rico	3.4	4.0	4.1	1,167.8	29,195.0	17	8	0.9%	75	11	70	79	71	
Saint Lucia	0.2	0.2	0.3	855.1	10,689.3	19	6	1.3%	56	17	71	72	48	$3,660

TABLE I-1   25

	Land Area 1000 (sq mi)	Population 2002 (Millions)	Population 2010 (Millions)	Population Density Arithmetic	Population Density Physiologic	Birth Rate	Death Rate	Natural Increase	Doubling Time (years)	Infant Mortality per 1,000 (births)	Life Expectancy Males (years)	Life Expectancy Females (years)	Percent Urban Pop	Per Capita GNP ($US)
St. Vincent and the Grenadines	0.2	0.1	0.1	682.8	1,796.7	19	7	1.2%	59	20	71	74	44	$2,560
Trinidad and Tobago	2.0	1.3	1.4	659.1	4,394.2	14	7	0.7%	103	16	68	73	72	$4,520
South America	**6763.3**	**356.8**	**400.1**	**52.8**		**23**	**6**	**1.7%**	**41**	**34**	**66**	**73**	**78**	**$4,270**
Argentina	1056.6	37.8	41.6	35.8	397.7	19	8	1.1%	62	19	70	77	90	$8,030
Bolivia	418.7	8.6	10.0	20.6	1,031.2	30	10	2.0%	34	67	59	62	62	$1,010
Brazil	3265.1	175.2	193.6	53.7	1,073.4	21	6	1.5%	45	38	64	71	78	$4,630
Chile	289.1	15.6	17.2	54.0	1,079.1	18	5	1.3%	54	11	72	78	85	$4,990
Colombia	401.0	41.6	48.3	103.8	2,594.5	26	6	2.0%	34	28	65	73	71	$2,470
Ecuador	106.9	13.1	15.0	122.9	2,047.8	27	6	2.1%	33	40	67	72	63	$1,520
French Guiana	34.0	0.2	0.3	6.2	616.8	27	3	2.4%	29	18	71	77	79	
Guyana	76.0	0.7	0.8	9.5	476.3	24	7	1.7%	40	63	63	69	36	$780
Paraguay	153.4	5.8	7.2	37.7	629.0	32	6	2.6%	26	27	68	72	52	$1,760
Peru	494.2	28.3	32.6	57.2	1,905.4	27	6	2.1%	32	43	66	71	72	$2,440
Suriname	60.2	0.4	0.4	6.9	689.9	26	7	1.9%	37	29	68	73	69	$1,660
Uruguay	67.5	3.3	3.6	49.5	618.5	16	10	0.6%	107	15	70	78	92	$6,070
Venezuela	340.6	25.2	29.0	73.9	1,848.0	25	5	2.0%	34	21	70	76	86	$3,530
North Africa/ Southwest Asia	**7655.8**	**526.6**	**620.5**	**52.3**		**28**	**8**	**2.0%**	**35**	**52**	**63**	**69**	**54**	
Afghanistan	251.8	28.1	36.0	111.4	928.4	43	18	2.5%	28	150	46	45	20	
Algeria	919.6	33.0	38.4	35.8	1,194.9	29	6	2.3%	29	44	68	70	49	$1,550
Bahrain	0.3	0.7	1.1	2,692.0	269,204.7	22	3	1.9%	37	8	68	71	88	$7,640
Djibouti	9.0	0.6	0.8	69.8		39	16	2.3%	30	115	47	50	83	
Egypt	384.3	71.1	81.6	184.9	9,245.3	26	6	2.0%	35	52	64	67	44	$1,290
Eritrea	39.0	4.3	6.0	111.5	929.4	43	13	3.0%	23	82	52	57	16	$200
Iran	631.7	69.4	78.0	109.9	1,099.2	21	6	1.5%	48	31	68	71	63	$1,650
Iraq	168.9	24.4	31.0	144.5	1,204.4	38	10	2.8%	25	127	58	60	68	
Israel	8.0	6.4	7.2	800.0	4,705.9	22	6	1.6%	45	6	76	80	90	$16,180
Jordan	34.3	5.4	6.8	157.1	3,928.3	33	5	2.8%	24	34	68	70	78	$1,150
Kazakhstan	1031.2	15.0	14.9	14.6	121.4	14	10	0.4%	161	21	59	70	63	$1,650
Kuwait	6.9	2.3	2.9	333.0		24	2	2.2%	32	13	72	73	100	
Kyrgyzstan	74.1	5.0	5.3	68.1	973.2	22	7	1.5%	47	26	63	71	34	$380
Lebanon	4.0	4.3	4.8	1,083.9	5,161.3	23	7	1.6%	43	35	68	73	88	$3,560
Libya	679.4	5.4	6.5	7.9	788.7	28	3	2.5%	28	33	73	77	86	
Morocco	172.3	29.8	33.6	172.9	823.2	23	6	1.7%	41	37	67	71	54	$1,240
Oman	82.0	2.6	3.6	31.5		43	5	3.8%	18	25	69	73	72	

	Land Area 1000 (sq mi)	Population 2002 (Millions)	Population 2010 (Millions)	Population Density Arithmetic	Population Density Physiologic	Birth Rate	Death Rate	Natural Increase	Doubling Time (years)	Infant Mortality per 1,000 (births)	Life Expectancy Males (years)	Life Expectancy Females (years)	Percent Urban Pop	Per Capita GNP ($US)
Palestinian Territ. (West Bank/Gaza)	2.4	3.3	5.0	1.4		41	5	3.6%	19	27	70	73		$1,560
Qatar	4.3	0.6	0.7	144.6	14,460.3	20	2	1.8%	38	20	70	75	91	
Saudi Arabia	830.0	22.9	29.7	27.6	1,380.4	35	5	3.0%	23	46	68	71	83	$6,910
Somalia	242.2	7.7	10.6	31.9	1,595.7	47	18	2.9%	24	126	45	48	24	
Sudan	917.4	30.8	37.0	33.5	670.4	33	12	2.1%	32	70	50	52	27	$290
Syria	71.0	17.4	21.2	245.1	875.4	33	6	2.7%	25	35	67	68	51	$1,020
Tajikistan	54.3	6.6	7.3	121.7	2,027.8	21	5	1.6%	43	28	66	71	27	$370
Tunisia	60.0	9.9	11.1	164.8	867.6	22	7	1.5%	44	35	67	70	61	$2,060
Turkey	297.2	67.3	75.6	226.4	707.4	22	7	1.5%	46	38	67	71	66	$3,160
Turkmenistan	181.4	5.4	5.9	29.5	984.4	21	6	1.5%	48	33	62	69	44	
United Arab Emirates	32.3	2.9	3.3	90.5		24	2	2.2%	32	16	73	76	84	$17,870
Uzbekistan	159.9	25.7	28.0	160.4	1,782.4	23	6	1.7%	40	22	66	72	38	$950
Western Sahara	102.7	0.3	0.4	3.1		46	18	2.8%	24	150	46	48		
Yemen	203.9	18.0	26.2	88.1	2,937.0	39	11	2.8%	25	75	58	61	26	$280
Subsaharan Africa	**7916.3**	**646.8**	**781.0**	**81.7**		**41**	**16**	**2.6%**	**27**	**93**	**48**	**50**	**29**	**$522**
Angola	481.4	13.7	18.2	28.4	1,418.7	48	19	2.9%	23	125	45	48	32	$380
Benin	42.7	6.8	8.7	158.4	1,218.4	45	17	2.8%	24	94	49	51	38	$380
Botswana	218.8	1.6	1.5	7.5	753.4	32	17	1.5%	45	57	38	40	49	$3,070
Burkina Faso	105.6	12.6	16.2	119.3	917.8	47	18	2.9%	24	105	47	47	15	$240
Burundi	9.9	6.4	8.0	646.0	1,468.3	42	17	2.5%	28	75	46	47	8	$140
Cameroon	179.7	16.2	19.6	90.0	692.6	37	12	2.5%	27	77	55	56	44	$610
Cape Verde Is.	1.6	0.4	0.5	264.2	2,401.8	37	9	2.8%	25	77	65	72	44	$1,200
Central African Republic	240.5	3.6	4.1	15.1	504.7	38	18	2.0%	34	97	43	46	39	$300
Chad	486.1	8.5	12.0	17.6	585.4	50	17	3.3%	21	110	46	51	22	$230
Comoros Is.	0.9	0.6	0.8	737.3	2,106.5	38	10	2.8%	25	77	57	62	29	$370
Congo	131.9	2.9	3.6	22.3	1,113.0	40	16	2.4%	29	109	45	50	41	$680
Congo, The	875.3	55.4	75.3	63.3	2,109.0	48	16	3.2%	22	109	47	50	29	$110
Equatorial Guinea	10.8	0.5	0.6	48.6	972.8	41	16	2.5%	28	108	48	52	37	$1,110
Ethiopia	386.1	67.2	86.3	174.1	1,450.7	45	21	2.4%	29	116	45	47	15	$100
Gabon	99.5	1.3	1.6	12.6	1,259.7	38	16	2.2%	32	87	51	54	73	$4,170
Gambia	3.9	1.4	1.7	349.5	1,941.8	43	19	2.4%	29	130	43	47	37	$340
Ghana	87.9	20.4	22.9	232.6	1,938.5	34	10	2.4%	29	56	56	59	37	$390

TABLE I-1   27

	Land Area 1000 (sq mi)	Population 2002 (Millions)	Population 2010 (Millions)	Population Density Arithmetic	Population Density Physiologic	Birth Rate	Death Rate	Natural Increase	Doubling Time (years)	Infant Mortality per 1,000 (births)	Life Expectancy Males (years)	Life Expectancy Females (years)	Percent Urban Pop	Per Capita GNP ($US)
Guinea	94.9	7.9	9.8	82.9	4,143.5	42	18	2.4%	29	98	43	47	26	$530
Guinea-Bissau	10.9	1.3	1.5	115.0	1,045.4	42	20	2.2%	31	130	47	44	22	$160
Ivory Coast	122.8	16.7	19.3	136.1	1,701.1	38	16	2.2%	32	112	45	48	46	$700
Kenya	219.8	31.6	32.7	143.7	2,052.9	35	14	2.1%	33	74	48	49	20	$350
Lesotho	11.7	2.2	2.3	186.7	1,697.6	33	13	2.0%	33	85	52	55	16	$570
Liberia	37.2	3.4	4.4	91.8	9,179.3	50	17	3.3%	21	139	49	52	45	
Madagascar	224.5	15.8	20.9	70.4	1,760.3	44	14	3.0%	24	96	51	53	22	$260
Malawi	36.3	10.8	11.5	297.5	1,652.7	41	22	1.9%	36	127	38	40	20	$210
Mali	471.1	11.9	15.7	25.3	1,263.5	47	16	3.1%	22	123	55	52	26	$250
Mauritania	395.8	2.9	3.6	7.2	720.9	41	13	2.8%	25	92	52	55	54	$410
Mauritius	0.8	1.2	1.3	1,569.4	3,202.8	17	7	1.0%	66	19	67	74	43	$3,730
Moçambique	302.7	19.9	20.2	65.9	1,647.6	41	19	2.2%	32	134	40	39	28	$210
Namibia	317.9	1.9	2.0	5.8	584.5	36	20	1.6%	42	68	47	45	27	$1,940
Niger	489.1	10.7	13.9	21.9	730.3	54	24	3.0%	23	123	41	41	17	$200
Nigeria	351.7	130.6	160.1	371.2	1,124.9	42	13	2.9%	24	77	52	53	36	$300
Réunion	1.0	0.7	0.8	743.5	4,373.3	20	5	1.5%	49	9	70	79	73	
Rwanda	9.5	7.5	7.7	793.2	2,266.2	43	20	2.3%	30	121	39	40	5	$230
São Tomé and Príncipe	0.3	0.2	0.2	737.3	36,867.4	43	9	3.4%	20	51	63	66	44	$270
Senegal	74.3	10.0	12.7	135.1	1,126.0	41	13	2.8%	25	68	51	54	41	$520
Seychelles	0.2	0.1	0.1	511.1	25,553.0	18	7	1.1%	65	9	67	73	59	$6,420
Sierra Leone	27.7	5.5	7.2	197.6	2,823.1	47	21	2.6%	26	157	42	47	37	$140
South Africa	471.4	44.5	40.8	94.5	944.8	25	12	1.3%	55	45	54	57	45	$3,310
Swaziland	6.6	1.0	1.3	157.3	1,430.2	41	22	1.9%	37	108	36	39	22	$1,400
Tanzania	341.1	37.4	46.3	109.6	3,652.6	42	13	2.9%	24	99	52	54	20	$220
Togo	21.0	5.3	6.2	253.1	666.0	42	11	3.1%	23	80	48	50	31	$330
Uganda	77.1	24.6	34.0	319.4	1,277.5	48	20	2.8%	24	81	42	43	15	$310
Zambia	287.0	10.0	11.7	34.7	496.2	42	23	1.9%	35	109	37	38	38	$330
Zimbabwe	149.4	11.5	10.7	77.2	1,102.2	30	20	1.0%	69	80	41	39	32	$620
South Asia	**1592.0**	**1376.7**	**1558.5**	**864.8**		**28**	**9**	**1.9%**	**36**	**74**	**60**	**61**	**28**	**$436**
Bangladesh	50.3	133.0	150.7	2,644.4	3,622.5	27	8	1.9%	38	82	59	58	20	$350
Bhutan	18.2	1.0	1.1	52.6	2,628.2	40	9	3.1%	22	71	66	66	15	$470
India	1148.0	1038.5	1168.3	904.6	1,615.4	27	9	1.8%	39	72	60	61	28	$440
Maldives	0.1	0.3	0.4	2,652.3	26,522.5	35	5	3.0%	23	27	71	72	25	$1,130
Nepal	52.8	25.1	30.3	475.6	2,797.5	36	11	2.5%	28	79	58	57	11	$210
Pakistan	297.6	159.2	186.3	534.8	1,980.7	39	11	2.8%	25	91	58	59	33	$470
Sri Lanka	25.0	19.7	21.4	786.5	5,618.2	18	6	1.2%	60	17	70	74	22	$810

	Land Area 1000 (sq mi)	Population 2002 (Millions)	Population 2010 (Millions)	Population Density Arithmetic	Population Density Physiologic	Birth Rate	Death Rate	Natural Increase	Doubling Time (years)	Infant Mortality per 1,000 (births)	Life Expectancy Males (years)	Life Expectancy Females (years)	Percent Urban Pop	Per Capita GNP ($US)
East Asia	**4450.1**	**1517.2**	**1574.2**	**340.9**		**15**	**7**	**0.8%**	**87**	**29**	**70**	**74**	**38**	**$3,880**
China	3705.8	1294.4	1349.0	349.3	3,594.2	15	6	0.9%	79	31	69	73	31	$750
Japan	145.4	127.2	124.7	874.5	7,950.1	9	8	0.1%	462	4	77	84	78	$32,350
Korea, North	46.5	22.3	23.7	479.8	3,427.3	21	7	1.4%	48	26	67	73	59	
Korea, South	38.1	48.2	50.2	1,263.9	6,652.2	14	5	0.9%	82	11	71	78	79	$8,600
Mongolia	604.8	2.6	2.9	4.2	424.2	20	7	1.3%	50	34	60	66	52	$380
Taiwan	14.0	22.6	23.7	1,615.2	6,730.1	13	6	0.7%	97	7	72	78	77	
Southeast Asia	**1735.4**	**546.1**	**614.5**	**314.7**		**24**	**7**	**1.7%**	**41**	**46**	**63**	**67**	**36**	**$1,240**
Brunei	2.0	0.3	0.4	156.7	15,667.3	25	3	2.2%	32	24	70	73	67	
Cambodia	68.2	12.7	16.1	186.8	1,436.7	38	12	2.6%	27	80	54	58	16	$260
East Timor	5.7	0.8	1.0	145.4		34	16	1.8%	39	143	45	47		
Indonesia	705.2	219.0	240.8	310.6	3,106.1	24	8	1.6%	44	46	62	66	39	$640
Laos	89.1	5.5	6.6	61.4	2,047.9	41	15	2.6%	26	104	50	52	17	$320
Malaysia	126.9	24.2	29.3	191.0	6,367.6	25	5	2.0%	34	8	70	75	57	$3,670
Myanmar/Burma	253.9	50.9	57.8	200.4	1,335.8	30	10	2.0%	35	83	53	56	26	
Philippines	115.1	83.9	97.2	728.7	3,835.2	29	7	2.2%	31	35	66	69	47	$1,050
Singapore	0.2	4.1	5.6	16,934.4	846,720.0	13	5	0.8%	84	3	76	80	100	$30,170
Thailand	197.3	63.1	66.7	319.9	941.0	16	7	0.9%	70	22	70	75	31	$2,160
Vietnam	125.7	80.9	92.5	643.7	3,786.7	20	6	1.4%	48	37	63	69	24	$350
Austral Realm	**3067.9**	**23.3**	**24.9**	**7.6**		**13**	**7**	**0.6%**	**109**	**5**	**76**	**82**	**85**	**$19,639**
Australia	2966.2	19.4	20.8	6.6	109.7	13	7	0.6%	110	5	76	82	85	$20,640
New Zealand	103.5	3.9	4.1	37.3	414.5	15	7	0.8%	89	5	74	80	85	$14,600
Pacific Realm	**207.7**	**7.5**	**9.0**	**36.3**		**29**	**8**	**2.1%**	**34**	**6**	**54**	**56**	**21**	**$912**
Federated States of Micronesia	0.3	0.1	0.1	389.9		33	7	2.6%	27	46	65	67	27	$1,800
Fiji	7.1	0.8	0.9	116.1	1,160.8	22	7	1.5%	46	13	65	69	46	$2,210
French Polynesia	1.4	0.2	0.2	147.5	14,746.5	21	5	1.6%	44	10	69	74	54	
Guam	0.2	0.2	0.2	998.6	9,078.6	28	4	2.4%	29	9	72	77	38	
Marshall Islands	0.1	0.1	0.1	1,492.1		26	4	2.2%	31	31	63	67	65	$1,540
New Caledonia	7.1	0.2	0.2	29.1		21	5	1.6%	42	7	69	77	59	
Papua New Guinea	174.9	5.0	6.1	28.8		34	10	2.4%	29	77	56	57	15	$890
Samoa	1.1	0.2	0.2	191.0	1,005.4	31	6	2.5%	28	25	65	72	21	$1,070
Solomon Is.	10.8	0.4	0.6	39.4	3,936.9	37	6	3.1%	23	25	69	74	13	$760
Vanuatu	4.7	0.2	0.2	45.0	2,248.5	35	7	2.8%	25	39	64	67	18	$1,260

1 / Europe

This Chapter's Media Highlights include:

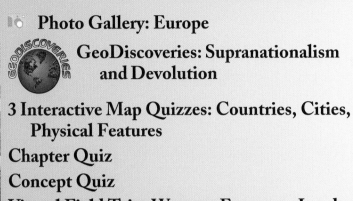

- **Photo Gallery: Europe**
- **GeoDiscoveries: Supranationalism and Devolution**
- **3 Interactive Map Quizzes: Countries, Cities, Physical Features**
- **Chapter Quiz**
- **Concept Quiz**
- **Virtual Field Trip: Western European Landscapes**

  www.wiley.com/college/deblij

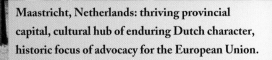

Maastricht, Netherlands: thriving provincial capital, cultural hub of enduring Dutch character, historic focus of advocacy for the European Union.

Relief

Meters		Feet
3050		10 000
1525		5000
610		2000
305		1000
152.5		500
0	Sea Level	0
152.5		500
1525		5000
3050		10 000

Sea Level

Below Sea Level

Scale 1: 16 000 000; one inch to 250 miles. Conic Projection

Elevations and depressions are given in feet

Longitude West of Greenwich   Longitude East of Greenwich

0	50	100		200		300	400		500 Miles
0	100	200		400		600		800 Kilometers	

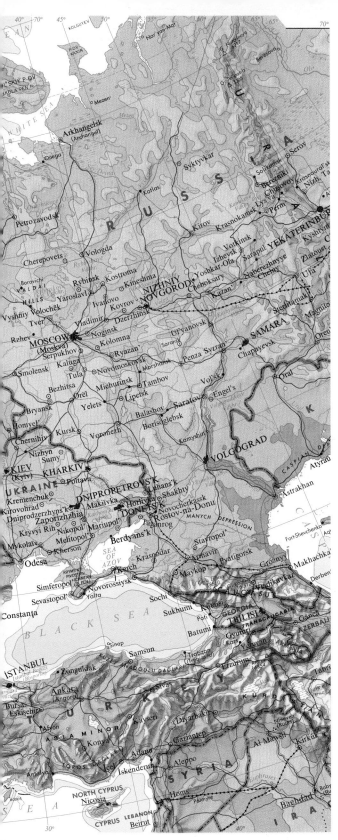

chapter 1 / Europe

CONCEPTS, IDEAS, AND TERMS

1	Land hemisphere	11	Indo-European languages	20	Regional state
2	Infrastructure	12	Complementarity	21	Site
3	Local functional specialization	13	Transferability	22	Situation
4	*The Isolated State*	14	Intervening opportunity	23	Conurbation
5	Model	15	Primate city	24	Landlocked location
6	Industrial Revolution	16	Metropolis	25	Break-of-bulk
7	Nation-state	17	Supranationalism	26	Entrepôt
8	Nation	18	Devolution	27	Shatter belt
9	Centrifugal forces	19	Four Motors of Europe	28	Balkanization
10	Centripetal forces			29	Exclave
				30	Irredentism

REGIONS

- ▷ **WESTERN EUROPE**
- ▷ **THE BRITISH ISLES**
- ▷ **NORTHERN (NORDIC) EUROPE**
- ▷ **MEDITERRANEAN EUROPE**
- ▷ **EASTERN EUROPE**

FIGURE 1-1

It is appropriate to begin our investigation of the world's geographic realms in Europe because over the past five centuries Europe and Europeans have influenced and changed the rest of the world more than any other realm or people has done. European empires spanned the globe and transformed societies far and near. European colonialism propelled the first wave of globalization. Millions of Europeans migrated from their homelands to the Old World as well as the New, changing (and sometimes nearly obliterating) traditional communities and creating new societies from Australia to North America. Colonial power and economic incentive combined to impel the movement of millions of imperial subjects from their ancestral homes to distant lands: Africans to the Americas, Indians to Africa, Chinese to Southeast Asia, Malays to South Africa's Cape, Native Americans from east to west. In agriculture, industry, politics, and other spheres, Europe generated revolutions—and then exported those revolutions across the world, thereby consolidating the European advantage.

But throughout much of that 500-year period of European hegemony, Europe also was a cauldron of conflict. Religious, territorial, and political disputes precipitated bitter wars that even spilled over into the colonies. And during the twentieth century, Europe twice plunged the world into war. The terrible, unprecedented toll of World War I (1914–1918) was not enough to stave off World War II (1939–1945), which ended with the first-ever use of nuclear weapons in Japan. In the aftermath of that war, Europe's weakened powers lost most of their colonial possessions and a new rivalry emerged: an ideological Cold War between the communist Soviet Union and the capitalist United States. This Cold War lowered an Iron Curtain across the heart of Europe, leaving most of the east under Soviet control and most of the west in the American camp. Western Europe proved resilient, overcoming the destruction of war and the loss of colonial power to regain economic strength. Meanwhile the Soviet communist experiment failed at home and abroad, and in 1990 the last vestiges of the Iron Curtain were lifted. Since then, a massive effort has been underway to reintegrate and reunify Europe from the Atlantic coast to the Russian border, the key geographic story of this chapter.

DEFINING THE REALM

As Figure 1-1 shows, Europe is a realm of peninsulas and islands on the western margin of the world's largest landmass, Eurasia. It is a realm of 583 million people and 39 countries, but it is territorially quite small. Yet despite its modest proportions it has had—and continues to have—a major impact on world affairs. For many centuries Europe has been a hearth of achievement, innovation, and invention.

◆ Major Geographic Qualities of Europe

1. The European realm lies on the western extremity of the Eurasian landmass, a locale of maximum efficiency for contact with the rest of the world.

2. Europe's lingering and resurgent world influence results largely from advantages accrued over centuries of global political and economic domination.

3. The European natural environment displays a wide range of topographic, climatic, and soil conditions and is endowed with many industrial resources.

4. Europe is marked by strong internal regional differentiation (cultural as well as physical), exhibits a high degree of functional specialization, and provides multiple exchange opportunities.

5. European economies are dominated by manufacturing, and the level of productivity has been high; levels of development generally decline from west to east.

6. Europe's nation-states emerged from durable power cores that formed the headquarters of world colonial empires. A number of those states are now plagued by internal separatist movements.

7. Europe's rapidly aging population is generally well off, highly urbanized, well educated, and enjoys long life expectancies.

8. A growing number of European countries are experiencing population declines; in many of these countries the natural decrease is partially offset by immigration.

9. Europe has made significant progress toward international economic integration and, to a lesser extent, political coordination.

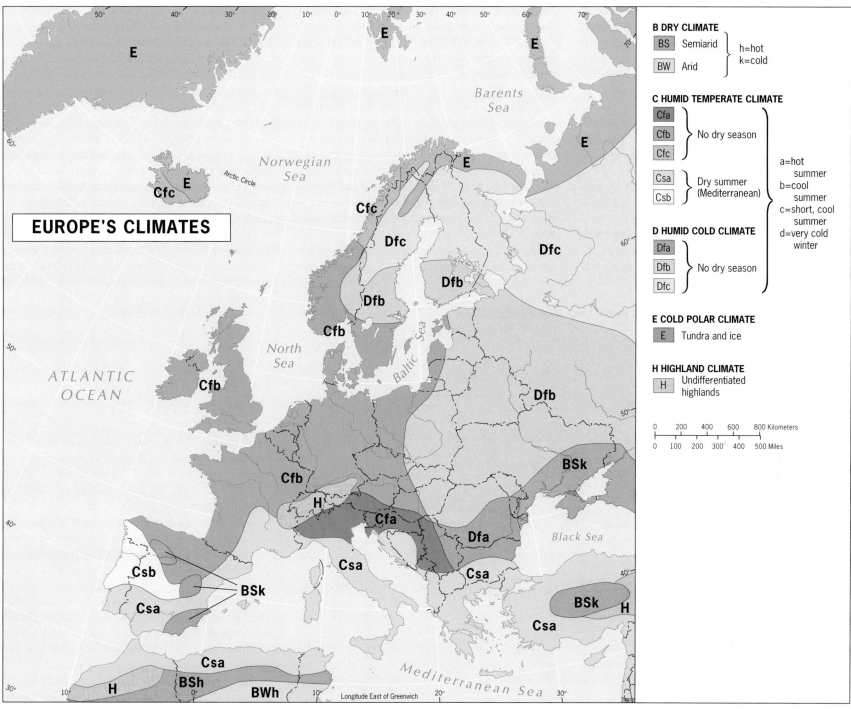

EUROPE'S CLIMATES

B DRY CLIMATE
BS Semiarid   } h=hot
BW Arid   } k=cold

C HUMID TEMPERATE CLIMATE
Cfa }
Cfb } No dry season
Cfc }

Csa } Dry summer
Csb } (Mediterranean)

a=hot summer
b=cool summer
c=short, cool summer
d=very cold winter

D HUMID COLD CLIMATE
Dfa }
Dfb } No dry season
Dfc }

E COLD POLAR CLIMATE
E Tundra and ice

H HIGHLAND CLIMATE
H Undifferentiated highlands

0   200   400   600   800 Kilometers
0   100 200 300 400 500 Miles

FIGURE 1-2

The European realm is bounded on the west, north, and south by Atlantic, Arctic, and Mediterranean waters, respectively. But where is Europe's eastern limit? Some scholars place it at the Ural Mountains, deep inside Russia, thereby recognizing a "European" Russia and, presumably, an "Asian" one as well. Our regional definition places Europe's eastern boundary between Russia and its numerous European neighbors to the west. This definition is based on several geographic factors including European-Russian contrasts in territorial dimensions, population size, cultural properties, and historic aspects, all discernible on the maps in Chapters 1 and 2.

Europe's peoples have benefited from a large and varied store of raw materials. Whenever the opportunity or need arose, the realm proved to contain what was required. Early on, these requirements included cultivable soils, rich fishing waters, and wild animals that could be domesticated; in addition, extensive forests provided wood for houses and boats. Later, mineral fuels and ores propelled industrialization.

From the balmy shores of the Mediterranean Sea to the icy peaks of the Alps, and from the moist woodlands and moors of the Atlantic fringe to the semiarid prairies north of the Black Sea, Europe presents an almost infinite range of natural environments (Fig. 1-2). Compare Western Europe and eastern North America on Figure I-8 (p. 12) and you will see the moderating influence of the warm ocean current known as the North Atlantic Drift and its onshore windflow.

The European realm is home to peoples of numerous cultural-linguistic stocks, including not only Latins, Germanics, and Slavs but also minorities such as Finns, Magyars (Hungarians), Basques, and Celts. This diversity of ancestries continues to be an asset as well as a liability. It has generated not only interaction and exchange, but also conflict and war.

Europe also has outstanding locational advantages. **1** Its *relative location*, at the heart of the **land hemisphere**, creates maximum efficiency for contact with the rest of the world (Fig. 1-3). A "peninsula of peninsulas," Europe is nowhere far from the ocean and its avenues of seaborne trade and conquest. Hundreds of miles of navigable rivers, augmented by an unmatched system of canals, open the interior of Europe to its neighboring seas and to the waterways of the world.

Also consider the scale of the maps of Europe in this chapter. Europe is a realm of moderate distances and close proximities. Short distances and large cultural differences make for intense interaction, the constant circulation of goods and ideas. That has been the hallmark of Europe's geography for over a millennium.

RELATIVE LOCATION: EUROPE IN THE LAND HEMISPHERE

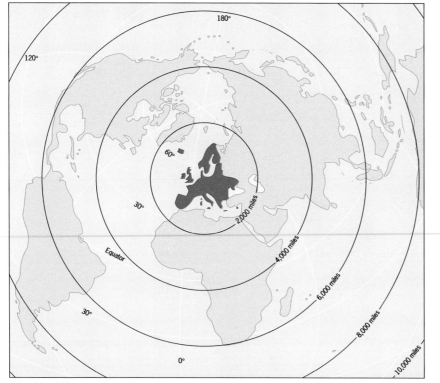

Azimuthal equidistant projection centered on Hamburg, Germany

FIGURE 1-3

LANDSCAPES AND OPPORTUNITIES

Europe's area may be small, but its landscapes are varied and complex. Regionally, we identify four broad units: the Central Uplands, the southern Alpine Mountains, the Western Uplands, and the North European Lowland (Fig. 1-4).

The *Central Uplands* form the heart of Europe. It is a region of hills and low plateaus loaded with raw materials whose farm villages grew into towns and cities when the Industrial Revolution transformed this realm.

The *Alpine Mountains*, a highland region named after the Alps, extend from the Pyrenees on the French-Spanish border to the Balkan Mountains near the Black Sea, and include Italy's Appennines and Eastern Europe's Carpathians.

The *Western Uplands*, geologically older, lower, and more stable than the Alpine Mountains, extend from Scandinavia through western Britain and Ireland to the heart of the Iberian Peninsula.

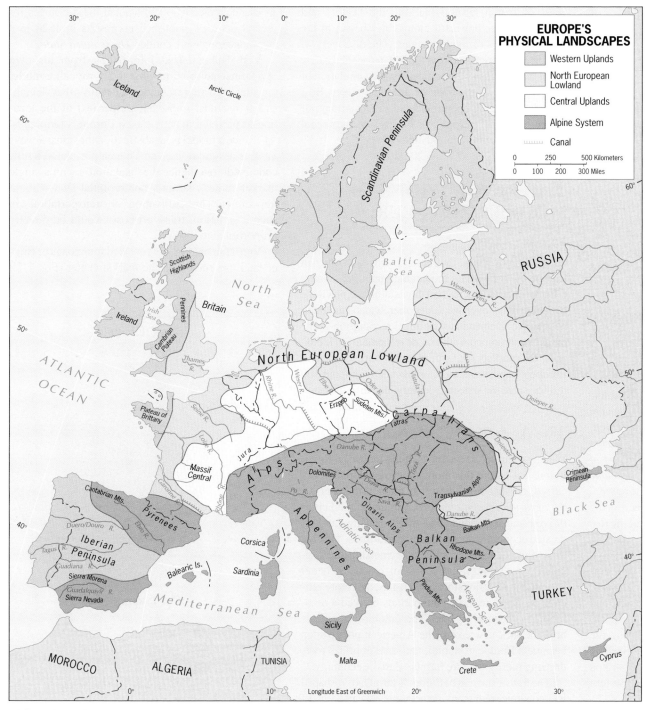

EUROPE'S PHYSICAL LANDSCAPES

Western Uplands

North European Lowland

Central Uplands

Alpine System

Canal

0   250   500 Kilometers

0   100   200   300 Miles

FIGURE 1-4

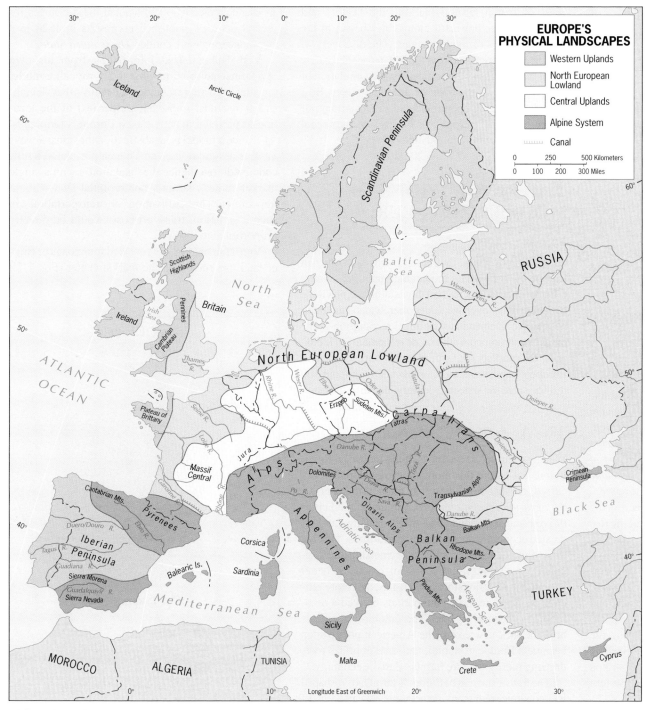 The *North European Lowland* extends in a lengthy arc from southeast Britain and central France across Germany and Denmark into Poland and Ukraine, from where it continues well into Russia. Also known as the Great European Plain, this has been an avenue for human migration time after time, so that complex cultural and economic mosaics developed here together with a jigsaw-like political map. As Figure 1-4 shows, many of Europe's major rivers and connecting waterways serve this populous region, where many of Europe's leading cities (London, Paris, Amsterdam, Copenhagen, Berlin, Warsaw) are located.

HISTORICAL GEOGRAPHY

Modern Europe was peopled in the wake of the Pleistocene's most recent glacial retreat and global warming—a gradual warming that caused tundra to give way to deciduous forest and ice-filled valleys to turn into grassy vales. On Mediterranean shores, Europe witnessed the rise of its first great civilizations, on the islands and peninsulas of Greece and later in what is today Italy.

Ancient Greece lay exposed to influences radiating from the advanced civilizations of Mesopotamia and the Nile Valley, and in their fragmented habitat the Greeks laid the foundations of European civilization. Their achievements in political science, philosophy, the arts, and other spheres have endured for 25 centuries. But the ancient Greeks never managed to unify their domain, and their persistent conflicts proved fatal when the Romans challenged them from the west. By 147 BC the last of the sovereign Greek intercity leagues (alliances) had fallen to the Roman conquerors.

The center of civilization and power now shifted to Rome in present-day Italy. Borrowing from Greek culture, the Romans created an empire that stretched from Britain to the Persian Gulf and from the Black Sea to Egypt; they made the Mediterranean Sea a Roman lake carrying armies to distant shores and goods to imperial Rome. With an urban population that probably exceeded 1 million, Rome was the first metropolitan-scale urban center in Europe.

The Romans founded numerous other cities throughout their empire and linked them to the capital through a vast system of highway and water routes, facilitating political control and enabling economic growth in their provinces. It was an unparalleled **2 infrastructure**, much of which long outlasted the empire itself.

Roman rule brought disparate, isolated peoples into the imperial political and economic sphere. By guiding (and often forcing) these groups to produce particular goods or materials, they launched Europe down a road **3** for which it would become famous: **local functional specialization**. The workers on Elba, a Mediterranean island, mined iron ore. Those near Cartagena in Spain produced silver and lead. Certain farmers were taught irrigation to produce specialty crops. Others raised livestock for meat or wool. The *production of particular goods by particular people in particular places* became and remained a hallmark of the realm.

The Romans also spread their language across the empire, setting the stage for the emergence of the *Romance* languages; they disseminated Christianity; and they established durable systems of education, administration, and commerce. But when their empire collapsed in the fifth century, disorder ensued, and massive migrations soon brought Germanic and Slavic peoples to their present positions on the European stage. Capitalizing on Europe's weakness, the Arab-Berber Moors from North Africa, energized by Islam, conquered most of Iberia and penetrated France. Later the Ottoman Turks invaded Eastern Europe and reached the outskirts of Vienna.

Europe's revival—its *Renaissance*—did not begin until the fifteenth century. After a thousand years of feudal turmoil marking the "Dark" and "Middle" Ages, powerful monarchies began to lay the foundations of modern states. The discovery of continents and riches across the oceans opened a new era of *mercantilism*, the competitive accumulation of wealth chiefly in the form of gold and silver. Best placed for this competition were the kingdoms of Western Europe. Europe was on its way to colonial expansion and world domination.

THE REVOLUTIONS OF MODERNIZING EUROPE

Even as Europe's rising powers reached for world domination overseas, they fought with each other in Europe itself. Powerful monarchies and land-owning ("landed") aristocracies had their status and privilege challenged by ever-wealthier merchants and businesspeople. Demands for political recognition grew; cities mushroomed with the development of industries; the markets for farm products burgeoned; and Europe's population, more or less stable at about 100 million since the sixteenth century, began to increase.

The Agrarian Revolution

We know Europe as the focus of the Industrial Revolution, but before this momentous development occurred, another revolution was already in progress: the *agrarian revolution*. Port cities and capital cities thrived and expanded, and their growing markets created economic opportunities for farmers. This led to revolutionary changes in land ownership and agricultural methods. Improved farm practices, better equipment, superior storage facilities, and more efficient transport to the urban markets marked a revolution in the countryside. The colonial merchants brought back new crops (the American potato soon became a European staple), and market prices rose, drawing more and more farmers into the economy.

The transformation of Europe's farmlands reshaped its economic geography, producing new patterns of land use and market links. The economic geographer Johann Heinrich von Thünen (1783–1850), himself an estate farmer who had studied these changes for several decades, published his results in 1826 in a **4** pioneering work entitled ***The Isolated State***.

Von Thünen used information from his own **5** farmstead to build what today we call a **model** (an idealized representation of reality that demonstrates its most important properties) of the location of productive activities in Europe's farmlands. ✳ Since a model is an abstraction that must always involve assumptions, von Thünen postulated a self-contained area (hence the "isolation") with a single market center, flat and uninterrupted land without impediments to cultivation or transportation. In such a situation, transport costs would be directly proportional to distance.

Von Thünen's model revealed four zones or rings of land use encircling the market center (Fig. 1-5).

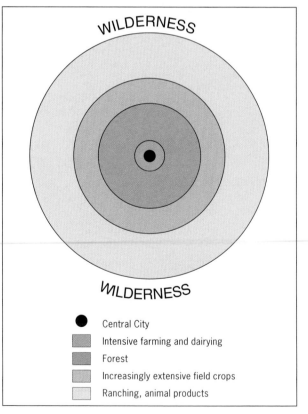

● Central City

▨ Intensive farming and dairying

▨ Forest

▨ Increasingly extensive field crops

▢ Ranching, animal products

FIGURE 1-5

Innermost and directly adjacent to the market would lie a zone of intensive farming and dairying, yielding the most perishable and highest-priced products. Immediately beyond lay a zone of forest used for timber and firewood (still a priority in von Thünen's time). Next there would be a ring of field crops, for example, grains or potatoes. A fourth zone would contain pastures and livestock. Beyond lay wildnerness, from where the costs of transport to market would become prohibitive.

In many ways, von Thünen's model was the first analysis in a field that would eventually become known as *location theory*. Von Thünen knew, of course, that the real Europe did not present the conditions postulated in his model. But it did demonstrate the economic-geographic forces that shaped the new Europe, which is why it is still being discussed today. More than a century after the publication of *The Isolated State*, geographers Samuel van Valkenburg and Colbert Held produced a map of twentieth-century European agricultural intensity, revealing a striking, ring-like concentricity focused on the vast urbanized area lining the North Sea—now the dominant market for a realmwide, macroscale "Thünian" agricultural system (Fig. 1-6).

The Industrial Revolution

6 The term **Industrial Revolution** suggests that an agrarian Europe was suddenly swept up in wholesale industrialization that changed the realm in a few decades. In reality, seventeenth- and eighteenth-century Europe had been industrializing in many spheres, long before the chain of events known as the Industrial Revolution began. From the textiles of England and Flanders to the iron farm implements of Saxony (in present-day Germany), from Scandinavian furniture to French linens, Europe had already entered a new era of local functional specialization. It would therefore be more appropriate to call what happened next the period of Europe's *industrial intensification*.

In the 1780s James Watt and others devised a steam-driven engine, which was soon adopted for numerous industrial uses. At about the same time, coal (converted into carbon-rich coke) was recognized as a vastly superior substitute for charcoal in smelting iron. These momentous innovations had a rapid effect. The power loom revolutionized the weaving industry. Iron smelters, long dependent on Europe's dwindling forests for fuel, could now be concentrated near coalfields. Engines could move locomotives as well as power looms. Ocean shipping entered a new age.

Britain had an enormous advantage, for the Industrial Revolution occurred when British influence reigned worldwide and the significant innovations were achieved in Britain itself. The British controlled the flow of raw materials, they held a monopoly over products that were in global demand, and they alone possessed the skills necessary to make the machines that manufactured the products. Soon the fruits of the Industrial Revolution were being exported, and the modern industrial spatial organization of Europe began to take shape. In Britain, manufacturing regions, densely populated and heavily urbanized, developed near coalfields in the English Midlands, at Newcastle to the northeast, in southern Wales, and along Scotland's Clyde River around Glasgow.

In mainland Europe, a belt of major coalfields extends from west to east, roughly along the southern margins of the North European Lowland, due eastward from southern England across northern France and Belgium, Germany (the Ruhr), western Bohemia in the Czech Republic, Silesia in southern Poland, and the Donets Basin (Donbas) in eastern Ukraine. Iron ore is found in a broadly similar belt, and the industrial map of Europe reflects the resulting concentrations of economic activity (Fig. 1-7). Another set of manufacturing

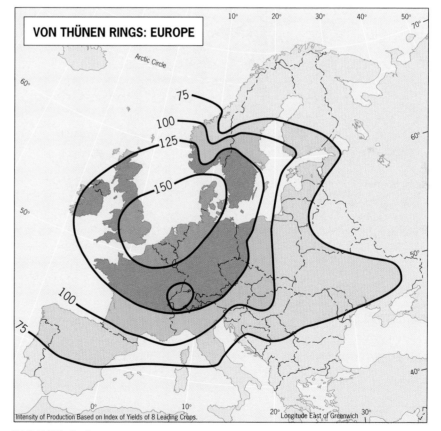

VON THÜNEN RINGS: EUROPE

Intensity of Production Based on Index of Yields of 8 Leading Crops.

FIGURE 1-6

 Photo Gallery ✦ More to Explore **20** Concepts, Ideas, and Terms GEODISCOVERIES Geodiscoveries Module

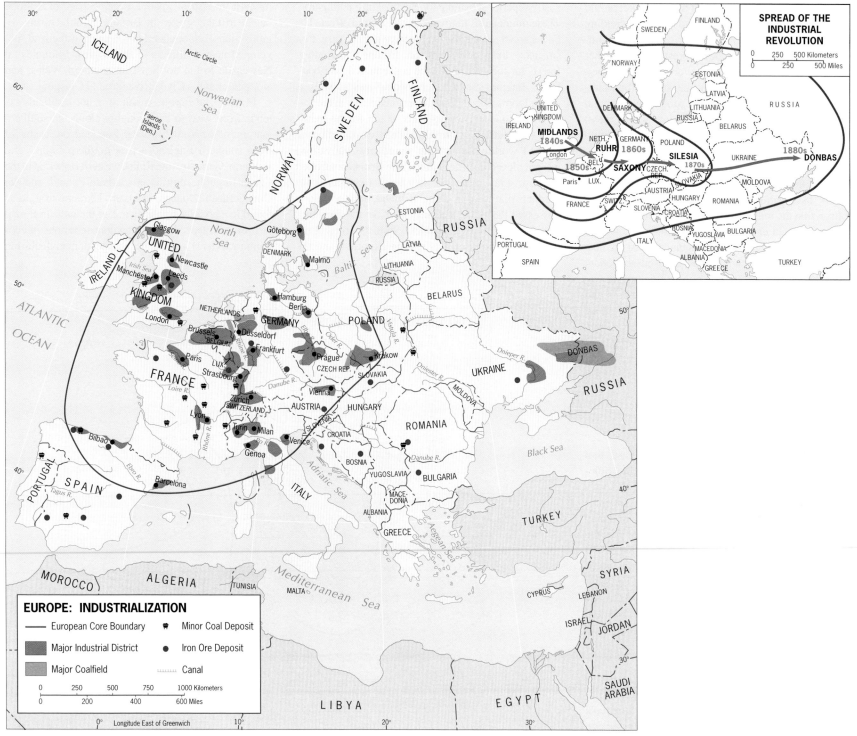

SPREAD OF THE INDUSTRIAL REVOLUTION

MIDLANDS 1840s
RUHR 1860s
1850s
SAXONY
SILESIA 1870s
1880s
DONBAS

EUROPE: INDUSTRIALIZATION

— European Core Boundary
⬛ Major Industrial District
⬛ Major Coalfield
⚒ Minor Coal Deposit
● Iron Ore Deposit
∷∷∷ Canal

0   250   500   750   1000 Kilometers
0   200   400   600 Miles

Longitude East of Greenwich

FIGURE 1-7

regions emerged in and near the growing urban centers of Europe, as the same map demonstrates. London—already Europe's leading urban focus and Britain's richest domestic market—was typical of these developments. Many local industries were established here, taking advantage of the large supply of labor, the ready avail-ability of capital, and the proximity of so great a number of potential buyers. Although the Industrial Revolution thrust other places into prominence, London did not lose its primacy: industries in and around the British capital multiplied.

The industrial transformation of Europe, like the agrarian revolution, became the focus of geographic research. One of the leaders in this area was the economic geographer Alfred Weber (1868–1958), who published a spatial analysis of the process titled *Concerning the Location of Industries* (1909). Unlike von Thünen, Weber focused on activities that take place at particular points rather than over large areas. His model, therefore, represented the factors of industrial location, the clustering or dispersal of places of intense manufacturing activity.

One of Weber's most interesting conclusions has to do with what he called *agglomerative* (concentrating) and *deglomerative* (dispersive) forces. It is often advantageous for certain industries to cluster together, sharing equipment, transport facilities, labor skills, and other assets of urban areas. This is what made London (as well as Paris and other cities that were not situated on rich deposits of industrial raw materials) attractive to many manufacturing plants that could benefit from agglomeration and from the large markets that these cities anchored. As Weber found, however, excessive agglomeration may lead to disadvantages such as competition for increasingly expensive space, congestion, overburdening of infrastructure, and environmental pollution. Manufacturers may then move away and deglomerative forces will increase.

The Industrial Revolution spread eastward from Britain onto the European mainland throughout the middle and late nineteenth century (see inset, Fig. 1-7). Population skyrocketed, emigration mushroomed, industrializing cities burst at the seams. European states already had acquired colonial empires before this revolution started; now colonialism gave Europe an unprecedented advantage in its dominance over the rest of the world.

Political Revolutions

Revolution in a third sphere—the political—had been going on in Europe even longer than the agrarian or industrial revolutions. Europe's *political revolution* took many different forms and affected diverse peoples and countries, but in general it headed toward parliamentary representation and democracy. Most dramatic was the French Revolution (1789–1795), but political transformation had come much earlier to the Netherlands, Britain, and Scandinavian countries. Other parts of Europe remained under the control of authoritarian (dictatorial) regimes headed by monarchs or despots. Europe's patchy political revolution lasted into the twentieth century, and by then *nationalism* (national spirit, pride, patriotism) had become a powerful force in European politics.

When you look at the political map of Europe, the question that arises is how did so small a geographic realm come to be divided into so many political entities? Europe's map is a legacy of its feudal and royal periods, when powerful kings, barons, dukes, and other rulers, rich enough to fund armies and powerful enough to exact taxes and tribute from their domains, created bounded territories in which they reigned supreme. Royal marriages, alliances, and conquests actually simplified Europe's political map. In the early nineteenth century there still were 39 German states; Germany as we know it today did not emerge until the 1870s.

Europe's political revolution produced a form of political-territorial organization known as the **7** **nation-state**, a state embodied by its culturally distinctive population. But what is a nation-state **8** and what is not? The term **nation** has multiple meanings. In one sense it refers to a people with a single language, a common history, a similar ethnic background. In the sense of *nationality* it relates to legal membership in the state, that is, citizenship. Very few states today are so homogeneous culturally that the culture is conterminous with the state. Europe's prominent nation-states of a century ago—France, Spain, the United Kingdom, Italy—have become multicultural societies, their nations defined more by an intangible "national spirit" and emotional commitment than by cultural or ethnic homogeneity. Today, Poland, Hungary, and Sweden are among the few states that still satisfy the definition of the nation-state in Europe.

Mercantilism and colonialism empowered the states of Western Europe; the United Kingdom (Britain) was the superpower of its day. But all countries, even Europe's nation-states in their heyday, are subject to divisive stresses. Political geographers use the **9** term **centrifugal forces** to identify and measure the strength of such division, which may result from religious, racial, linguistic, political, or regional factors. In the United States, racial issues form a centrifugal force; during the Vietnam (Indochina) War (1964–1975) politics created strong and dangerous ✳ disunity.

10 Centrifugal forces are measured against **centripetal forces**, the binding, unifying glue of the state. General satisfaction with the system of government and administration, legal institutions, and other functions of the state (notably including its treatment of minorities) can ensure stability and continuity when centrifugal forces threaten. In the recent case of Yugoslavia, the centrifugal forces unleashed after the end of the Cold War exceeded the weak centripetal forces in that relatively young state, and it disintegrated.

Europe's political revolution continues. Today a ✳ growing group of European states is trying to create a realmwide union, what might some day become a European superstate.

CONTEMPORARY EUROPE

Europe has been a regional laboratory of political revolution and evolution, and some of its nation-states were among the first of their kind to emerge on the world stage. Enriched and empowered by colonialism, European states competed and fought with each other, but Europe's nations survived and prospered. Strong cultural identities and historic durability gave European peoples a confidence that continues to mark the realm today, long after Europe's global empires collapsed.

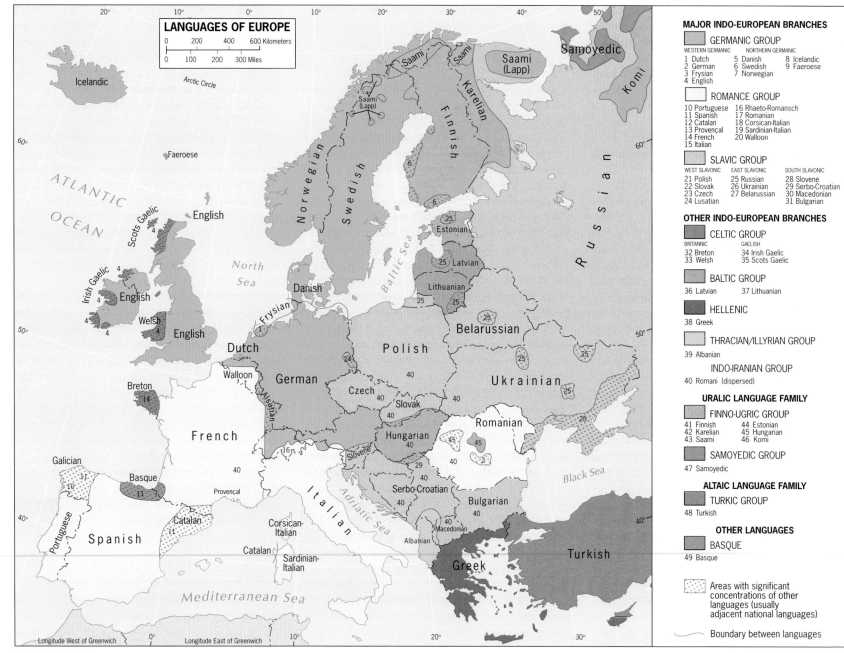

FIGURE 1-8

The European realm exhibits only limited geographic homogeneity, which is a challenge that confronts those leading states that want to create a more unified European system. As Figure 1-8 shows, most **11** Europeans speak **Indo-European languages**, but in fact Europe remains a veritable Tower of Babel. Not only are many of those Indo-European languages not mutually understandable, but peoples such as the Hungarians and the Finns have other linguistic sources. English has become Europe's *lingua franca*, but generally with declining effectiveness from west to east.

Christian religious traditions, another factor that might serve as a unifier, have instead been the source of endless conflict. Shared Christian values, for example, have done little to bring peace to Northern Ireland, where longstanding sectarian strife continues. Religion and politics remain closely connected, and some political parties still have religious names, such as Germany's Christian Democratic Union and the Dutch Katholieke Volkspartij (Catholic People's Party).

The name "European" is sometimes taken to refer not just to someone residing in Europe, but as a racial reference, describing a common ancestry. But here again, Europe's purported homogeneity is more apparent than real. In terms of physical characteristics, Europeans, from Swede to Spaniard and from Scot to Sicilian, are as varied as any of the world's other geographic realms.

Spatial Interaction

If not culture, what does unify Europe? The answer lies in this realm's outstanding opportunities for productive contact and profitable interaction. The ancient Romans knew it, but they centered their system on the imperial capital. Modern Europeans have seized the same opportunities to create a regionwide structure of *spatial interaction* that links regions, countries, and places in countless ways. The American geographer Edward Ullman conceptualized this process around three operating principles: (1) complementarity, (2) transferability, and (3) intervening opportunity.

12 Complementarity occurs when one area has a surplus of a commodity required by another area. The mere existence of a particular resource or product is no guarantee of trade: it must be needed elsewhere. One of Europe's countless examples (at various levels of scale) involves Italy, leader among Mediterranean countries in economic development but lacking adequate coal supplies. Italy imports coal from Western Europe, and in turn Italy exports to Western Europe its citrus fruits, olives, and grapes—which are in high demand on Western European markets. This is a case of *double complementarity*, and even the physical barrier of the Alps has not restricted this two-way trade by rail and road.

13 Transferability refers to the ease with which a commodity is transported between producer and consumer. Sheer distance, in terms of cost and time, may make it economically impractical to transfer a product. This is not a problem in modestly sized Europe, where distances are short and transport systems efficient.

14 Intervening opportunity, the third of Ullman's spatial interaction principles, holds that potential trade between two places, even if they are in a position of complementarity and do not have problems of transferability, will develop only in the absence of a closer, intervening source of supply. Using our current example, if a major coal reserve were discovered in southern Switzerland, Italy would avail itself of that (intervening) opportunity, reducing or eliminating its imports from Western Europe.

Europe's internal spatial interaction is facilitated by what in many respects is the world's most effective network of railroads, highways, waterways, and air routes. This network is continuously improving as tunnels, bridges, high-speed rail lines, and augmented airports are built. The continent's burgeoning cities and their environs, however, are increasingly troubled by severe congestion.

An Urbanized Realm

Overall, 73 percent of Europe's population resides in cities and towns, but this average is exceeded by far in the west and not attained in the east.

Large cities are the crucibles of their nations' cultures. In his 1939 study of the pivotal role of great cities in the development of national cultures, American geographer Mark Jefferson postulated the law **15** of the **primate city**, which stated that "a country's leading city is always disproportionately large and exceptionally expressive of national capacity and feeling." Though imprecise, this "law" can readily be demonstrated using European examples. Certainly Paris personifies France in countless ways, and nothing in England rivals London. In both of these primate cities, the culture and history of a nation and empire are indelibly etched in the urban landscape. Similarly, Vienna is a microcosm of Austria, Warsaw is the heart of Poland, Stockholm typifies Sweden, and Athens is Greece. Today, each of these (together with the other primate cities of Europe) sits atop a hierarchy of urban centers that has captured the lion's share of its national population growth since World War II.

Primate cities tend to be old, and in general the European cityscape looks quite different from the American. Seemingly haphazard inner-city street systems impede traffic; central cities may be picturesque, but they are also cramped. The urban layout of the London region (Fig. 1-9) reveals much about the internal spatial structure of the European **16** **metropolis** (the central city and its suburban ring). The metropolitan area remains focused on the large city at its center, especially the downtown *central business district* (*CBD*), which is the oldest part of the urban agglomeration and contains the region's largest concentration of business, government, and shopping facilities as well as its wealthiest and most prestigious residences. Wide residential sectors radiate outward from the CBD across the rest of the central city, each one home to a particular income group. Beyond the central city lies a sizeable suburban ring, but residential densities are much higher here than in the United States because the European tradition is one of setting aside recreational spaces (in "greenbelts") and living in apartments rather than in detached single-family houses. There is also a greater reliance on public transportation, which further concentrates the suburban development pattern. That

METROPOLITAN LONDON

- Built-up area
- Greenbelt
- Central Business District
- Road
- Railroad

0   5   10   15   20 Kilometers

0   5   10 Miles

FIGURE 1-9

Meanwhile, *immigration* is partially offsetting the losses European countries face. Millions of Turks (mainly to Germany), Algerians (France), Moroccans (Spain), West Africans (Britain), and Indonesians (Netherlands) are changing the social fabric of what once were unicultural nation-states.

EUROPE'S TRANSFORMATION

At the end of World War II, much of Europe lay shattered, its cities and towns devastated, its infrastructure wrecked, its economies devastated. The Soviet Union had taken control over the bulk of Eastern Europe, and communist parties seemed poised to dominate the political life of major Western European countries.

In 1947 U.S. Secretary of State George C. Marshall proposed a European Recovery Program designed to counter all this dislocation and to create stable political conditions in which democracy would survive. Over the next four years, the United States provided about $13 billion in assistance to Europe (almost $100 billion in today's money). Because the Soviet Union refused U.S. aid and forced its Eastern European satellites to do the same, the Marshall Plan applied solely to 16 European countries, including defeated (West) Germany, and Turkey.

The Marshall Plan did far more than stimulate European economies. It showed European leaders that their countries needed a joint economic-administrative structure not only to coordinate the financial assistance, but also to ease the flow of resources and products across Europe's mosaic of boundaries, to lower restrictive trade tariffs, and to seek ways to effect political cooperation.

For all these needs Europe's governments had some guidelines. While in exile in Britain, three small countries—Belgium, the Netherlands, and Luxembourg—had been discussing an association of this kind even before the end of the war. There, in 1944, they formulated and signed the Benelux Agreement, intended to achieve total economic inte-

has allowed many nonresidential activities to suburbanize as well, and today ultramodern outlying business centers increasingly compete with the CBD in many parts of urban Europe.

A Changing Population Complexion

When a population urbanizes, average family size declines, and so does the overall rate of natural increase. There was a time when Europe's population was (in the terminology of population geographers) exploding, sending millions to the New World and the colonies and still growing at home. But today Europe's indigenous population, unlike most of the rest of the world's, is actually shrinking. To keep a popula-

tion from declining, the (statistically) average woman must bear 2.1 children. For Europe as a whole, that figure was 1.4 in 2002. Seven European countries, including Italy and Spain, recorded below 1.3—the lowest ever seen in any human population.

Such *negative population growth* poses serious challenges for any nation. When the population pyramid becomes top-heavy, the number of workers whose taxes pay for the social services of the aged goes down, leading to reduced pensions and dwindling funds for health care. Governments that impose tax increases endanger the business climate; their options are limited. Europe, and especially Western Europe, is experiencing a *population implosion* that will be a formidable challenge in decades to come.

gration. When the Marshall Plan was launched, the Benelux precedent helped speed the creation of the Organization for European Economic Cooperation ✳ (OEEC), which was established to coordinate the investment of America's aid (see box below).

Soon the economic steps led to greater political cooperation as well. In 1949 the participating governments created the Council of Europe, the beginnings of what was to become a European Parliament meeting in Strasbourg, France. Europe was embarked on still another political revolution, the formation of a multinational union involving a growing number of European states. Geographers **17** define **supranationalism** as the voluntary association in economic, political, or cultural spheres of three or more independent states willing to yield

some measure of sovereignty for their mutual benefit. In later chapters we will encounter other supranational organizations, including the North American Free Trade Agreement (NAFTA), but none had reached the plateau achieved by the European ✳ Union (EU).

In Europe, the key initiatives arose from the Marshall Plan and lay in the economic arena; political integration came more haltingly. By the Treaty of Rome, six countries joined to become the European Economic Community (EEC) in 1957, also called the "Common Market." In 1973 the United Kingdom, Ireland, and Denmark joined, and the renamed European Community (EC) had nine members. As Figure 1-10 shows, membership reached 15 countries in 1995, after the organization had been re-

named yet one more time to become the European Union (EU).

The European Union is not just a paper organization for bankers and manufacturers. It has a major impact on the daily lives of its member countries' citizens in countless ways. (You will see one of these ways when you arrive at an EU airport and find that EU-passport holders move through inspection at a fast pace, while non-EU citizens wait in long immigration lines.) Some of the EU's conditions are not universally popular: farmers in particular repeatedly object to EU regulations on agriculture. One major plan is to replace all national currencies such as France's franc and Germany's mark with the euro, but in 2002 three of the 15 EU members, including Britain, had not agreed to this move. Still, use of the

Supranationalism in Europe

1944   Benelux Agreement signed.

1947   Marshall Plan created (effective 1948–1952).

1948   Organization for European Economic Cooperation (OEEC) established.

1949   Council of Europe created.

1951   European Coal and Steel Community (ECSC) Agreement signed (effective 1952).

1957   Treaty of Rome signed, establishing European Economic Community (EEC) (effective 1958), also known as the Common Market and "The Six." European Atomic Energy Community (EURATOM) Treaty signed (effective 1958).

1959   European Free Trade Association (EFTA) Treaty signed (effective 1960).

1961   United Kingdom, Ireland, Denmark, and Norway apply for EEC membership.

1963   France vetoes UK EEC membership; Ireland, Denmark, and Norway withdraw applications.

1965   EEC-ECSC-EURATOM Merger Treaty signed (effective 1967).

1967   European Community (EC) inaugurated.

1968   All customs duties removed for intra-EC trade; common external tariff established.

1973   United Kingdom, Denmark, and Ireland admitted as members of EC, creating "The Nine." Norway rejects membership in the EC by referendum.

1979   First general elections for a European Parliament held; new 410-member legislature meets in Strasbourg. European Monetary System established.

1981   Greece admitted as member of EC, creating "The Ten."

1985   Greenland, acting independently of Denmark, withdraws from EC.

1986   Spain and Portugal admitted as members of EC, creating "The Twelve." Single European Act ratified, targeting a functioning European Union in the 1990s.

1987   Turkey and Morocco make first application to join EC. Morocco is rejected; Turkey is told that discussions will continue.

1990   Charter of Paris signed by 34 members of the Conference on Security and Cooperation in Europe (CSCE). Former East Germany, as part of newly reunified Germany, incorporated into EC.

1991   Maastricht meeting charts European Union (EU) course for the 1990s.

1993   Single European Market goes into effect. Modified European Union Treaty ratified, transforming EC into EU.

1995   Austria, Finland, and Sweden admitted into EU, creating "The Fifteen."

1999   European Monetary Union (EMU) goes into effect. Helsinki summit discusses fast-track negotiations with six prospective members and applications from six others; prospects for Turkey considered in longer term.

2001   Denmark's voters reject EMU participation by 53 to 47 percent.

2002   The euro is introduced as historic national currencies disappear in 12 countries.

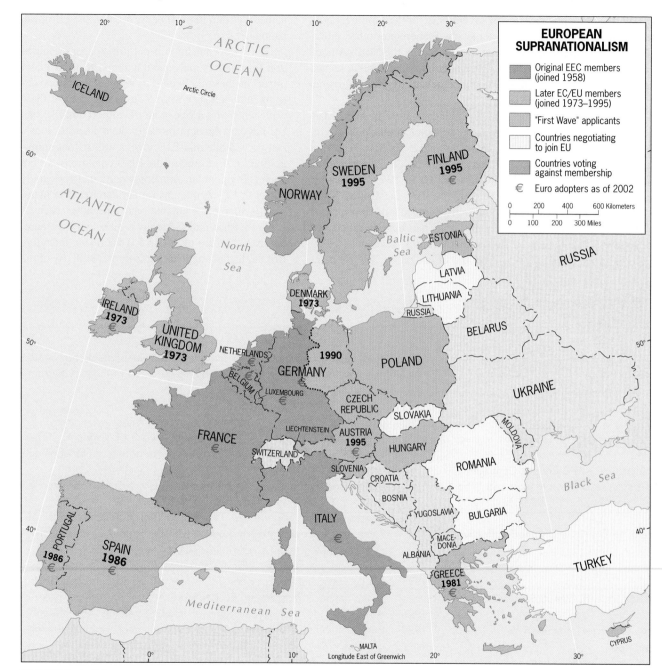

FIGURE 1-10

EUROPEAN SUPRANATIONALISM

- Original EEC members (joined 1958)
- Later EC/EU members (joined 1973–1995)
- "First Wave" applicants
- Countries negotiating to join EU
- Countries voting against membership
- € Euro adopters as of 2002

0   200   400   600 Kilometers
0 100   200   300 Miles

euro took effect in 12 EU countries in March 2002, and the old currencies disappeared.

Another EU objective is further expansion (Fig. 1-10). Even with 15 members, the EU includes well under half of Europe's countries. As the map shows, 13 potential member states are under discussion, and even a non-European state, Turkey, is under consideration. But expansion entails major risks and problems. States must meet certain economic and social criteria to join the EU, and many aspirant members cannot meet them. So the EU must either exclude certain countries (Romania, for example), or it must relax its rules. Neither option is attractive to the architects of European supranationalism.

Such problems notwithstanding, the European Union is a huge achievement in a fractious, fragmented realm. The influence of the European Parliament is growing, and border-crossing formalities are being eliminated. As of 2002, the 15 member-states represent 380 million people who constitute one of the world's richest markets and who produce about 40 percent of all exports. (Germany, in many ways the EU's leader, alone is the world's number two exporter by value.) Some European leaders continue to dream of a United States of Europe, a federal superpower that would constitute a major force in world affairs during the twenty-first **GEODISCOVERIES** century.

Centrifugal Forces

Europe has been described as a realm of geographic contradictions. Its current progress toward unification should be seen against a history of conflict, division, and self-destruction. Will supra-

EUROPE: FOCI OF DEVOLUTIONARY PRESSURES, 2002

FIGURE 1-11

nationalism finally overcome the centrifugal forces that have so long and so frequently afflicted this part of the world?

Even as Europe's states have been working to join forces in the EU, many of those same states are confronting severe centrifugal stresses. **18** The term **devolution** has come into use to describe the powerful centrifugal forces whereby regions or peoples within a state, through negotiation or active rebellion, demand and gain political strength and sometimes autonomy at the expense of the center. Most states exhibit some level of internal regionalism, but the process of devolution is set into motion when a key centripetal, binding force—the nationally accepted idea of what a country stands for—erodes to the point that a regional drive for autonomy, or for outright secession, is launched.

As Figure 1-11 shows, numerous European countries are affected by devolution. States large and small, young and old, EU members and non-EU members must deal with the problem. Even the long-stable United Kingdom is affected. England, the historic core area of the British Isles, dominates the UK in terms of population as well as political and economic power. The country's three other entities—Scotland, Wales, and Northern Ireland—were acquired over several centuries and attached to England (hence the "United" Kingdom). But neither time nor representative democratic government was enough to eliminate all latent regionalism in these three components of the UK. During the 1960s and 1970s the British government confronted a virtual civil war in Northern Ireland and

2 / Russia

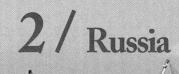

This Chapter's Media Highlights include:

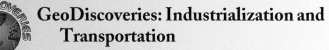 **Photo Gallery: Russia**

GeoDiscoveries: Industrialization and Transportation

Chapter Quiz

Concept Quiz

Virtual Field Trip: Journey Across Post-Soviet Russia www.wiley.com/college/deblij

St. Petersburg: Petrograd, Leningrad, St. Petersburg again: window on the Baltic and Europe as czarist Russia's capital, center of communist revolution, Soviet citadel, and now the economic symbol of the new Russia.

chapter 2 / Russia

Scale 1:20 000 000; one inch to 315 miles.
Lambert's Azimuthal, Equal Area Projection
Elevations and depressions are given in feet

FIGURE 2-1

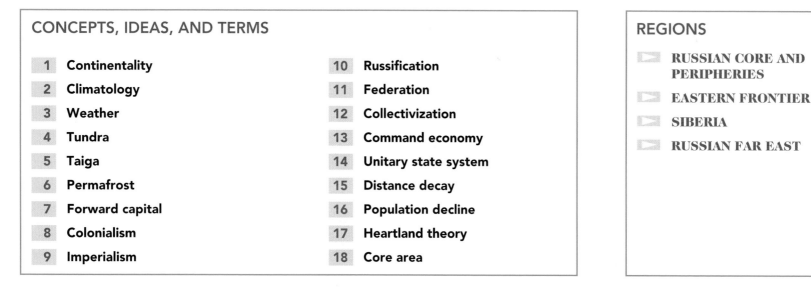

The Russian state constitutes a world geographic realm because of it territorial size, relative location, and substantial population. Turn back one page and consider the map: Russia borders Norway as well as North Korea and ten countries in between; it is about three times as large as the contiguous United States; and it extends across 11 time zones from its eastern frontier (the volcano-studded Kamchatka Peninsula) to its western port of St. Peters-burg. The western half of Russia is bisected by the north-south trending Ural Mountains, which extend from the island of Novaya Zemlya in the Arctic Ocean to the border with Kazakhstan. The vast majority of all Russians live to the west of the Urals, which has given rise to the false notion of a "European" Russia and, consequently, a non-European Russia farther east. But the cultural landscapes of Russia continue from Moscow to Novosibirsk to Vladivos-tok on the Pacific coast. The Urals bound the Siberian Lowland, but they do not limit "European" Russia.

Russia's predominantly Slavic population of over 140 million may not match those of China or India, but it is much larger than that of any European country. The demographic as well as the political map confirms Russia's status as a geographic realm. As we will see, environmental, historic, and ideological factors combine to strengthen Russia's geographic identity.

DEFINING THE REALM

Although the geographic realm under discussion is dominated by the giant Russian state, it also includes three small countries in the area between the Black Sea and the Caspian Sea: Georgia, Armenia, and GEODISCOVERIES Azerbaijan. For centuries Russia was a powerful empire whose influence reached far beyond its borders into Transcaucasia, Turkestan, Mongolia, and even China. Today, Russia retains strong links with all three of these Transcaucasian republics, placing the realm boundary along the borders with Turkey and Iran (Fig. I-2).

Russia is a land of vast distances, remote frontiers, bitter cold, and isolated outposts. Its core area lies west of the Urals, centered on Moscow. To the north, south, and east lie distant, problematic peripheries where economies falter and political systems fail. During the early twentieth century, Russia embarked on an immense social experiment with the goal of an egalitarian communist state that would be a model for the rest of the world. The experiment failed, but not before Russia and its Eurasian empire had been

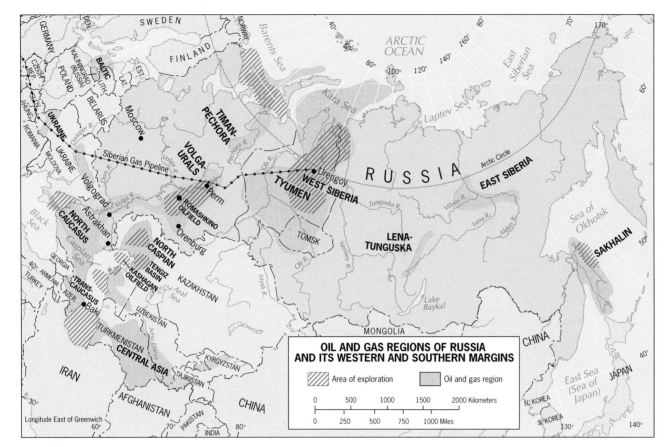

OIL AND GAS REGIONS OF RUSSIA
AND ITS WESTERN AND SOUTHERN MARGINS

Area of exploration   Oil and gas region

0   500   1000   1500   2000 Kilometers

0   250   500   750   1000 Miles

FIGURE 2-12

The External Southern Periphery

Beyond Russia's tier of internal republics lies still another periphery, legally outside the Russian sphere but closely tied to it nevertheless. This is *Transcaucasia*, historically a battleground for Christians and Muslims, Russians and Turks, Armenians and Persians. Today it is a subregion containing three former Soviet Socialist Republics: landlocked Armenia, coastal Georgia on the Black Sea, and Azerbaijan on the land-encircled Caspian Sea. Although these three countries are independent states today, they are so closely bound up with Russia that they are, functionally, part of the Russian realm (Fig. 2-13).

Armenia

As Figure 2-1 shows, landlocked Armenia (population: 3.8 million) occupies some of the most rugged and mountainous terrain in the earthquake-prone Transcaucasus. The Armenians are an embattled people who adopted Christianity 17 centuries ago and for more than a millennium sought to secure their ancient homeland here on the margins of the Muslim world. During World War I, the Ottoman Turks massacred much of the Christian Armenian minority and drove the survivors from eastern Anatolia and what is now Iraq into the Transcaucasus. At the end of that war in 1918, an independent Armenia arose, but its autonomy lasted only two years. In 1920, Armenia was taken over by the Soviets; in 1936, it became one of the 15 constituent republics of the Soviet Union. The collapse of the Soviet Empire gave Armenia what it had lost three generations earlier: independence.

Or so it seemed. Soon afterward, the Armenians found themselves at war with neighboring Azerbaijan over the fate of some 150,000 Armenians living

crucial oil-industry center, pipeline junction, and service hub during the Soviet era.

But Chechnya also contains a sizeable Muslim population, and fiercely independent Muslim Chechens used Caucasus mountain hideouts to resist Russian colonization during the nineteenth century. Accused of collaboration with the Nazis during World War II, on Stalin's orders the Soviets exiled the entire Chechen population to Central Asia, with much loss of life. Rehabilitated and allowed to return by Khrushchev in the 1950s, the Chechens seized their opportunity in 1991 after the collapse of the Soviet Union and fought the Russian army to a stalemate. But Moscow never granted Chechnya independence (one quarter of the population of 1.2 million was Russian), and an uneasy standoff contin-

ued. Meanwhile, Chechen hit-and-run attacks on targets in neighboring republics continued, and in 1999 three apartment buildings in Moscow were bombed, resulting in 230 deaths. Russian authorities blamed Chechnyan terrorists for these bombings and ordered a full-scale attack on Chechnya's Muslim holdouts. Groznyy, already severely damaged in the earlier conflict, was totally devastated and the rebels were driven into the mountains, but the Russian armed forces, still taking substantial losses, were unable to establish unchallenged control of all of the Republic.

Russia's costly problems in Chechnya illustrate the fractious nature of its internal southern periphery, from Buddhist Kalmykiya to also-Muslim Ingushetiya. Here the still-evolving federal framework will be put to the test for many years to come.

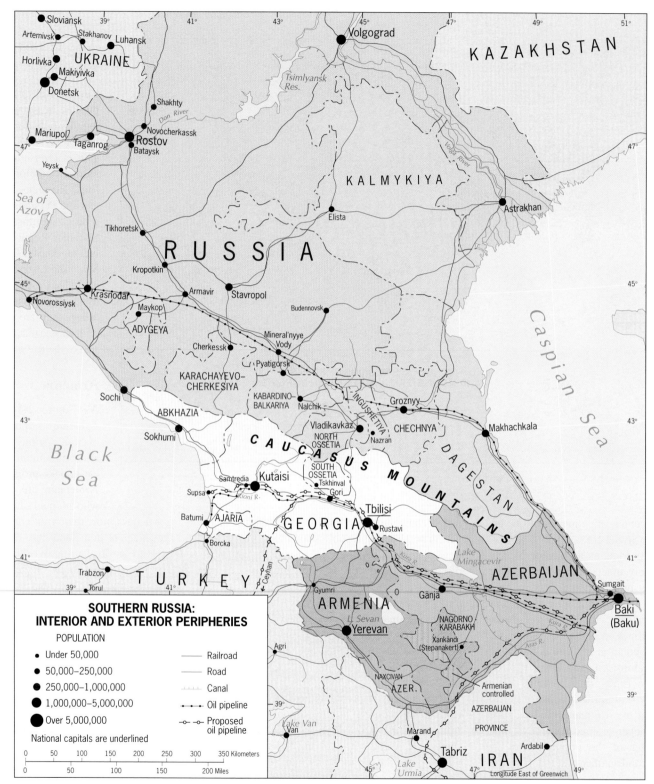

SOUTHERN RUSSIA:
INTERIOR AND EXTERIOR PERIPHERIES

POPULATION

• Under 50,000
• 50,000–250,000
● 250,000–1,000,000
● 1,000,000–5,000,000
● Over 5,000,000

National capitals are underlined

——— Railroad
——— Road
++++++ Canal
•–•–• Oil pipeline
o–o–o Proposed oil pipeline

0   50   100   150   200   250   300   350 Kilometers
0   50   100   150   200 Miles

in Nagorno-Karabakh, a pocket of territory surrounded by Azerbaijan. Such a separated territory is called an *exclave*, and this one had been created by Soviet sociopolitical planners who, while acknowledging the cultural (Christian) distinctiveness of this cluster of Armenians, nevertheless gave (Muslim) Azerbaijan jurisdiction over it.

That was a recipe for trouble: the arrangement was made to work under authoritarian Soviet rule, but once this rule ended, the Christian Armenians encircled by Muslim Azerbaijan felt insecure and appealed to Armenia for help. In the ensuing conflict, Armenian troops entered Azerbaijan and gained control over the exclave, even ousting Azerbaijanis from the zone between the main body of Armenia and Nagorno-Karabakh (Fig. 2-13). The international community, however, has not recognized Armenia's occupation, and officially the territory remains a part of Azerbaijan. After the turn of the twenty-first century, the matter still remained unresolved.

Georgia

Of the three former Soviet Republics in Transcaucasia, only Georgia has a Black Sea coast and thus an outlet to the wider world. Smaller than South Carolina, Georgia is a country of high mountains and fertile valleys. Its social and political geographies are complicated. The population of 5.5 million is more than 70 percent Georgian but also includes Armenians (8%), Russians (6%), Ossetians (3%), and Abkhazians (2%). The Georgian Orthodox Church dominates the religious community, but about 10 percent of the people are Muslims, most of them concentrated in Ajaria in the southwest.

Unlike Armenia and Azerbaijan, Georgia has no exclaves, but its political geography is problematic nonetheless (Fig. 2-13). Within Georgia's borders lie three minority-based autonomous entities: the Abkhazian and Ajarian Autonomous Republics, and the South Ossetian Autonomous Region.

Sakartvelos, as the Georgians call their country, has a long and turbulent history. Tbilisi, the capital for 15 centuries, lay at the core of an empire around the turn of the thirteenth century, but the Mongol invasion ended this era. Next, the Christian Georgians found themselves in the path of wars between Islamic Turks and Persians. Turning northward for protection, the Georgians were annexed by the Russians, who were looking for warm-water ports, in 1800. Like other peoples overpowered by the czars, the Georgians took advantage of the Russian Revolution to reassert their independence; but the Soviets reincorporated Georgia in 1921 and proclaimed a Georgian Soviet Socialist Republic in 1936. Josef Stalin, the communist dictator who succeeded Lenin, was a Georgian.

Georgia is renowned for its scenic beauty, warm and favorable climates, agriculture (especially tea), timber, manganese, and other products. Georgian wines, tobacco, and citrus fruits are much in demand. The diversified economy could support a viable state.

Unfortunately, Georgia's political geography is loaded with centrifugal forces. After Georgia declared its independence in 1991, factional fighting destroyed its first elected government. But worse was to come. In the Autonomous Region of South Ossetia, a movement for union with (Russian) North Ossetia created havoc; Abkhazia proclaimed independence; and Muslim-infused Ajaria asserted itself by ignoring a government preoccupied with problems elsewhere. The price of all this instability was Russian intervention, and today Georgia remains a dysfunctional state.

Azerbaijan

Azerbaijan is the name of an independent state *and* of a province in neighboring Iran. The Azeris (short for Azerbaijanis) on both sides of the border have the same ancestry: they are a Turkish people divided between the (then) Russian and Persian Empires by a treaty signed in 1828. By that time, the Azeris had become Shi'ite Muslims, and when the Soviet communists laid out their grand design for the USSR, they awarded the Azeris their own republic. On the Persian side, the Azeris were assimilated into the Persian Empire, and their domain became a province. Today the former Soviet Socialist Republic is the independent state of Azerbaijan (population: 7.8 million), and the 10 million Azeris to the south live in the Iranian province.

During the brief transition to independence and at the height of their war with the Armenians, the dominantly Muslim Azeris tended to look southward, toward Iran. But geographic realities dictate a more practical orientation. Azerbaijan possesses huge reserves of oil and natural gas; under the Soviets it was one of Moscow's chief regional sources of fuels. The center of the oil industry is Baki (Baku), the capital on the shore of the Caspian Sea—but the Caspian Sea is a lake. To export its oil, Azerbaijan needs pipelines, but those of Soviet vintage link Baki to Russia's Black Sea terminal of Novorossiysk. During the mid-1990s, when Azerbaijan's leaders announced plans to build new pipelines via Georgia or Iran, Russia objected and even meddled in Azeri politics to stymie such plans.

A look at the map (Fig. 2-13) shows Azerbaijan's options. A pipeline could be routed from Baki through Georgia to the Black Sea coast near Batumi, avoiding Russian territory altogether. Another route could run through Georgia and across Turkey to the Mediterranean terminal at Ceyhan. Still another option would run the pipeline across Iran to a terminal on the Persian Gulf. Moscow objects to such alternatives, but Azerbaijan now has powerful international allies. American, French, British, and Japanese oil companies are developing Azerbaijan's offshore Caspian reserves, and the U.S. government prefers the Turkish route.

All these alternate routes are in the future, however; for the present, Azerbaijan's oil must flow across Russian territory, and Azerbaijan must cooperate with Moscow. Given its Muslim roots and its location, the time may come when Azerbaijan turns toward the Islamic world. For the moment, however, this country is inextricably bound up with its neighbors in Transcaucasia and the north.

◻▷ THE EASTERN FRONTIER

From the eastern flanks of the Ural Mountains to the headwaters of the Amur River, and from the latitude of Tyumen to the northern zone of neighboring

Kazakhstan, lies Russia's vast Eastern Frontier Region, product of a gigantic experiment in the eastward extension of the Russian Core (Fig. 2-10). As the maps of cities and surface communications suggest, this eastern frontier is more densely peopled and more fully developed in the west than in the east; at the longitude of Lake Baykal, settlement has become linear, marked by ribbons and clusters along the east-west railroads. Two subregions dominate the geography: the Kuznetsk Basin in the west and the Lake Baykal area in the east.

The Kuznetsk Basin (Kuzbas)

Some 900 miles (1450 km) east of the Urals lies another of Russia's primary regions of heavy manufacturing resulting from the communist period's national planning: the Kuznetsk Basin, or *Kuzbas* (Fig. 2-11). In the 1930s, it was opened up as a supplier of raw materials (especially coal) to the Urals, but that function became less important as local industrialization accelerated. The original plan was to move coal from the Kuzbas west to the Urals and allow the returning trains to carry iron ore east to the coalfields. However, good-quality iron ore deposits were subsequently discovered near the Kuznetsk Basin itself. As the new resource-based Kuzbas industries grew, so did its urban centers. The leading city, located just outside the region, is Novosibirsk, which stands at the intersection of the Trans-Siberian Railroad and the Ob River as the symbol of Russian enterprise in the vast eastern interior. To the northeast lies Tomsk, one of the oldest Russian towns in all of Siberia, founded in the seventeenth century and now caught up in the modern development of the Kuzbas Region. Southeast of Novosibirsk lies Novokuznetsk, a city that produces steel for the region's machine and metal-working plants and aluminum products from Urals bauxite.

The Lake Baykal Area (Baykaliya)

East of the Kuzbas, development becomes more insular, and distance becomes a stronger adversary. North of the Tyva Republic and eastward around Lake Baykal, larger and smaller settlements cluster along the two railroads to the Pacific coast (Fig. 2-11). West

of the lake, these rail corridors lie in the headwater zone of the Yenisey River and its tributaries. A number of dams and hydroelectric projects serve the valley of the Angara River, particularly the city of Bratsk. Mining, lumbering, and some farming sustain life here, but isolation dominates it. The city of Irkutsk, near the southern end of Lake Baykal, is the principal service center for a vast Siberian region to the north and for a lengthy east-west stretch of southeastern Russia.

Beyond Lake Baykal, the Eastern Frontier really lives up to its name: this is southern Russia's most rugged, remote, forbidding country. Settlements are rare, many being mere camps. The Buryat Republic (Fig. 2-7) is part of this zone; the territory bordering it to the east was taken from China by the czars and may become an issue in the future. Where the Russian-Chinese boundary turns southward, along the Amur River, the region called the Eastern Frontier ends and Russia's Far East begins.

▷ SIBERIA

Before we assess the potential of Russia's Pacific Rim, we should remember that the ribbons of settlement just discussed hug the southern perimeter of this giant country, avoiding the vast Siberian region to the north (Fig. 2-10). Siberia extends from the Ural Mountains to the Kamchatka Peninsula—a vast, bleak, frigid, forbidding land. Larger than the conterminous United States but inhabited by only an estimated 15 million people, Siberia quintessentially symbolizes the Russian environmental plight: vast distances, cold temperatures worsened by strong Arctic winds, difficult terrain, poor soils, and limited options for survival.

But Siberia also has resources. From the days of the first Russian explorers and Cossack adventurers, Siberia's riches have beckoned. Gold, diamonds, and other precious minerals were found. Later, metallic ores including iron and bauxite were discovered. Still more recently, the Siberian interior proved to contain sizeable quantities of oil and natural gas (Fig. 2-12) and began to contribute significantly to Russia's energy supply.

As the physiographic map (Fig. 2-2) shows, major rivers—the Ob, Yenisey, and Lena—flow

gently northward across Siberia and the Arctic Lowland into the Arctic Ocean. Hydroelectric power development in the basins of these rivers has generated electricity used to extract and refine local ores, and in the lumber mills that have been set up to exploit the vast Siberian forests.

The human geography of Siberia is fragmented, and much of the region is virtually uninhabited (Fig. 2-4). Ribbons of Russian settlement have developed; the Yenisey River, for instance, can be traced on this map of Soviet peoples (a series of small settlements north of Krasnoyarsk), and the upper Lena Valley is similarly fringed by ethnic Russian settlement. Yet hundreds of miles of empty territory separate these ribbons and other islands of habitation.

The political geography of eastern Siberia is marked by the growing identity of Sakha (the Yakut Republic). As additional resources are discovered here (including oil and natural gas), this Republic, centered on the capital, Yakutsk, will become more important.

Siberia, Russia's freezer, is stocked with goods that may become mainstays of future national development. Already, precious metals and mineral fuels are bolstering the Russian economy. In time, we may expect Siberian resources to play a growing role in the economic development of the Eastern Frontier and the Russian Far East as well. One step in that process was already taken during Soviet times: the completion of the BAM (Baykal-Amur Mainline) Railroad in the 1980s. This route, lying north of and parallel to the old Trans-Siberian Railroad, extends 2200 miles (3540 km) eastward from Tayshet (near the important center of Krasnoyarsk) directly to the Far East city of Komsomolsk (Fig. 2-6). In the post-Soviet era, the BAM Railroad has been beset by equipment breakdowns and workers' strikes. Nonetheless, it is a key element of the infrastructure that will serve the Eastern Frontier's economic growth in the twenty-first century.

▷ THE RUSSIAN FAR EAST

Imagine this: a country with 5000 miles (8000 km) of Pacific coastline, two major ports, major interior cities nearby, huge reserves of resources ranging from minerals to fuels to timber, directly across from one of the world's largest economies—all this at a time when the Asian Pacific Rim was the world's fastest-growing economic region. Would not that country have burgeoning cities, busy harbors, growing industries, and expanding trade?

In the Russian Far East (Fig. 2-10), the answer is—no. Activity in the port of Vladivostok is a shadow of what it was during the Soviet era, when it was the communists' key naval base. The nearby container terminal at Nakhodka suffers from breakdowns and inefficiencies. The railroad to western Russia carries just a fraction of the trade it did during the 1970s and 1980s. Cross-border trade with China is minimal. Trade with Japan is inconsequential. The region's cities are grimy, drab, moribund. Utilities are shut off for hours at a time because of fuel shortages and system breakdowns. Outdated factories are shut, their workers dismissed. Political relations with Moscow are poor. There is potential here, but little of it has been realized.

As a region, the Russian Far East consists of two parts: the mainland area extending from Vladivostok to the Stanovoy Mountains and the large island of Sakhalin (Figs. 2-1; 2-14). This is cold country: icebreakers have to keep the ports of Vladivostok and Nakhodka open throughout the winter. Winters here are long and bitterly cold; summers are brief and cool. Although the population is small (about 7 million), food must be imported because not much can be grown. Most of the region is rugged, forested, and remote. Vladivostok, Khabarovsk, and Komsomolsk are the only cities of any size. Nakhodka and the newer railroad terminal at Vanino are smaller towns; the population of the whole island of Sakhalin is about 700,000 (on an island

the size of Caribbean Hispaniola [15.4 million]). Offshore lie productive fishing grounds, and Russian fleets from Vladivostok and points north catch salmon, herring, cod, and mackerel to be frozen or canned and shipped to local and distant markets.

The Soviet regime rewarded people willing to move to this region with housing and subsidies. The communists, like the czars before them, realized the importance of this frontier (Vladivostok means "We Own the East"), and they used every possible incentive to develop it and link it ever closer to Russia's distant western core. Freight rates on the Trans-Siberian Railroad, for example, were about 10 percent of their real costs; the trains were always loaded in both directions. Vladivostok was a military base and a city closed to foreigners, and Moscow invested heavily in its infrastructure. Komsomolsk in the north and Khabarovsk near the region's center were endowed

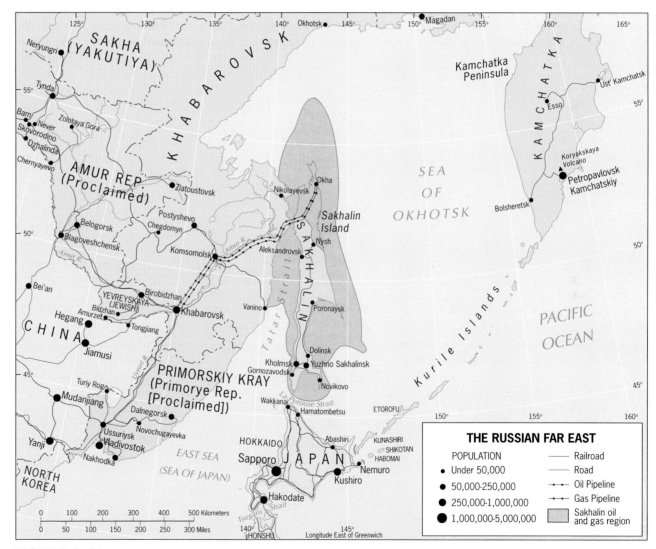

FIGURE 2-14

with state-owned industries using local resources: iron ore from Komsomolsk, oil from Sakhalin, timber from the ubiquitous forests. The steel, chemical, and furniture industries sent their products westward by train, and they received food and other consumer goods from the Russian heartland.

For several reasons, the post-Soviet transition has been especially difficult here in the Far East. The new economic order has canceled the region's communist-era advantages: the Trans-Siberian Railroad now must charge the real cost of transporting products from the Eastern Frontier to the Russian Core. State-subsidized industries must compete on market principles, their subsidies having ended. The decline of Russia's armed forces has hit Vladivostok hard. The fleet lies rusting in port; service industries have lost their military markets; the shipbuilding industry has no government contracts. Coal miners in the Bureya River Valley (a tributary of the Amur) go unpaid for months and go on strike; coal-fired power plants do not receive fuel shipments, and cities and towns go dark.

Locals put much of the blame for their region's failure on Moscow, and with reason. As Figure 2-7 shows, the Far East contains only five administrative regions: Primorskiy Kray, Khabarovsk Kray, Amur Oblast, Sakhalin Oblast, and Yevreyskaya, originally the Jewish Autonomous Region. This does not add up to much political clout, which is the way Moscow appears to want it. Compare Figures 2-8 and 2-14 and you will note that the city of Khabarovsk, not the Primorskiy Kray capital of Vladivostok, has been made the headquarters of the Far Eastern Region as defined by Moscow.

For all its stagnation, Russia's Far East will figure prominently in Russian (and probably world) affairs. Here Russia meets China on land and Japan at sea (an unresolved issue between Russia and Japan involves several Kurile islands). Here lie vast resources ranging from Sakhalin's fuels to Siberia's lumber, Sakha's gold to Khabarovsk's metals. Here Russia has a foothold on the Pacific Rim, a window on the ocean on whose shores the world is being transformed.

This Chapter's Media Highlights include:

Photo Gallery: North America

GeoDiscoveries: Urbanization and Sprawl

3 Interactive Map Quizzes: States/Provinces, Cities, Physical Features

Chapter Quiz

Concept Quiz

Virtual Field Trip: California to St. Louis Via Route 66 www.wiley.com/college/deblij

Boston, Massachusetts: the cultural landscape of a bygone era survives in the heart of a burgeoning metropolis.

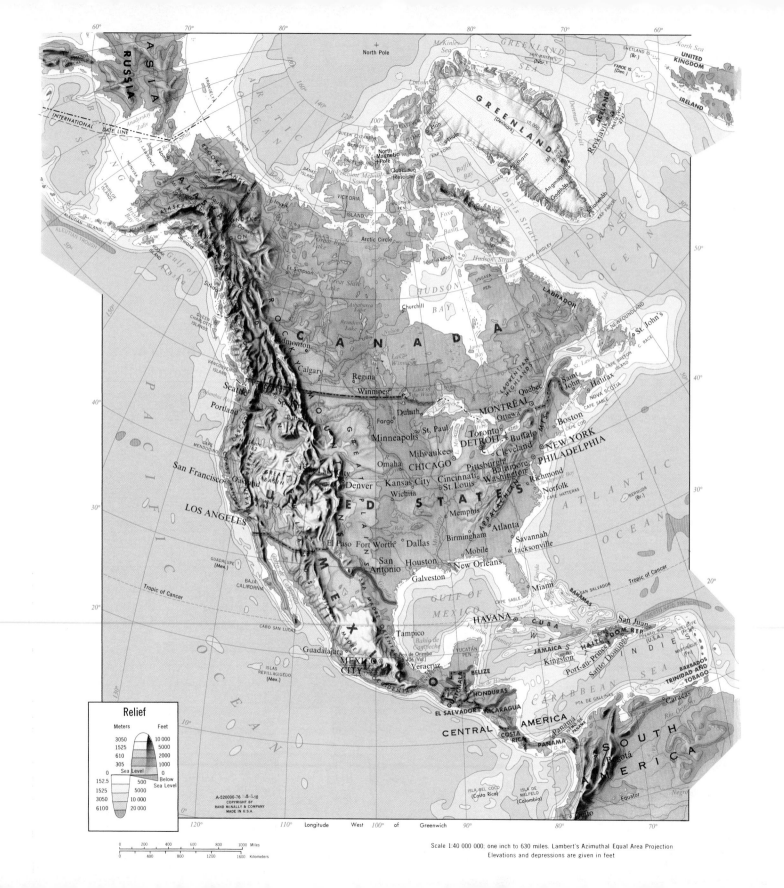

RUSSIA

ASIA

UNITED KINGDOM

IRELAND

GREENLAND
(Denmark)

North Pole

ARCTIC OCEAN

ALASKA

BROOKS RANGE

BAFFIN ISLAND

CANADA

HUDSON BAY

LABRADOR

St. John's

PACIFIC OCEAN

Edmonton

Calgary

Regina

Winnipeg

Seattle

Portland

Spokane

Duluth

Fargo

Minneapolis

St. Paul

Milwaukee

CHICAGO

Omaha

MONTRÉAL

Ottawa

Québec

Toronto

DETROIT

Cleveland

Buffalo

Boston

NEW YORK

PHILADELPHIA

Saint John

Halifax

NOVA SCOTIA

San Francisco

Oakland

Salt Lake City

Denver

Kansas City

St. Louis

Cincinnati

Pittsburgh

Baltimore

Washington

Richmond

Norfolk

UNITED STATES

Wichita

Memphis

APPALACHIAN MTS.

ATLANTIC OCEAN

BERMUDA
(Br.)

LOS ANGELES

Birmingham

Atlanta

Savannah

Jacksonville

El Paso

Fort Worth

Dallas

San Antonio

Houston

Galveston

New Orleans

Mobile

Miami

BAHAMAS

GUADALUPE
(Mex.)

BAJA CALIFORNIA

Tropic of Cancer

M E X I C O

Guadalajara

MEXICO CITY

Tampico

Veracruz

GULF OF MEXICO

YUCATÁN PEN.

HAVANA

CUBA

San Salvador

HAITI

DOM. REP.

San Juan

PUERTO RICO
(U.S.A.)

JAMAICA

Kingston

Port-au-Prince

Santo Domingo

W E S T   I N D I E S

CARIBBEAN SEA

BELIZE

GUATEMALA

HONDURAS

EL SALVADOR

NICARAGUA

CENTRAL   AMERICA

COSTA RICA

PANAMA

BARBADOS

TRINIDAD AND TOBAGO

Caracas

SOUTH AMERICA

Bogotá

Equator

ISLA DEL COCO
(Costa Rica)

ISLA DE MALPELO
(Colombia)

International Date Line

Arctic Circle

North Magnetic Pole

Tropic of Cancer

Longitude   West   of   Greenwich

Relief

Meters		Feet
3050		10 000
1525		5000
610		2000
305		1000
0	Sea Level	0
152.5		Below Sea Level
1525		500
3050		5000
6100		10 000
		20 000

A-520000-76 -5--18
COPYRIGHT BY
RAND McNALLY & COMPANY
MADE IN U.S.A.

0   200   400   600   800   1000 Miles

0   400   800   1200   1600 Kilometers

Scale 1:40 000 000; one inch to 630 miles. Lambert's Azimuthal Equal Area Projection
Elevations and depressions are given in feet

chapter 3 / North America

CONCEPTS, IDEAS, AND TERMS

1 Cultural pluralism

2 Physiographic province

3 Rain shadow effect

4 Migration

5 Push/pull factors

6 Sunbelt

7 American Manufacturing Belt

8 Ghetto

9 Outer city

10 Suburban downtown

11 Urban realms model

12 Mosaic culture

13 Productive activities

14 Fossil fuels

15 Economies of scale

16 Postindustrialism

17 Technopole

18 Ecumene

19 Regional state

20 Pacific Rim

REGIONS

NORTH AMERICAN CORE

MARITIME NORTHEAST

FRENCH CANADA

CONTINENTAL INTERIOR

SOUTH

SOUTHWEST

WESTERN FRONTIER

NORTHERN FRONTIER

PACIFIC HINGE

The North American realm consists of two countries, the United States and Canada, which are alike in many ways. Culturally, most (but not all) of the people trace their ancestries to various European countries, and the realm is often called *Anglo-America*, with English serving as the dominant language of the United States and officially sharing equal status with French in Canada. Economically, both rank among the world's most highly advanced by every measure of national development, and they continue to benefit from a still-intensifying trade relationship that functions across the longest open international boundary on Earth.

With more than 75 percent of the population of the United States and Canada residing in towns and cities,* North America's societies have become the world's most highly urbanized. Indeed, nothing symbolizes them as strongly as the skyscrapered

*The population and activity agglomerations we call cities are changing throughout the world as the urban areas they anchor continue to grow. The term *city* applies to the older core or central city; the term *metropolitan* (also *metropolis*) refers to the entire urban complex consisting of the core city and its ring of newer surrounding suburbs. As indicated in the Preface, all urban population data provided in this book are metropolitan-area totals unless otherwise specified.

panoramas of New York, Toronto, or Los Angeles—and, increasingly, the booming new suburban business complexes that are transforming their outer metropolitan rings. North Americans also are hypermobile, with networks of superhighways, commercial air routes, and state-of-the-art telecommunications media efficiently interconnecting the realm's far-flung cities and regions.

At the opening of the twenty-first century, North America has entered a new age, the third since Columbus arrived in the New World more than 500 years ago. Agriculture and rural life dominated the first four centuries; the second age—industrial

◆Major Geographic Qualities of North America

1. North America encompasses two of the world's biggest states territorially (Canada is the second largest in size; the United States is third).

2. Both Canada and the United States are federal states, but their systems differ. Canada's is adapted from the British parliamentary system and is divided into ten provinces and three territories. The United States separates its executive and legislative branches of government, and it consists of 50 States, the Commonwealth of Puerto Rico, and a number of island territories under U.S. jurisdiction in the Caribbean Sea and the Pacific Ocean.

3. Both Canada and the United States are plural societies. Although ethnicity is increasingly important, Canada's pluralism is most strongly expressed in regional bilingualism. In the United States, major divisions occur along racial lines.

4. A large number of Quebec's French-speaking citizens supports a movement that seeks independence for the province. The movement's high-water mark may have been reached in the 1995 referendum in which (minority) non-French speakers were the difference in the narrow defeat of separation. The prospects for a break-up of the Canadian state have diminished considerably since 2000.

5. North America's population, not large by international standards, is the world's most highly urbanized and mobile. Largely propelled by a continuing wave of immigration, the realm's population total is expected to grow by more than 40 percent over the next half-century.

6. By world standards, this is a rich realm where high incomes and high rates of consumption prevail. North America possesses a highly diversified resource base, but nonrenewable fuel and mineral deposits are consumed prodigiously.

7. North America is home to one of the world's great manufacturing complexes. The realm's industrialization generated its unparalleled urban growth, but a new postindustrial society and economy are rapidly maturing in both countries.

8. The two countries heavily depend on each other for supplies of critical raw materials (e.g., Canada is the leading source of U.S. energy imports) and have long been each other's chief trading partners. Today, the North American Free Trade Agreement (NAFTA), which also includes Mexico, is linking all three economies ever more tightly as barriers to international trade and investment flow are steadily being dismantled.

9. North Americans are the world's most mobile people. Although plagued by recurrent congestion problems, the realm's networks of highways, commercial air routes, and cutting-edge telecommunications are the most efficient on Earth.

urbanization—spanned the now-ended twentieth century. In its aftermath, the United States and Canada are today experiencing the maturation of a *postindustrial society and economy*, which is dominated by the production and manipulation of information, skilled services, and high-technology manufactures, and operates within an ever more globalized framework of business interactions.

Although Canada and the United States share many historical, cultural, and economic qualities, they also differ in significant ways, as can be seen on the map. The United States, somewhat smaller territorially than Canada, occupies the heart of the North American continent and, as a result, encompasses a greater environmental range. The U.S. population is dispersed across most of the country, forming major concentrations along both the (north-south-trending) Atlantic and Pacific coasts. The overwhelming majority of Canadians, however, live in an interrupted east-west corridor that lies across southern Canada, mainly within 200 miles (320 km) of the U.S. border.

Differences between the two countries also become apparent when we examine population characteristics that form the basis for important internal social contrasts. The population of the United States in 2002 was 285.4 million; Canada's was 31.0 million, just over one-tenth as large. Although comparatively small, Canada's population is divided by culture and tradition, and this division has a pronounced regional expression. About 84 percent of Canada's citizens speak English, 31 percent speak French, and as many as 10 percent other languages (multilinguality affects the percentage total, with 17 percent speaking both English and French). *First Nations* (indigenous peoples of Amerindian descent, whose U.S. counterparts are called Native Americans) and *Inuit* (peoples of the Arctic zone, formerly called Eskimos) make up approximately 2 percent of the total.

The strong spatial clustering of *Francophones* (French speakers) in Quebec, the second most populous of the country's ten provinces, accentuates Canada's social division along ethnic and linguistic lines. Nearly 85 percent of this province's population is French Canadian, and Quebec is the historic, traditional, and emotional focus of French culture in Canada. Over the past few decades, a strong nation-

alist movement has emerged in Quebec together with demands for outright separation from the rest of Canada—a movement that peaked in the mid-1990s and is now declining.

No multilingual divisions affect the unity of the **1** realm's other federation, but **cultural pluralism** of another kind prevails in the United States. More persistent in the U.S. cultural mosaic than language or ethnicity is the division between peoples of European descent (more than 75 percent of the popula-

tion) and those of African origin (12 percent). Despite the significant progress of the modern civil rights movement, which decidedly weakened *de jure* racial segregation in the public domain, whites so rarely share their living space with blacks that *de facto* residential segregation is all but universal. Deprivation, too, is surprisingly widespread, with notable spatial concentrations of poverty inside the inner-ring slums of central cities and on rural reservations where Native Americans or Inuit reside.

DEFINING THE REALM

NORTH AMERICA'S PHYSICAL GEOGRAPHY

Before we examine the realm's human geography, we need to consider its varied natural environment (Fig. 3-1) that covers a vast zone extending from the near-tropical latitudes of southern Florida and Texas to subpolar Alaska and Canada's far-flung northern periphery.

Physiography

North America's physiography is characterized by its clear, well-defined division into physically **2** homogeneous regions called **physiographic provinces**. Each region is marked by considerable uniformity in relief, climate, vegetation, soils, and other environmental conditions, resulting in a scenic sameness that comes readily to mind. For example, we identify such regions when we refer to the Rocky Mountains or the Great Plains.

Figure 3-2 maps the realm's physiography and includes a cross-sectional terrain profile along the 40th parallel. The most obvious aspect of this map of North America's physiographic provinces is the north-south alignment of the great mountain backbone formed by the Rocky Mountains, whose rugged topography dominates the western segment of the realm from 📷 Alaska to New Mexico. The major feature of

eastern North America is another, much lower chain of mountain ranges called the Appalachian Highlands, which also trend approximately north-south and extend from Canada's Atlantic Provinces to Alabama. The orientation of the Rockies and Appalachians is important because, unlike Europe's Alps, they do not form a topographic barrier to polar or tropical airmasses flowing across the continent's interior.

Between the Rocky Mountains and the Appalachians lie North America's vast interior plains, which extend from the Mackenzie Delta on the Arctic Ocean to the Gulf of Mexico. We can subdivide these into several provinces: (1) the great Canadian Shield, which is the geologic core area containing North America's oldest rocks; (2) the Interior Lowlands, covered largely by glacial debris laid down by meltwater and wind during the Pleistocene glaciation; and (3) the Great Plains, the extensive sedimentary surface that slowly rises westward toward the Rocky Mountains. Along the southern margin, these interior plainlands merge into the Gulf-Atlantic Coastal Plain, which stretches from southern Texas along the seaward margin of the Appalachian Highlands and the neighboring Piedmont until it ends at Long Island just to the east of New York City.

On the western side of the Rocky Mountains lies the zone of Intermontane Basins and Plateaus. Within the conterminous United States, this physiographic province includes the Colorado Plateau in

the south, the lava-covered Columbia Plateau in the north, and the central Basin-and-Range country of Nevada and Utah. To the west of the intermontane province, the entire west coast of North America is dominated by an almost unbroken corridor of high mountain ranges that originated from contact between the North American and Pacific Plates (Fig. I-4). The major components of this coastal mountain belt include California's Sierra Nevada, the Cascades of Oregon and Washington, and the long chain of highland massifs that line the British Columbia and southern Alaska coasts. Three broad valleys—which contain dense populations—are the only noteworthy interruptions: California's Central (San Joaquin-Sacramento) Valley; the Puget Sound lowland of Washington State that merges southward into western Oregon's Willamette Valley; and the lower Fraser Valley, which slices through southern British Columbia's coast range.

Climate

The world climate map (Fig. I-8) clearly depicts the various climatic regimes and regions of North America. In general, temperature varies latitudinally—the farther north one goes, the cooler it gets. Regional land-and-water-heating differentials, however, distort this broad pattern. Because land surfaces heat and cool far more rapidly than water bodies, yearly temperature ranges are much larger where *continentality*

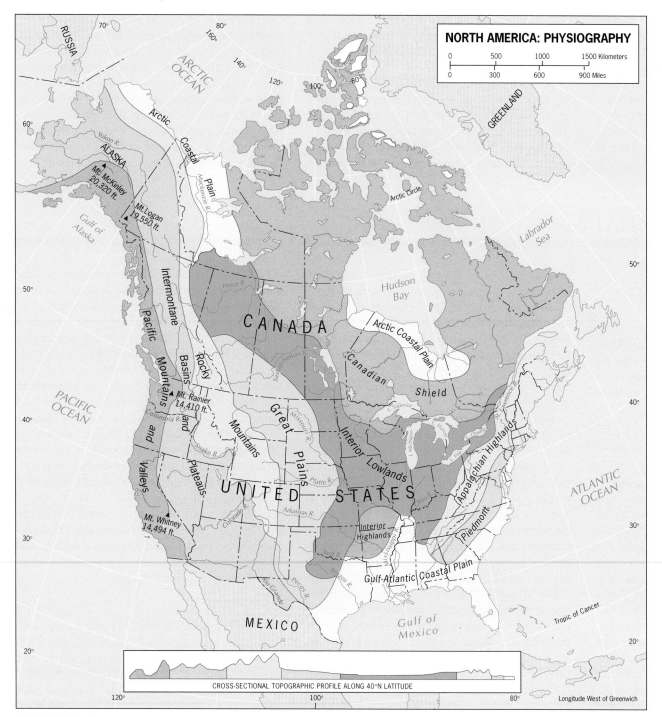

NORTH AMERICA: PHYSIOGRAPHY

FIGURE 3-2

CROSS-SECTIONAL TOPOGRAPHIC PROFILE ALONG 40°N LATITUDE

Longitude West of Greenwich

(interior remoteness from the sea) is greatest.

Precipitation generally tends to decline toward the west (except for the Pacific coastal strip itself) as a result of **3** the **rain shadow effect**. This occurs because Pacific airmasses, driven by prevailing winds, carry their moisture onshore but soon collide with the Sierra Nevada–Cascades wall, forcing them to rise—and cool—in order to crest these mountain ranges. Such cooling is accompanied by major condensation and precipitation, so that by the time these airmasses descend and warm along the eastern slopes to begin their journey across the continent's interior, they have deposited much of their moisture. Thus the mountains produce a downwind "shadow" of dryness, which is reinforced whenever eastward-moving air must surmount other ranges farther inland, especially the massive Rockies. Indeed, this semiarid (and in places truly arid) environment extends so deeply into the central United States that a broad division can be made between Arid (western) and Humid (eastern) America, which face each other along a fuzzy boundary that is best viewed as a wide transitional zone. Although the separating criterion of 20 inches (50 cm) of annual precipitation is easily mapped (see Fig. I-7), that generally north-south *isohyet* (the line connecting all places receiving exactly 20 inches per year) can and does swing widely across the drought-prone Great Plains from year to year because highly variable warm-season rains from the Gulf of Mexico come and go in unpredictable fashion.

Precipitation in Humid America is far more regular. The prevailing winds that blow from west to east, which normally come up dry for the large zone west of the 100th meridian, pick up considerable moisture over the Interior Lowlands and distribute it throughout eastern North America. Many storms develop here on the active weather front between tropical Gulf air to the south and polar air to the north. Even if major storms do not materialize, local weather disturbances created by sharply contrasting temperature differences are always a danger. Moreover, in winter the northern half of this region receives large amounts of snow, particularly around the Great Lakes.

The broad environmental partitioning into Humid and Arid America is also reflected in the distribution of the realm's soils and vegetation. For farming purposes there is usually sufficient soil moisture to support crops where annual precipitation exceeds the critical 20 inches. Where the yearly total is less, soils may still be fertile (especially in the Great Plains), but irrigation is often necessary to achieve their full agricultural potential. As for vegetation, the Humid/Arid America dichotomy is again a valid generalization: the natural vegetation of areas receiving more than 20 inches of water annually is *forest*, whereas the drier climates give rise to a *grassland* cover.

Hydrography (Surface Water)

Surface water patterns in North America are dominated by the two major drainage systems that lie between the Rockies and the Appalachians: (1) the five Great Lakes (Superior, Michigan, Huron, Erie, and Ontario) that drain into the St. Lawrence River, and (2) the mighty Mississippi-Missouri River network, fed by such major tributaries as the Ohio, Tennessee, and Arkansas rivers. Both are products of the last episode of Pleistocene glaciation, and together they amount to nothing less than the best natural inland waterway system in the world.

Elsewhere, the northern east coast of the continent is well served by a number of short rivers leading inland from the Atlantic. In fact, a number of northeastern seaboard cities, including Washington, D.C. and Philadelphia, are located at the waterfalls that marked the limit to tidewater navigation (hence their designation as *Fall Line cities*). Rivers in the Southeast and west of the Rockies at first offered little practical value because of their orientation and the difficulty of navigating them. In the Far West, however, the Colorado and Columbia rivers have become important as

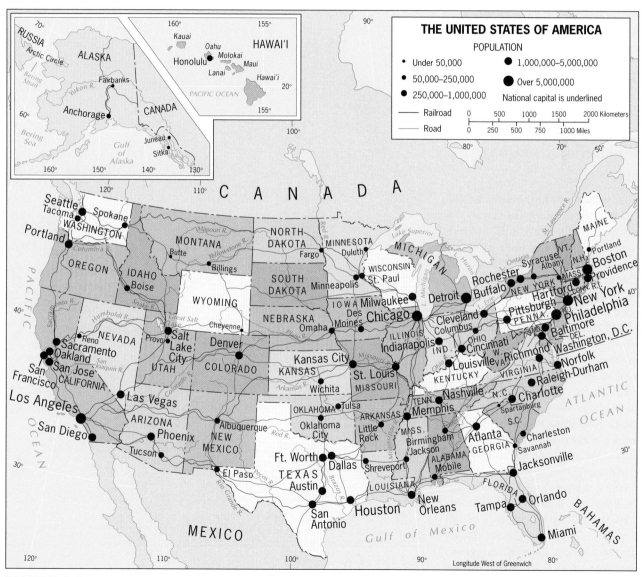

FIGURE 3-3

suppliers of drinking and irrigation water as well as hydroelectric power.

THE UNITED STATES

The broad outline of North America we have just sketched will be useful in delineating the realm's internal regional organization, but first we need to examine the changing human geography of each country. We begin with the United States, and it may be helpful to take a few moments to review the map of its basic contents (Fig. 3-3).

Population in Time and Space

The population distribution of the United States is shown in Figure 3-4, a map that is the latest "still" in a motion picture that has been unreeling for nearly four centuries since the founding of the first permanent European settlements on the northeastern coast. Slowly at first and then with accelerating speed after 1800, Americans (and Canadians) took charge of their remarkable continent and pushed the settlement frontier westward to the Pacific. The swiftness of this expansion was dramatic, but North Americans have long been the world's most mobile people.

To understand the current population map, we need to review the major forces that have shaped, and continue to shape, the distribution of Americans and their activities. Since its earliest days, the United States has attracted a steady influx of immigrants who were rapidly assimilated into the societal mainstream. Within the coun-

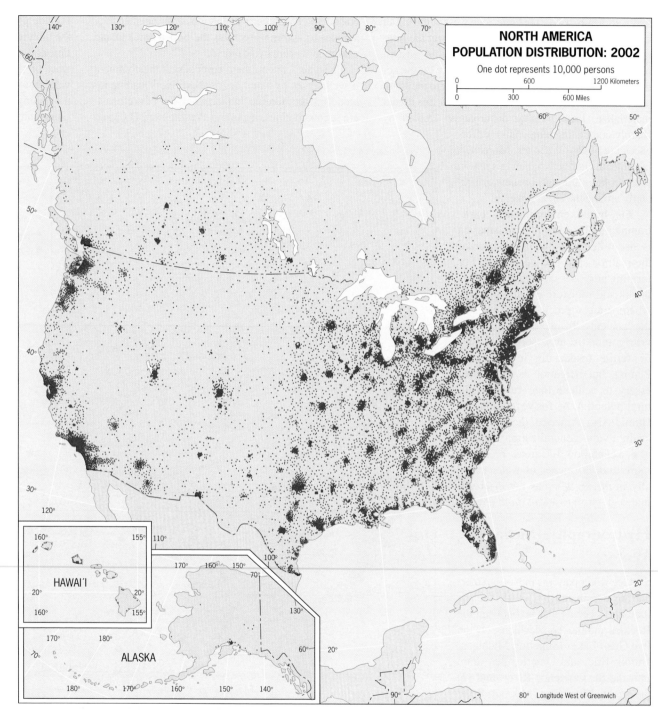

NORTH AMERICA POPULATION DISTRIBUTION: 2002

One dot represents 10,000 persons

FIGURE 3-4

try, people have sorted themselves out to maximize their proximity to existing economic opportunities, and they have shown little resistance to relocating as the nation's evolving economic geography has successively favored different sets of places over time. These movements continue to redistribute millions of Americans, and before we trace the evolution of the population map we need to acquaint ourselves with the process of migration.

The Role of Migration

4 **Migration** refers to a change in residential location intended to be permanent, and it has long played a key role in the human geography of this realm. After thousands of years of native settlement, European explorers reached North American shores beginning with Columbus in 1492 (and probably centuries before that). The first permanent colonies were established in the early 1600s, and from them evolved the modern United States of America. The Europeanization of North America doomed the aboriginal societies, but this was only one of many areas around the world where local cultures and foreign in-

vaders came face to face. Between 1835 and 1935, perhaps as many as 75 million Europeans departed for distant shores—most of them bound for the Americas (Fig. 3-5). Some sought religious freedom; others escaped poverty and famine; still others simply hoped for a better life.

Studies of the *migration decision* indicate that migration flows vary in size with: (1) the perceived degree of difference between one's home, or source, and the destination; (2) the effectiveness of flows of information about the destination which migrants sent to those who stayed behind waiting to decide; and (3) the distance between the source and the destination. More than a century ago, the British social scientist Ernst Georg Ravenstein studied the migration process, and many of his conclusions remain valid today. For example, every migration stream from source to destination produces a counter-stream of returning migrants who cannot adjust. Studies of migration also conclude that several factors are at **5** work in the migration process. **Push factors** motivate people to move away; **pull factors** attract them to new destinations. To those early Europeans the United States was a new frontier, a place where one

might acquire a piece of land, and the opportunities were reported to be unlimited. That perception has never changed, and international immigration continues to significantly shape the demographic complexion of the United States in the twenty-first century (with Middle America and eastern Asia now replacing Europe as the dominant source).

Within the United States today, the leading migration flow that continues to transform the population map is the persistent drift of people and livelihoods toward the West and South—the so-called **6** **Sunbelt**. In addition, a pair of lesser but longstanding migratory flows still play a major role: (1) the rapid growth of metropolitan areas, first triggered by the late-nineteenth-century Industrial Revolution, which since the 1960s has been largely rechanneled from the central cities to the suburban ring; and (2) the movement of African Americans from the rural South to the urban North, which since the 1970s has become a stronger return flow.

Let us now look more closely at the historical geography of these changing population patterns, considering first the initial rural influence and then the decisive impacts of industrial urbanization.

Pre-Twentieth-Century Population Patterns

Indigenous North America   When the first Europeans set foot on its soil, the North American realm was occupied by millions of people whose ancestors had reached the Americas from Asia, via Alaska and probably also across the Pacific, more than 13,000 years (and possibly as long as 30,000 years) ago. In North America these Native Americans or First Nations, as they are now called in the United States and Canada, respectively, had organized themselves into hundreds of nations with a rich mosaic of languages and a great diversity of cultures (Fig. 3-6).

The eastern nations first bore the brunt of the European invasion. By

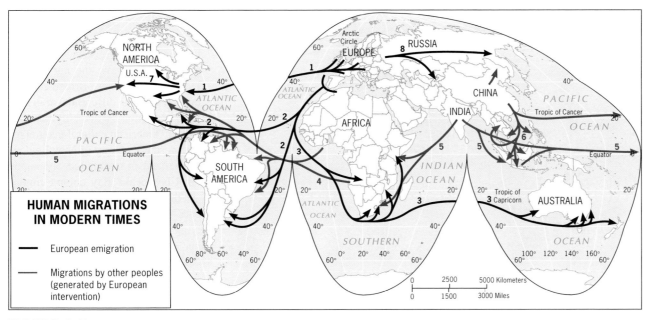

HUMAN MIGRATIONS IN MODERN TIMES

—— European emigration

—— Migrations by other peoples (generated by European intervention)

FIGURE 3-5

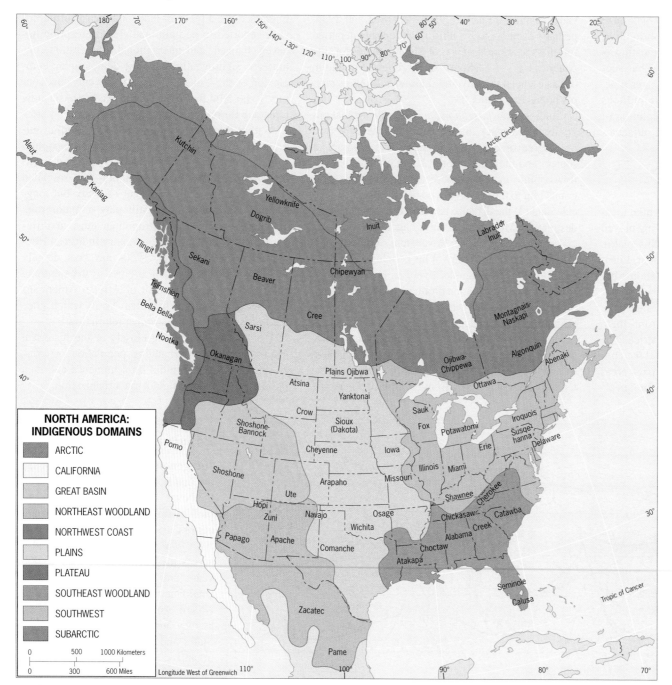

**NORTH AMERICA:
INDIGENOUS DOMAINS**

- ARCTIC
- CALIFORNIA
- GREAT BASIN
- NORTHEAST WOODLAND
- NORTHWEST COAST
- PLAINS
- PLATEAU
- SOUTHEAST WOODLAND
- SOUTHWEST
- SUBARCTIC

0   500   1000 Kilometers
0   300   600 Miles

Longitude West of Greenwich

FIGURE 3-6

the end of the eighteenth century, land-hungry settlers had driven most of the Native American peoples living along the Atlantic and Gulf coasts from their homes and lands, beginning a westward push that was to devastate indigenous society. As early as 1789, the U.S. Congress proclaimed that "Indian . . . land and property shall never be taken from them without their consent," but in fact this is just what happened as the European settlement frontier marched steadily toward the Pacific. By the end of the nineteenth century, decades of war left what remained of North America's native peoples with about 4 percent of U.S. territory in the form of mostly impoverished reservations.

Colonial and Nineteenth-Century Development The current distribution of the U.S. population is rooted in the colonial era of the seventeenth and eighteenth centuries that was dominated by England and France. The French sought mainly to organize a lucrative fur-trading network, while the English established settlements along what is now the northeastern U.S. seaboard. These British colonies quickly became differentiated in their local economies, a diversity that later shaped American cultural geography. The northern colony of New England specialized in commerce; the southern Chesapeake Bay colony emphasized the plantation farming of tobacco; and the Middle Atlantic area lying in between was home to a number of smaller, independent-farmer colonies.

These neighboring colonies soon thrived and yearned to expand, but the British government responded by closing the inland frontier and tightening economic controls. By 1783 this had led to colonial unification, British defeat in the Revolutionary War, and independence for the newly formed United States of America. The western frontier of the fledgling nation now swung open, and the zone north of the Ohio River was promptly settled following the discovery that the soils of the Interior Lowlands were more favorable for farming than those of the Atlantic seaboard. This triggered the rapid growth of trans-Appalachian agriculture and coastal-interior trading ties as U.S. spatial organization assumed national-scale proportions.

By the time the westward-moving frontier swept across the Mississippi Valley in the 1820s, the three former seaboard colonies (**A**, **B**, and **C** in Fig. 3-7) had become separate *culture hearths*—primary source areas and innovation centers from which migrants carried cultural traditions into the central United States (as the arrows in Fig. 3-7 show). The northern half of this vast interior space soon became well unified as its infrastructure steadily improved following the introduction of the railroad in the 1830s. The American South, however, did not wish to integrate itself economically with the North, preferring to export tobacco and cotton from its plantations to overseas markets. Its insistence on preserving slavery to support this system soon led the South into secession, the disastrous Civil War (1861–1865), and a dismal aftermath that took a full century to overcome.

The second half of the nineteenth century saw the frontier cross the western United States, and by 1869 agriculturally booming California was linked to the rest of the nation by transcontinental railroad (these same steel tracks also opened up the bypassed, semi-arid Great Plains). When the American frontier closed in the 1890s, today's rural settlement pattern was firmly in place, anchored to a set of enduring national agricultural regions (discussed later in this chapter). By then, however, the exodus of rural Americans toward the burgeoning cities had begun in response to the Industrial Revolution that had taken hold after 1870.

Post–1900 Industrial Urbanization

The Industrial Revolution occurred almost a century later in the United States than in Europe, but when it finally did cross the Atlantic in the 1870s, it took hold so successfully and advanced so robustly that only 50 years later America was surpassing Europe as the world's mightiest industrial power. The impact of industrial urbanization occurred simultaneously at two levels of generalization. At the national or *macroscale*, a system of new cities swiftly emerged, specializing in the collection, processing, and distribution of raw materials and manufactured goods, linked together by an efficient web of railroad lines. Within that urban network, at the local or *microscale*, individual cities prospered in their new roles as manufacturing centers, generating an internal structure that still forms the geographic framework of most of the central cities in America's large metropolitan areas. We now examine the urban trend at both of these scales.

Evolution of the U.S. Urban System

The rise of the national urban system

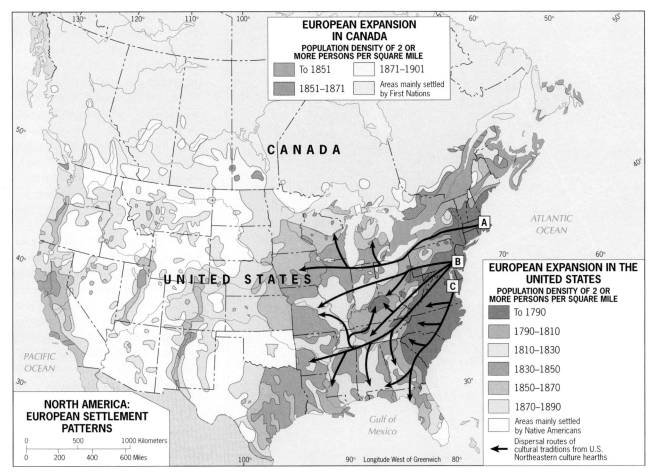

FIGURE 3-7

was based on the traditional external role of cities: providing goods and services for their hinterlands in exchange for raw materials. Because people, commercial activities, investment capital, and transport facilities were already agglomerated in existing cities, new industrialization favored such locations. Their growing incomes permitted industrially intensifying cities to develop their infrastructures, public services, and housing—and thereby convert each round of industrial expansion into a new stage of urban growth.

The evolution of the national urban system has been studied by John Borchert, who identified five epochs of metropolitan evolution based on transportation technology and industrial energy. (1) *The Sail-Wagon Epoch* (1790–1830), marked by primitive overland and waterway circulation; the leading cities were the northeastern ports oriented to the European overseas trade. (2) *The Iron Horse Epoch* (1830–1870), dominated by the spread of the steam-powered railroad; a nationwide transport system appeared and the national urban system took shape, with New York emerging as the primate city. (3) *The Steel-Rail Epoch* (1870–1920), spanning the Industrial Revolution; new forces shaping growth were the increasing scale of manufacturing and the introduction of steel rails that enabled trains to travel faster and haul heavier cargoes. (4) *The Auto-Air-Amenity Epoch* (1920–1970), encompassing the later stage of U.S. industrial urbanization; key elements were the automobile and the airplane, the expansion of white-collar services jobs, and the growing locational pull of *amenities* (pleasant environments) that favored suburbs and certain Sunbelt locales. (5) *The Satellite-Electronic-Jet Propulsion Epoch* (1970–   ), shaped by advancements in information management, computer technologies, global communications, and intercontinental travel; favors globally oriented metropolises, particularly those functioning as international gateways.

Industrialization and the accompanying growth of the urban system reconfigured the realm's economic landscape. The most notable regional transformation was the emergence of the North American Core, or **7 American Manufacturing Belt**, which contained the lion's share of industrial activity in both the United States and Canada. As Figure 3-8 shows, the geographic form of the Core Region—which includes southern Ontario—was a near rectangle whose four corners were Boston, Milwaukee, St. Louis, and Baltimore. Within this region, manufacturing is heavily concentrated into a dozen districts centered on the cities mapped in Figure 3-8.

At the subregional scale, as transportation breakthroughs permitted progressive urban decentralization and *megalopolitan growth*, the expanding peripheries of major cities soon coalesced to form a number of conurbations. The most important of these by far is the *Atlantic Seaboard Megalopolis* (Fig. 3-9), the 600-mile (1000-km) urbanized northeastern coastal strip extending from southern Maine to Virginia that contains metropolitan Boston, New York, Philadelphia, Baltimore, and Washington. This was the economic heartland of the Core; the seat of U.S. government, business, and

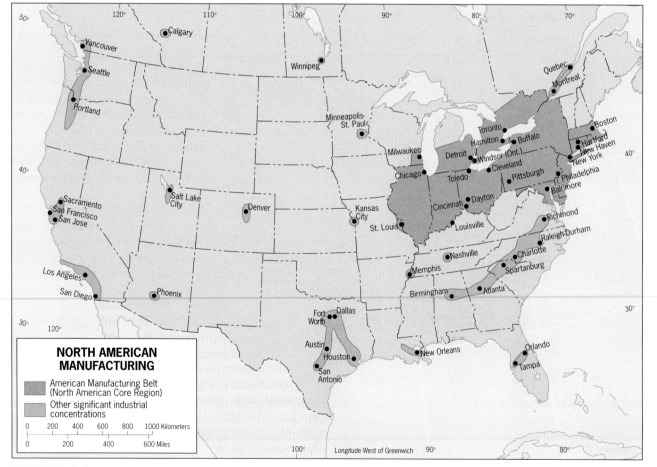

NORTH AMERICAN MANUFACTURING

 American Manufacturing Belt (North American Core Region)

 Other significant industrial concentrations

0   200   400   600   800   1000 Kilometers

0   200   400   600 Miles

FIGURE 3-8

FIGURE 3-9

This growth process was conceptualized into a four-stage model by John Adams, who identified the four eras of intraurban structural evolution that are diagrammed in Figure 3-10. Stage I, prior to 1888, was the *Walking-Horsecar Era*, which produced a compact pedestrian city in which everything had to be within a 30-minute walk, a layout only slightly augmented when horse-drawn trolleys began to operate after 1850. The 1888 invention of the electric traction motor launched Stage II, the *Electric Streetcar Era* (1888–1920); higher speeds enabled the 30-minute travel radius and the urbanized area to expand considerably along new outlying trolley corridors; in the older core city, the central business district (CBD), industrial, and residential land uses differentiated into their modern form. Stage III, the *Recreational Automobile Era* (1920–1945), was marked by the initial impact of cars and highways that steadily improved the accessibility of the outer metropolitan ring, thereby launching a wave of mass suburbanization that further extended the urban frontier. During this era, the still-dominant central city experienced its economic peak and the partitioning of its residential space into neighborhoods sharply defined by income, ethnicity, and race. Stage IV, the *Freeway Era* (under way since 1945), saw the full impact of automobiles, with the metropolis turning inside-out as expressways pushed suburban development more than 30 miles (50 km) from the CBD.

The social geography of the evolving industrial metropolis was marked by the development of a residential mosaic of ever-more-specialized groups. The electric streetcar, which introduced affordable transit for all, allowed the immigrant-dominated city population to sort itself into ethnically uniform neighborhoods. When the United States sharply curtailed immigration in the 1920s, industrial managers

culture; and the trans-Atlantic trading interface between much of North America and Europe. Six other primary conurbations have also emerged: *Lower Great Lakes* (Chicago-Detroit-Cleveland-Pittsburgh), *Piedmont* (Atlanta-Charlotte-Raleigh/Durham), *Florida* (Jacksonville-Tampa-Orlando-Miami), *Texas* (Houston-Dallas/Fort Worth-San Antonio), *California* (San Diego-Los Angeles-San Francisco), and the *Pacific Northwest* (Portland-Seattle-Vancouver). Note that the last spills across the border into Canada, which has also spawned its own nationally predominant conurbation—*Main Street* (Windsor-Toronto-Montreal-Quebec City).

The Changing Structure of the U.S. Metropolis

The internal structure of the metropolis reflected the same mixture of forces that shaped the national urban system. Rails—in this case, lighter street rail lines—again shaped spatial organization as horse-drawn trolleys were succeeded by electric streetcars in the late nineteenth century. The mass introduction of the automobile after World War I changed all that, and America increasingly turned from building compact cities to widely dispersed metropolises. By 1970, the new intraurban expressway network had equalized location costs throughout the metropolis, setting the stage for the suburban ring to transform itself from a residential preserve into a complete outer city. Newly urbanized suburbs now began to capture major economic activities, and their rise came at the expense of the central city which saw its ✴ status diminish to that of coequal.

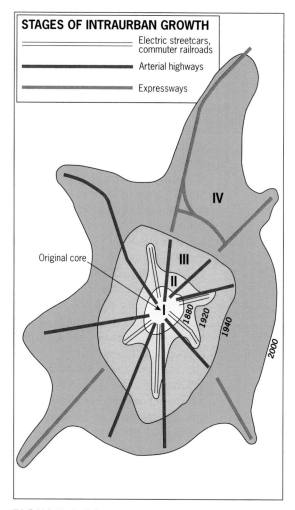

STAGES OF INTRAURBAN GROWTH

——————— Electric streetcars, commuter railroads

——————— Arterial highways

——————— Expressways

Original core

IV

III

II

I

1880

1920

1940

2000

FIGURE 3-10

As ties to the central city loosened, the suburban **9** ring was transformed into a full-fledged **outer city**. Its independence was accelerated by the rise of major new suburban nuclei, multipurpose activity nodes that grew up around large shopping centers with prestigious images that attracted industrial parks, office campuses and high-rises, entertainment facilities, and even major league sports stadiums and **10** arenas. These burgeoning new **suburban downtowns**, in fact, are nothing less than an automobile-age version of the CBD. The newest spatial elements of the contemporary urban complex are assembled in the model displayed in Figure 3-11. The outer city today anchors a *multicentered* metropolis consisting of the traditional CBD and a constellation of coequal suburban downtowns, with each activity center serving a discrete and self-sufficient surrounding area. James Vance defined these tributary areas as **11** **urban realms**, recognizing in his studies that each such realm maintains a separate, distinct eco-

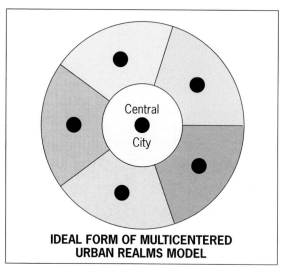

Central City

IDEAL FORM OF MULTICENTERED URBAN REALMS MODEL

FIGURE 3-11

discovered the large African American population of the rural South and began to recruit these workers by the thousands for the factories of Manufacturing Belt cities. This influx had an immediate impact on the social geography of the industrial city: whites were unwilling to share their living space, and the result was the segregation of these newest migrants into geographically separate, all-black areas. By the 1950s, these mostly inner-city areas became large **8** expanding **ghettos**, speeding the departure of many white central-city communities and reinforcing the trend toward a racially divided urban society.

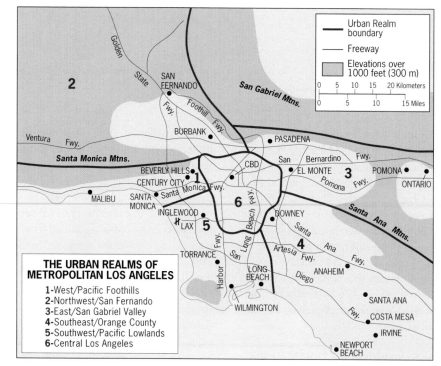

Urban Realm boundary

Freeway

Elevations over 1000 feet (300 m)

0   5   10   15   20 Kilometers

0   5   10   15 Miles

THE URBAN REALMS OF METROPOLITAN LOS ANGELES

1-West/Pacific Foothills
2-Northwest/San Fernando
3-East/San Gabriel Valley
4-Southeast/Orange County
5-Southwest/Pacific Lowlands
6-Central Los Angeles

FIGURE 3-12

nomic, social, and political significance and strength. Figure 3-12 applies the urban realms model to Los GEODISCOVERIES Angeles.

The position of the central city within the new multinodal metropolis of realms is eroding. The CBD increasingly serves the less affluent residents of the innermost realm and those working downtown. As manufacturing employment declined, many large cities adapted by shifting to service industries, embodied by downtown commercial revitalization. Residential reinvestment has also occurred in many downtown-area neighborhoods, but beyond the CBD the vast inner city remains the problem-ridden domain of low- and moderate-income people, with most forced to reside in ghettos.

Cultural Geography

In the United States over the past two centuries, the contributions of a wide spectrum of immigrant groups have shaped—and continue to shape—a rich and varied cultural complex. Great numbers of these newcomers were willing to set aside their original cultural baggage in favor of assimilation into the emerging culture of their adopted homeland, which itself was a hybrid nurtured by constant infusions of new influences.

Language and Religion

Although linguistic variations play a far more important role in Canada, more than one-eighth of the U.S. population speaks a primary language other than English. Differences in English usage are also evident at the subnational level in the United States, where regional variations (*dialects*) can still be noted despite the recent trend toward a truly national society.

North America's Christian-dominated kaleidoscope of religious faiths contains important spatial variations. Many major Protestant denominations are clustered in particular regions, with Southern Baptists localized in the southeastern quadrant of the United States, Lutherans in the Upper Midwest, and Mormons focused on Utah. Roman Catholics are most visibly concentrated in Manufacturing Belt metropolises and the Mexican borderland zone. Jews are most heavily clustered in the suburbs of Megalopolis, Southern California, South Florida, and the Midwest.

Ethnic Patterns

Ethnicity (national ancestry) has always played a key role in American cultural geography. Today, whites of European background no longer dominate the increasingly diverse U.S. ethnic tapestry, with ethnics of color and non-European origin comprising a steadily expanding proportion that will surpass 50 percent by 2050. African Americans still constitute the largest single minority group (12 percent of the U.S. total), but Hispanics (11 percent) are growing at a rate that will make them the nation's leading minority by 2005. The spatial distribution of the four largest ethnic minorities is mapped in Figure 3-13.

Immigration has long influenced the ethnic complexion of the United States, and at the end of the 1990s nearly 1 million immigrants annually entered the country. The source areas, however, have changed dramatically over the past half-century. During the 1950s, just over 50 percent came from Europe, 25 percent from Middle and South America, and 15 percent from Canada; today,

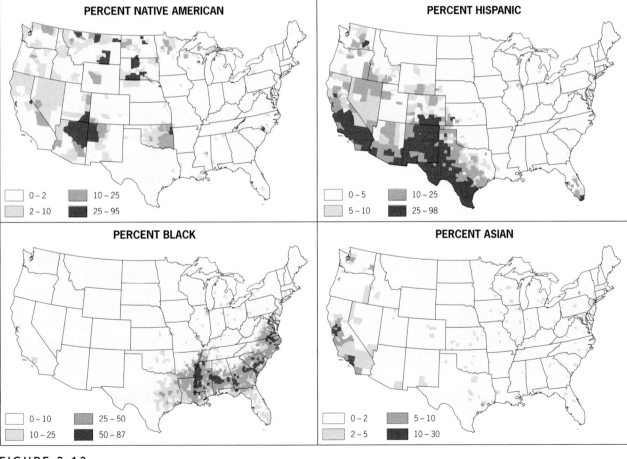

FIGURE 3-13

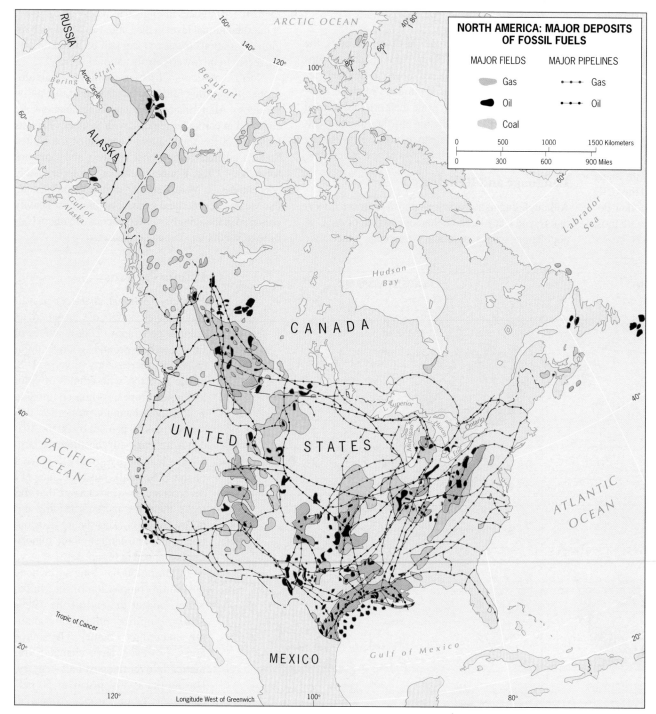

NORTH AMERICA: MAJOR DEPOSITS
OF FOSSIL FUELS

MAJOR FIELDS	MAJOR PIPELINES
Gas	Gas
Oil	Oil
Coal	

FIGURE 3-14

approximately 50 percent come from Middle and South America, 30 percent from Asia, and only about 15 percent from Europe and Canada.

The Emerging Mosaic Culture

American cultural geography continues to evolve. What is now taking place is a new fragmentation into the **12** emerging nationwide **mosaic culture**, an increasingly heterogeneous complex of separate, uniform "tiles" that cater to more specialized groups than ever before. No longer based solely on such broad divisions as income, race, and ethnicity, today's residential communities of interest are also forming along the dimensions of age, occupational status, and especially lifestyle.

The Changing Geography of Economic Activity

The economic geography of the United States today is the product of all of the foregoing, with people and activities overcoming the tyranny of distance to organize a continentwide spatial economy that took full advantage of agricultural, industrial, and urban development opportunities. Despite these past achievements, today American economic geography is again in the throes of restructuring as the transition is completed from industrial to postindustrial society.

Major Components of the Spatial Economy

Economic geography is heavily concerned with the locational analysis

13 of **productive activities**. Four major sets may be identified:

- **Primary activity**: the extractive sector of the economy in which workers and the environment come into direct contact, especially in *mining* and *agriculture*.

- **Secondary activity**: the *manufacturing* sector, in which raw materials are transformed into finished industrial products.

- **Tertiary activity**: the *services* sector, including a wide range of activities from retailing to finance to education to routine office-based jobs.

- **Quaternary activity**: today's dominant sector, involving the collection, processing, and manipulation of *information*; a subset, sometimes referred to as **quinary activity**, is managerial decision-making in large organizations.

Historically, each of these activities has successively dominated the U.S. labor force for a period over the past 200 years, with the quaternary sector now dominant. The approximate current breakdown by major sector of employment is agriculture, 2 percent; manufacturing, 15 percent; services, 18 percent; and quaternary, 65 percent (with about 10 percent in the quinary sector). We now review these major productive components of the spatial economy in the following coverage of resource use, agriculture, manufacturing, and the postindustrial revolution.

Resource Use

The United States (and Canada) was blessed with abundant deposits of mineral and energy resources. North America's mineral resources are localized in three zones: the Canadian Shield north of the Great Lakes, the Appalachian Highlands, and scattered areas across the mountain ranges of the West. The Shield's most noteworthy minerals are iron ore, nickel, uranium, and copper. Besides vast deposits of coal, the Appalachian region also contains iron ore in central Alabama. The western mountain zone contains significant deposits of coal, copper, lead, zinc, and gold.

The realm's most vital energy resources are its petroleum (oil), natural gas, and coal deposits 14 (mapped in Fig. 3-14); these are the **fossil fuels**, so named because they were formed by the geologic compression of tiny organisms that lived hundreds of millions of years ago. The leading *oil*-production areas of the United States are located along and offshore from the Texas-Louisiana Gulf Coast; in the Midcontinent district, extending through western Texas-Oklahoma-eastern Kansas; and along Alaska's North Slope facing the Arctic Ocean. (Canada's leading oilfields lie in a crescent curving southeastward from northern Alberta to southern Manitoba.) The distribution of *natural gas* supplies resembles the geography of oilfields because both fuels are usually found in the floors of ancient shallow seas. The realm's *coal* production zones, which rank among the greatest on Earth, are found in Appalachia, the northern U.S. Great Plains/southern Alberta, and southern Illinois/western Kentucky.

Agriculture

Despite the post–1900 emphasis on developing the nonprimary sectors of the spatial economy, agriculture

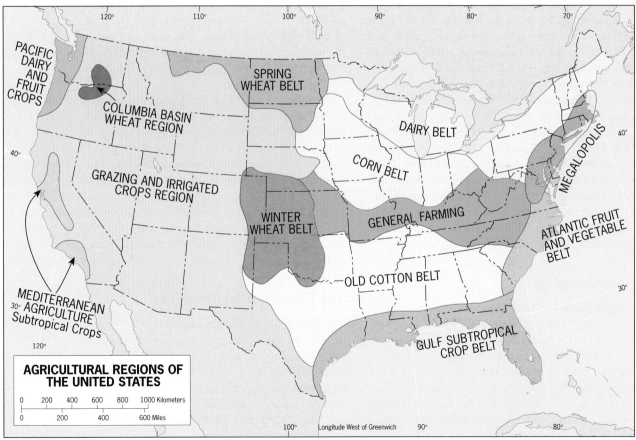

FIGURE 3-15

remains an important element in America's human geography. Vast expanses of the U.S. landscape are clothed with fields of grain or support great herds of livestock that are sustained by pastures and fodder crops. In recent decades, high-technology mechanization has revolutionized farming and has been accompanied by a sharp reduction in the agricultural workforce (today only about 1.5 percent of the U.S. population still resides on farms).

The regionalization of U.S. agricultural production is shown in Figure 3-15, its spatial organization developed largely within the framework of the *von Thünen model* (see pp. 36-37). As in Europe (Fig. 1-6), the early-nineteenth-century, original-scale model of town and hinterland expanded outward (driven by constantly improving transportation technology) from a locally "isolated state" to encompass the entire continent by 1900. The "supercity" anchoring this macro-Thünian regional system was the northeastern Megalopolis, already the dominant food market and transport focus of the entire country. Although the circular rings of the model are not apparent in Figure 3-15, many spatial regularities can be observed in this real-world application. Most significant is the sequence of farming regions as distance from the national market increased, especially westward from Megalopolis toward central California (the main directional thrust of the interior penetration of the United States). The Atlantic Fruit and Vegetable Belt, Dairy Belt, Corn Belt, Wheat Belts, and Grazing Region are indeed consistent with the model's logical structure, each zone successively farther inland astride the main transcontinental routeway.

Manufacturing

The geography of North America's industrial production has long been dominated by the Manufacturing Belt (Fig. 3-8). The emergence of this region was propelled by (1) superior access to the Megalopolis national market that formed its eastern edge, and (2) proximity to industrial resources, particularly iron ore and coal for the pivotal steel industry that arose in its western half. We noted earlier that manufacturers had a strong locational

affinity for cities, and the internal structure of the Belt became organized around a dozen urban-industrial districts interconnected by a dense transportation network. As these industrial centers **15** expanded, they swiftly achieved **economies of scale**, savings accruing from large-scale production in which the cost of manufacturing a single item was further reduced as factories mechanized assembly lines, specialized their workforces, and purchased raw materials in massive quantities.

This production pattern served the nation well throughout the remainder of the industrial age. Today, however, much of the Manufacturing Belt is aging and the distribution of American industry is changing. As transportation costs equalize among U.S. regions, as energy costs now favor the south-central oil- and gas-producing States, as high-technology manufacturing advances reduce the need for lesser skilled labor, and as locational decision-making intensifies its attachment to noneconomic factors, industrial management has increasingly demonstrated its willingness to relocate to regions it perceives as more desirable in the South and West.

Nonetheless, parts of the Manufacturing Belt have been resisting this trend. This is particularly true of the industrial Midwest, whose prospects have greatly improved since the "Rustbelt" days of the 1970s. Shedding that label in recent years, Midwestern manufacturers are successfully reinventing their operations by rooting out inefficiencies, investing in cutting-edge factories and technologies, and not only beating back foreign competition in the United States but significantly expanding their exports.

The Postindustrial Revolution

16 The signs of **postindustrialism** are visible throughout the United States today, and they are popularly grouped under such labels as "the computer age" or "the new economy." High-technology, white-collar, office-based activities are the leading growth industries of the postindustrial economy. Most are relatively footloose and are therefore responsive to such noneconomic loca-

tional forces as geographic prestige, local amenities, and proximity to recreational opportunities. Northern California's *Silicon Valley*—the world's leading center for computer research and development and the headquarters of the U.S. microprocessor industry—epitomizes the blend of locational qualities that attract a critical mass of high-tech companies to a given locality. These include: (1) a world-class research university (Stanford); (2) a large pool of highly skilled labor; (3) proximity to a cosmopolitan urban center (San Francisco); (4) abundant venture capital; (5) a local entrepreneurial culture that supports risk-taking; (6) a locally based network of global business linkages; and (7) a high-amenity environment in the form of pleasant weather and year-round recreational opportunities.

The development of Silicon Valley is so significant to the new postindustrial era that planners Manuel Castells and Peter Hall have conceptualized **17** it as the first **technopole**. Technopoles are planned techno-industrial complexes that innovate, promote, and manufacture the hardware and software products of the new informational economy. On the landscape of the outer suburban city, where almost all of these complexes are located, the signature of a technopole is a low-density cluster of ultramodern buildings laid out as a campus. From Silicon Valley, technopoles have spread in all directions—from the Route 128 corridor around Boston to North Carolina's Research Triangle to the lakeside suburbs of Seattle—and many will be noted in the regional section of this chapter. Importantly, technopoles are also becoming a global phenomenon as they spring up in other geographic realms; several are discussed in other chapters.

CANADA

Like the United States, Canada is a federal state, but it is organized differently. Canada is divided into ten provinces and three territories (Fig. 3-16).

The provinces—where almost all Canadians live—range in territorial size from tiny, Delaware-sized Prince Edward Island to sprawling Quebec,

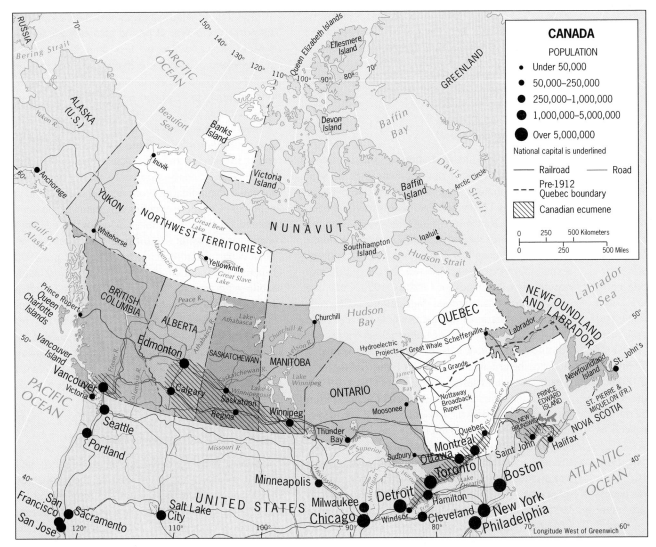

FIGURE 3-16

merly called Eskimos) and the federal government, and encompasses all of Canada's eastern Arctic.

In population size, Ontario (11.9 million) and Quebec (7.4 million) are again the leaders; British Columbia ranks third with 4.2 million; next come the three Prairie Provinces with a combined total of 5.4 million; the Atlantic Provinces are the smallest, together containing 2.4 million. Canada's total population of 31.0 million is only slightly larger than one-tenth the size of the U.S. population.

Population in Time and Space

The map showing the distribution of Canada's population (Fig. 3-4, p.102) reveals that only about one-eighth of this enormous country can be classified **18** as its **ecumene**—the inhabitable zone of permanent settlement. As Figure 3-16 indicates, the Canadian ecumene is dominated by a discontinuous strip of population clusters that lines the U.S. border. We can identify four such clusters on the map, the largest by far being Main Street (home to more than six out of every ten Canadians). As noted earlier (Fig. 3-9), *Main Street* is the conurbation that stretches across southernmost Quebec and Ontario, from Quebec City on the lower St. Lawrence River southwest through Montreal and Toronto to Windsor on the Detroit River. The three lesser clusters are: (1) the Saint John-Halifax crescent in central New Brunswick and Nova Scotia; (2) the prairies of southern Alberta, Saskatchewan, and Manitoba; and (3) the southwestern corner of British Columbia, focused on Canada's third-largest metropolis, Vancouver.

more than twice the area of Texas. Beginning in the east, the four Atlantic Provinces are Nova Scotia, New Brunswick, Prince Edward Island, and Newfoundland and Labrador. To their west lie Quebec and Ontario, Canada's two biggest provinces. Most of western Canada is covered by the three Prairie Provinces—Manitoba, Saskatchewan, and Alberta. In the far west, facing the Pacific, lies British Columbia.

The three territories—Yukon, the Northwest Territories, and Nunavut—together occupy a massive area half the size of Australia but are inhabited by only about 100,000 people. Nunavut is the newest addition to Canada's political map and deserves special mention. Created in 1999, this new territory is the outcome of a major aboriginal land claim agreement between the Inuit people (for-

Pre-Twentieth-Century Canada

As Figure 3-7 shows, compared to European expansion in the United States, penetration of the Canadian interior lagged well behind. In terms of political geography, Canada did not unify before the last third of the nineteenth century—and then mainly because of fears the United States was about to expand in a northerly direction.

The evolution of modern Canada is deeply rooted in the bicultural division discussed at the outset of this chapter. Its origin lies in the fact that it was the French, not the British, who were the first European colonizers of present-day Canada, with *New France* growing to encompass the St. Lawrence Basin, the Great Lakes region, and the Mississippi Valley. A series of wars between the English and French subsequently ended in France's defeat and the cession of New France to Britain in 1763. By the time London took control of its new possession, the French had made considerable progress in their North American domain. The British, anxious to avoid a war of suppression and preoccupied with problems in their other American colonies, gave former French Quebec—the territory extending from the Great Lakes to the mouth of the St. Lawrence—the right to retain its legal and land-tenure systems as well as freedom of religion.

After the American War for Independence, London was left with a region it called British North America but whose cultural imprint still was decidedly French. The Revolutionary War drove many thousands of English refugees northward, and soon difficulties arose between them and the French. In 1791, heeding appeals by these new settlers, the British Parliament divided Quebec into two provinces: Upper Canada, the region upstream from Montreal centered on the north shore of Lake Ontario, and Lower Canada, the valley of the St. Lawrence. Upper and Lower Canada became, respectively, the provinces of (English-speaking) Ontario and (Francophone) Quebec (Fig. 3-16). This earliest cultural division did not work well and in 1867 finally led to the British North America Act, which established the Canadian federation (consisting initially of Upper and Lower Canada, New Brunswick, and Nova Sco-

tia, later to be joined by the other provinces and territories). Under this Act, Ontario and Quebec were again separated, but this time Quebec was given important guarantees: the French civil code was left unchanged, and the French language was protected in Parliament and in the courts.

Canada Since 1900

By 1900, the Canadian federation was making major strides toward regional development and the spatial integration of a continentwide economy. The transcontinental Canadian Pacific Railway had been completed to Vancouver, along the way spawning the settlement of the fertile Prairie Provinces whose wheat-raising economy expanded steadily as immigrants arrived from the east and abroad. Industrialization also began to stir, and by 1920 Canadian manufacturing had surpassed agriculture as the leading source of national income. As noted earlier (Fig. 3-8), the dominant zone of industrial activity is the Toronto-Hamilton-Windsor corridor of southern Ontario, the crucible of Canada's Industrial Revolution that took hold during World War I (1914–1918).

As in the United States, industrial intensification was accompanied by the rise of a national urban system. Along the lines of Borchert's epochs of American metropolitan evolution, Maurice Yeates constructed a similar multistage model of development that divides the telescoped Canadian experience into three eras. The initial *Frontier-Staples Era* (prior to 1935) encompasses the century-long transition from a frontier-mercantile economy to one oriented to staples (production of raw materials and agricultural goods for export), with increasing manufacturing activity in the budding industrial heartland. By 1930, Montreal and Toronto had emerged as the two leading cities atop the national urban hierarchy (thus Canada has no single primate city).

Next came the *Era of Industrial Capitalism* (1935–1975), during which Canada achieved U.S.-style prosperity. A major stimulus was the investment of U.S. corporations in Canadian branch-plant construction, especially in the automobile industry in Ontario near the automakers' Detroit-area headquarters. In western Canada, the growth of oil and natural

gas production fueled Alberta's urban development. The post–World War II period also saw the ascent of Main Street, which on less than 2 percent of Canada's land quickly came to contain more than 60 percent of its people, contributed two-thirds of its national income, and claimed nearly 75 percent of its manufacturing jobs.

The third stage, ongoing since 1975, is the *Era of Global Capitalism*, signifying the rise of additional foreign investment from the Asian Pacific Rim and Europe. This, of course, is also the era of transformation into a postindustrial economy and society, and in the process Canada is experiencing many of the same upheavals as the United States. Most of this development is occurring in the form of new suburbanization, a departure from the past because the pre-1990 Canadian metropolis had experienced far less automobile-generated deconcentration than the United States. But today Canada's large cities are turning inside-out, and the new intraurban geography is symbolized by the suburban downtowns that anchor the ultramodern business complexes lining the Highway 401 freeway north of Toronto and Alberta's West Edmonton megamall, the world's biggest shopping center.

Cultural/Political Geography

The historic cleavage between Canada's French- and English-speakers has resurfaced in the past three decades to dominate the country's cultural and political geography. By the time the Canadian federation observed its centennial in 1967, it had become evident that Quebeckers regarded themselves as second-class citizens; they believed that bilingualism meant that French-speakers had to learn English but not vice versa; and they perceived that Quebec was not getting its fair share of the country's wealth. Since the 1960s, the intensity of ethnic feelings in Quebec has risen in surges despite the federal government's efforts to satisfy the province's demands. During the 1970s, while a separatist political party came to power in Quebec, a new federal constitution was drawn up in Ottawa. In 1980, Quebec's voters solidly rejected independence when given that choice in a referendum. But the new constitution did *not* satisfy the Quebeckers, and throughout the 1980s and early 1990s the Ottawa government struggled unsuccessfully to devise a plan, acceptable to all the provinces, that would keep Quebec in the Canadian federation.

By 1995, with Canada's interest in constitutional reform exhausted, a second referendum on Quebec's sovereignty could no longer be put off. With the reenergized separatist party again leading the way, the Francophone-dominated electorate very nearly approved independence. Subsequently, despite calls by many separatists for a follow-up vote that might turn narrow defeat into victory, opinion polls in Quebec have persistently shown a small but critical erosion in public support for secession. In 2001, this led to the unexpected resignation from politics of Lucien Bouchard, the province's charismatic premier and the one separatist leader with the best chance of reversing the results of the 1995 referendum.

After what many Canadians still call their "near-death experience," the federal government seized the initiative in late 1995 by asking Canada's Supreme Court to review the constitutionality of Quebec's separation. The court ruled in 1998 that if a clear majority of Quebec's electorate voted to secede, the Ottawa government and the other provinces would be obliged to negotiate the terms of separation as if the matter were a new amendment to Canada's constitution. Armed by this legal interpretation, the federal government passed legislation that formalized the court's rulings—and makes the process of secession for a breakaway province far more cumbersome and costly.

Among the issues that an exiting Quebec would have to negotiate, most significant from a geographic standpoint is that the French linguistic region does *not* coincide with the province's territorial boundaries. Thus, as the distribution of the "no" vote on separation in the 1995 referendum strongly suggests (Fig. 3-17; areas colored pink), there are numerous non-French communities located within the Francophone region. Dozens of English-speaking municipalities in the southern periphery of Quebec have already spearheaded a partitionist movement by declaring their intention to stay in Canada, arguing that they have the same right to secede from Quebec that Quebec has to secede from Canada. Even more important is an identical movement among the First Nations of Quebec's northern frontier, the Cree, whose historic domain covers more than half of the province. As Figure 3-16 shows, the territory of the Cree is no unproductive wilderness: it contains vital facilities of the James Bay Hydroelectric Project, a massive scheme of dams and artificial lakes that generates electrical power for a huge market within and outside the province—an enterprise that would be an

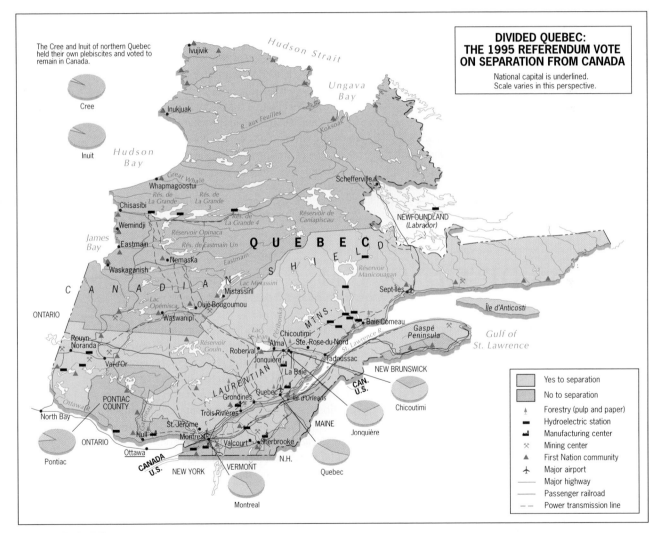

FIGURE 3-17

economic cornerstone in any plan for an independent Quebec.

Finally, the events of the past quarter-century have increasingly impacted Canada's political landscape, and not only in Quebec. Regionalism has also intensified in the west, whose leaders oppose federal concessions to Quebec. The most recent federal elections clearly reveal the emerging fault lines that surround Quebec and set the western provinces off from the rest of the country. Even the remaining Ontario-led center and the eastern bloc of Atlantic Provinces voted divergently, and today the politico-geographical hypothesis of "Four Canadas" may well be on its way toward becoming reality. Thus, with its national unity under siege, Canada today confronts the coalescing forces of *devolution* that threaten to transform the new fault lines into permanent fractures.

Economic Geography

As in the United States, the growth of Canada's spatial economy has been supported by a diversified *resource base*. We noted earlier that the Canadian Shield is endowed with major mineral deposits and that oil and natural gas are extracted in sizeable quantities in Alberta. Canada has long been a leading *agricultural* pro-

ducer and exporter, especially of wheat and other grains from its breadbasket in the Prairie Provinces. Postindustrialization has caused substantial employment decline in the *manufacturing sector*, with Southern Ontario's industrial heartland being the most adversely affected. On the other hand, Canada's robust *tertiary and quaternary sectors* (which today employ more than 70 percent of the total workforce) are creating a host of new economic opportunities.

Canada's economic future is also going to be strongly affected by the continuing development of its trading relationships. These include the 1989 United States-Canada Free Trade Agreement (today more than four-fifths of Canada's exports go to the United States, from which it also derives about two-thirds of its imports), and the 1994 North American Free Trade Agreement (NAFTA) which added Mexico to the trading partnership.

Because these free-trade agreements increasingly impact the Canadian spatial economy, they are likely to weaken domestic east-west linkages and strengthen international north-south ties. Since many local cross-border linkages built on geographical and historical commonalities are already well developed, they can be expected to intensify in the future: the Atlantic Provinces with neighboring New England; Quebec

with New York State; Ontario with Michigan; the Prairie Provinces with the Upper Midwest; and British Columbia with the (U.S.) Pacific Northwest. Such functional reorientations, of course, constitute yet another set of powerful devolutionary forces confronting the Ottawa government because most of these potential economic fault lines coincide with those that politically demarcate the "Four Canadas."

The rising importance of this framework of transnational regions straddling the U.S.-Canadian border recalls Kenichi Ohmae's regional state concept [19] introduced in Chapter 1. A **regional state** is a "natural economic zone" that defies old borders, and is shaped by the global economy of which it is a part; its leaders deal directly with foreign partners and negotiate the best terms they can with the national governments under which they operate. Writing about Canada, Ohmae identified a Pacific Northwest (the Seattle-Vancouver axis) and a Great Lakes regional state (the intertwined Ontario-Michigan industrial complex), and warned that the manner in which Ottawa's leaders dealt with these new economic entities was critical to the survival of the Canadian state. Indeed, these growing international interactions are increasingly evident in the regional configuration of North America, to which we now turn.

REGIONS OF THE REALM

The ongoing transformation of North America's human geography is fully reflected in its internal regional organization. We now examine that spatial arrangement within a framework of nine regions (Fig. 3-18).

☐ THE NORTH AMERICAN CORE

The Core Region (Fig. 3-18)—synonymous with the American Manufacturing Belt—was introduced earlier in the chapter. This region was the workshop for

the linked spatial economies of the United States and Canada during the century (1870–1970) dominated by industry. In the postindustrial era, however, that linchpin regional role is diminishing as strengthening challengers to the south and west siphon key functions away from the Core. But make no mistake: this is still the geographic heart of North America. Here, in each country, one finds over one-third of the population, the capital, and the largest city—as well as the leading financial markets, corporate headquarters, media centers, cultural facilities, and busiest transportation facilities.

Although manufacturing remains highly important within the transformed American economy, pro-

ductivity and obsolescence problems in the Core Region have erased many of its competitive advantages over newly emerging industrial areas in the southern and western United States. In parts of the Core, employment has dropped sharply in the smokestack industries as factories have closed or relocated, speeding the economic decline of surrounding communities. But other areas have fought back, none more successfully than the Midwest portion of the Manufacturing Belt, which reinvented itself by aggressively pursuing the high-tech upgrading of its aged industrial base and becoming far more competitive in the international marketplace. The recent renaissance of its key automobile industry is a prime

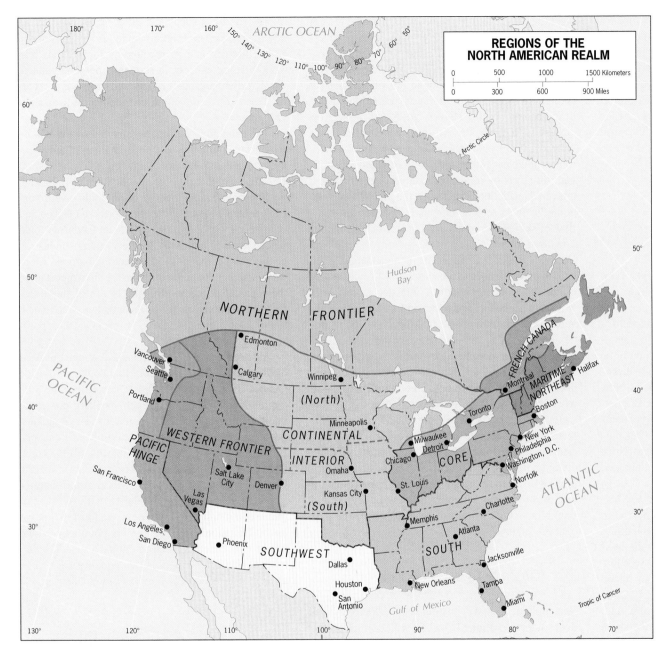

FIGURE 3-18

example, propelled by Toyota's investments in the Interstate-64 corridor between St. Louis and northern West Virginia, which became the axis of a state-of-the-art automaking complex centered on the Lexington, Kentucky area. Detroit's Big Three (GM, Ford, and Daimler-Chrysler) were forced to keep pace by modernizing fabrication technologies and dispersing many operations away from their traditional base in Michigan to locations elsewhere in the Midwest and as far away as Mexico.

Postindustrial development has also spawned a number of new growth centers in the Core. The major metropolitan complexes of the northeastern Megalopolis are adjusting fairly well. The Boston area, richly endowed with research facilities, is again attracting innovative high-tech businesses to the Route 128 freeway corridor that girdles the central city (North America's pioneering technopole of the 1950s); and Greater New York remains the national leader in finance, advertising, and corporate decision-making activity. But the Core Region metropolis that has gained the most from postindustrialization is Washington, D.C. As the quaternary and quinary sectors have blossomed, and as the U.S. federal government extended its ties to the private sector, Washington and its surrounding outer city of affluent Maryland and Virginia suburbs (interconnected by the 66-mile [105-km] Beltway encircling the capital city) has amassed an enormous complex of office, research, high-tech, lobbying, and consulting firms. The most important development has occurred along the outer-suburban expressway corridor between Tysons Corner on the Beltway and booming Dulles

International Airport. Since 1995, a large number of telecommunications and Internet companies have been attracted to this part of northern Virginia, one of the country's leading fiber-optic cable hubs thanks to its local cluster of federal intelligence agencies as well as the corporate headquarters of America Online. This major, still-developing technopole is already being called the "technological capital" of the United States, because more technical workers are now employed here than in Silicon Valley.

THE MARITIME NORTHEAST

The Maritime Northeast consists of upper New England (Vermont, New Hampshire, Maine) and the neighboring Atlantic Provinces of easternmost Canada (Fig. 3-18). New England, one of the realm's historic culture hearths, has retained a strong regional identity for almost 400 years. Even with the urbanized southern half (Massachusetts, Connecticut, Rhode Island) lying in the Core, the six New England States still share many common characteristics. The Maritime Northeast region also extends northeastward to encompass nearly all of Canada's four Atlantic Provinces of New Brunswick, Nova Scotia, Prince Edward Island, and Newfoundland and Labrador.

A long association based on economic and cultural similarities has tied northern New England to Atlantic Canada. Both have a strong maritime orientation, are rural in character, possess difficult environments with limited land resources, and were historically bypassed in favor of more fertile inland areas. Thus economic growth in upper New England has always lagged behind the rest of the realm, with development centered on fishing (once) rich offshore waters, forestry in the uplands, and farming in the few fertile valleys available. Recreation and tourism have boosted the regional economy in recent times, with scenic coasts and mountains attracting millions from the neighboring Core Region.

Since 1980, New England's roller-coaster economy has experienced prosperity, hard times, and, most recently, cautious recovery. To achieve greater stability, New England is developing a more diversified economic base built around telecommunications, financial services, health care, and biotechnology. Nonetheless, the economic revival has not yet taken hold in much of northern New England. Most of the benefits of the recovery are found in the zone closest to metropolitan Boston, which includes a wide swath of fringe areas spilling over into southern Maine and New Hampshire. New England's continued growth is by no means assured: apart from the uneven distribution of recovery, the region's costs of doing business remain high, and the continuing outflow of skilled workers raises concerns about the future labor force.

The Atlantic Provinces have also experienced hard economic times in recent years. Most adversely affected was the groundfish industry as offshore stocks of flounder, haddock, and cod became severely depleted through overfishing. New opportunities, never easy to come by here, are most promising in the provinces peripheral to more heavily populated Nova Scotia and New Brunswick. Economic prospects are brightest in remote Newfoundland Island, where major offshore oil deposits (see Fig. 3-14) were discovered in the 1990s just as government-mandated fisheries restrictions were inflicting disaster on the seafood industry. The construction and opening of seabed drilling platforms and coastal support facilities has been swift, and the hope is that oil can transform the economy of Newfoundland and Labrador along the lines of Norway's over the past quarter-century.

FRENCH CANADA

Francophone Canada constitutes the effectively settled, southern portion of Quebec, which straddles the central and lower St. Lawrence Valley from where that river crosses the Ontario-Quebec border just upstream from Montreal to its mouth in the Gulf of St. Lawrence. Also included is a sizeable concentration of French speakers, known as the Acadians, who reside just beyond Quebec's provincial boundary in neighboring New Brunswick (Fig. 3-18). The Old World charm of Quebec's cities is matched by an equally unique rural settlement landscape introduced by the French: narrow rectangular farms, known as *long lots*, are laid out in sequence perpendicular to the St. Lawrence, other rivers, and the roads that parallel them, thus allowing each farm access to an adjacent routeway.

The economy of French Canada, however, is no longer rural (although dairying remains a leading agricultural pursuit) and exhibits urbanization rates similar to those of the rest of the country. Industrialization is widespread, supported by cheap hydroelectric energy generated at huge dams in northern Quebec, but relatively little of the region's manufacturing could be classified as high-tech. Tertiary and postindustrial commercial activities are concentrated around Montreal, and tourism and recreation are also important to the regional economy. But the health of these sectors—now improving as the independence movement has stalled—is tied to the resolution of Quebec's political status within Canada.

As noted above, French Canada also includes Acadia, Canada's largest cluster of Francophones outside Quebec. Here in northernmost New Brunswick, the approximately 250,000 French-speakers constitute one-third of the province's population. The Acadians, however, not only shun the notion of independence for themselves but actively promote all efforts to keep Quebec within the Canadian federation. Unlike the Quebeckers, the Acadians have devoted their energies in recent years to accommodation with the Anglophone community. Today, a new relationship (known as "cohabitation") has been worked out between the two groups, based as much on mutual respect for each other's languages as the strict equality of a new, government-mandated system of bilingualism.

THE CONTINENTAL INTERIOR

The Continental Interior extends across the center of both the conterminous United States and the southern tier of Canada (Fig. 3-18). With few exceptions—most notably in the region's northeastern corner where

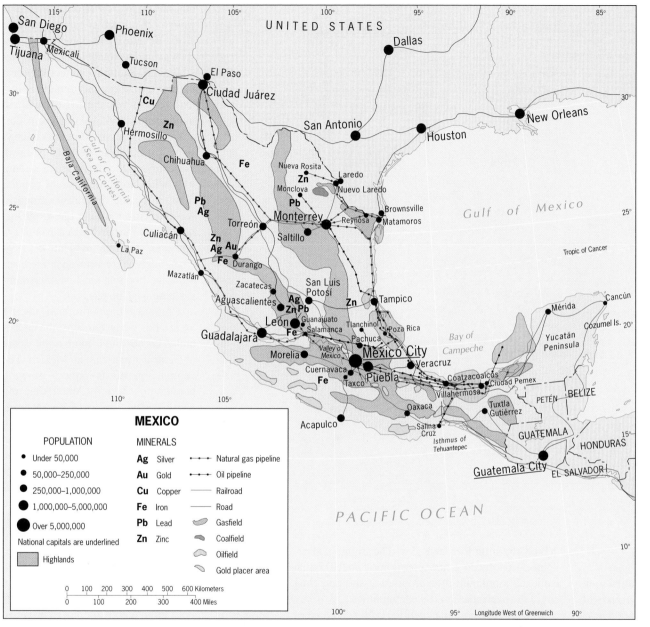

FIGURE 4-9

lion (making it the largest urban concentration on Earth) and is home to 26 percent of the national population. ✳ Among the other leading cities are Guadalajara, Puebla, and León in the central population corridor, and Monterrey, Ciudad Juarez, and Tijuana in the northern U.S. border zone (Fig. 4-10). Urbanization rates at the other end of Mexico, however, are at their lowest in those remote uplands where Amerindian society has been least touched by modernization.

Nationally, the Amerindian imprint on Mexican culture remains quite strong. Today, 60 percent of all Mexicans are mestizos, 20 percent are predominantly Amerindian, and about 10 percent are full-blooded Amerindians; only 9 percent are Europeans. Certainly the Mexican Amerindian has been Europeanized, but the Amerindianization of modern Mexican society is so powerful that it would be inappropriate here to speak of one-way, **9** European-dominated **acculturation**. Instead, what took place in **10** Mexico is **transculturation**—the two-way exchange of culture traits between societies in close contact. In the southeastern periphery (Fig. 4-4), several hundred thousand Mexicans still speak only an Amerindian language, and millions more still use these languages in everyday conversation even though they also speak Mexican Spanish. The latter has been strongly shaped by Amerindian influences, as have Mexican modes of dress, foods and cuisine, sculpture and painting, architectural styles, and folkways. This fusion of heritages, which makes Mexico unique, is the product of an upheaval that began to reshape the country nearly a century ago.

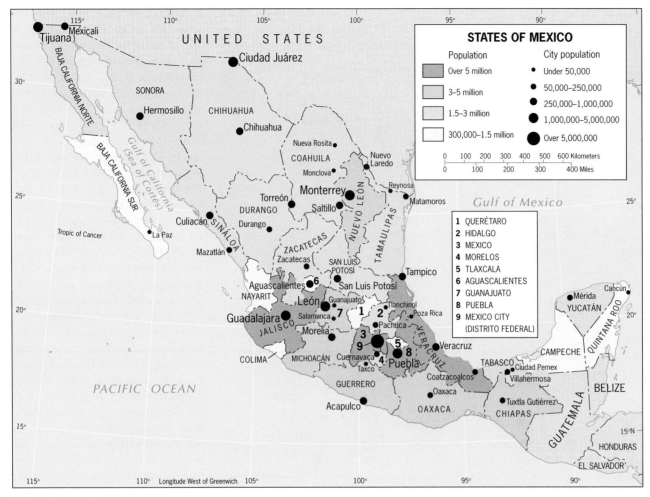

FIGURE 4-10

but the newly created landholdings were too finely fragmented to provide a real development opportunity. Despite an understandable temporary decline in agricultural productivity during the transition, the miracle is that land reform was carried off without major dislocation. Although considerable poverty persisted in the countryside, it was recognized that Mexico, alone among Middle America's countries with large Amerindian populations, had made significant strides toward solving the land question. Just how much farther Mexico has to go, however, became evident in 1994 when the issue resurfaced at the heart of a rebellion that broke out in southeasternmost Chiapas State.

Chiapas, the poorest of the 31 States, lies wedged against the Guatemalan border and is more reminiscent of Central America than Mexico. The complexion of its rapidly growing population is heavily Amerindian and dominated by families of peasant farmers who eke out a precarious existence by cultivating tiny plots of land in the rainforested hills. For centuries, the better valley soils have been incorporated into the estates of the large landholders, a system that endures here virtually unaffected by the land redistribution that reshaped so much of

Revolution and Its Aftermath

Modern Mexico was forged in a revolution that began in 1910 and set into motion events that are still unfolding today. At its heart, this revolution was about the redistribution of land, an issue that had not been resolved after Mexico freed itself from Spanish colonial control in the early nineteenth century. As late as 1900, more than 8000 haciendas blanketed virtually all of Mexico's good farmland, and about 95 percent of all rural families owned no land whatsoever and toiled as *peones* (landless, constantly in-

debted serfs) on the haciendas. The triumphant revolution produced a new constitution in 1917 that launched a program of expropriation and parceling out of the haciendas to rural communities.

Since 1917, more than half the cultivated land of Mexico has been redistributed, mostly to peasant communities consisting of 20 families or more. On such farmlands (known as *ejidos*) the government holds title to the land, and use rights are parceled out to villages and then individuals for cultivation. Most of these *ejidos* lie in central and southern Mexico, where Amerindian agricultural traditions survived,

rural Mexico. In response to this lack of development, during the 1980s the Mexican government had pledged to introduce new services and programs in Chiapas, but its main effort was an ineffective coffee-raising scheme and a feeble attempt to privatize Amerindian lands. This only intensified the longstanding bitterness of the Chiapans, and a radical group of Mayan peasant farmers began to organize to resume the historic struggle of Amerindian *peones* to gain land and fair treatment.

On January 1, 1994, this organization, now calling itself the Zapatista National Liberation Army

(ZNLA), ignited a guerrilla war with coordinated attacks on several Chiapan towns. By timing this uprising to coincide with the birth date of NAFTA and by taking the name of Zapata (a legendary leader of the 1910 revolution), the ZNLA ensured maximum publicity for its agenda, which included land reform, access to greater economic opportunity, heightened cultural identity, and local autonomy. The Mexican military reacted to this rebellion with a heavy hand but quickly called off its offensive when the international press exposed a number of human rights violations. Ever since, the army has maintained a major Chiapan presence—but always leaving sufficient maneuvering room for the unrepentant guerrillas to regroup in the mountains, where they have remained a formidable political, if not a military, force. Although negotiations soon replaced the initial armed conflict, peace talks collapsed in 1996 and gave way to a stalemate that has dragged into the twenty-first century.

The Chiapas conflict raises a number of broader issues that have important implications for Mexico's future. First, it emphasizes that certain areas of the country do not participate in the ongoing development thrust (in spatial terms, the Chiapas situation is a classic case of core-periphery confrontation). A second issue, of course, is devolution. The ZNLA demand for "autonomy" is modeled after Spain's Autonomous Communities (see p. 59) and involves the decentralization of powers from the federal to the State government, which allows the latter more local control, particularly over cultural affairs. A third issue is especially vital to Mexico's social geography: the ZNLA crusade elicited wide sympathy among indigenous populations in other southern States and could well spark a nationwide civil rights movement for Amerindians. As in Caribbean societies, darkness of skin color is directly related to a person's social status, a linkage that in Mexico further extends to an individual's degree of "Indianness." Clearly, large numbers of predominantly Amerindian Mexicans (a group now totaling 20 percent of the national population) made social gains during the past century. But the remaining 10 percent who are full-blooded Amerindians, and choose to preserve their pre-Hispanic cultural traditions, have been shunned by a racist mainstream society whose discrimination continues to render them second-class citizens.

The Changing Geography of Economic Activity

In recent decades, Mexico has made important progress in several productive sectors. During the early 1990s, its economy was further transformed by the boom preceding the implementation of NAFTA, which over the next few years will bind the economies of Mexico, the United States, and Canada into a single free-trade zone and market of more than 400 million people. Mexico will gain the most because this affiliation is expected to narrow the economic gap with its wealthy northern neighbor—which still produces landscapes marked by striking contrasts along their 1936-mile (3115-km)-long border.

The launching of NAFTA in 1994 was followed by a number of unexpected shocks to the Mexican political establishment and economic system. The most serious shock was a substantial devaluation of the peso in late 1994, which plunged Mexico into an economic recession that lingered through the end of the 1990s. The recession sharply challenged the government to correct the problems and restart the boom, an effort blunted by the political upheaval that accompanied the post–1997 disintegration of the power of the PRI Party, which had ruled Mexico for 71 consecutive years and was finally ousted in the election of 2000. We now review Mexico's changing economic geography against the background of these events.

Agriculture

Although traditional subsistence agriculture and the output of the inefficient *ejidos* have not changed a great deal in the poorer areas of rural Mexico, larger-scale commercial agriculture has diversified during the past three decades and made major gains with respect to both domestic and export markets. The country's arid northern tier has led the way as major irrigation projects have been built on streams flowing down from the interior highlands. Along the booming northwest coast of the mainland, which lies within a day's drive of Southern California, mechanized large-scale cotton production now supplies an increasingly profitable export trade. Here, too, wheat and winter vegetables are grown, with fruit and vegetable cultivation attracting foreign investors.

Energy Resources

While Mexico's metal mining industries are less important today than they once were, since 1970 the country has enjoyed the advantages and suffered the problems of being a major petroleum producer. Huge oilfields centered on the southern Gulf Coast's Bay of Campeche around Villahermosa in Tabasco State brought Mexico abundant revenues when the world oil price was high in the late 1970s and serious economic difficulties when the price fell after 1980. These discoveries of massive oil and natural gas reserves have made Mexico self-sufficient in energy, adding to already substantial reserves located in oilfields along the Gulf Coast to the northwest and inner Yucatán to the northeast (Fig. 4-9). The high petroleum prices of the 1970s stimulated the beginnings of Mexico's economic-geographic transformation, but the oil crash of the 1980s thwarted its momentum for a decade. This crash occurred because the government found itself without the expected oil revenues it needed to pay off the huge foreign loans it had taken to finance domestic development programs. Such a disaster could recur if a future oil boom again tempts Mexican leaders to take such risks.

Industrialization

Manufacturing is the centerpiece of Mexico's latest development episode, but this economic sector actually got its start exactly a century ago. Blessed with a wide range of raw materials (Fig. 4-9), the country began to industrialize in 1903 with the completion of an iron and steel plant in the northern city of Monterrey. A second steel complex was built at nearby Monclova in the 1950s, a period that saw the spreading of factories across many parts of central Mexico. Since that time, the industrial sector has grown steadily.

The most significant recent development in Mexico's manufacturing geography is the growth of **11** maquiladora plants in the U.S. border zone. **Maquiladoras** are factories that assemble imported, duty-free components and raw materials into finished products. At least 80 percent of these goods are then reexported to the United States, whose import tariffs are due to be phased out in 2003. Although foreign owners benefit from Mexico's lower wage rates, this industrial system also offers many economic advantages for Mexico. These advantages include the expansion of job opportunities, increased foreign investment, and the transfer of new technologies into the country. Unfortunately, there is a downside too: factory workers are usually exploited through long work weeks, receive minimal fringe benefits, and most can only afford to reside in the squalid slums and shacktowns GEODISCOVERIES that often surround the plants.

Although this development program began in the 1960s, it grew only modestly for the next two decades. Suddenly, however, the maquiladoras took flight, and by the early 1990s about 1800 plants were employing some 500,000 workers to assemble such goods as electronic equipment, electrical appliances, and auto parts. Today, approximately 4000 maquiladoras with over 1.2 million employees operate all along the northern border zone in the urban areas mapped in red in Figure 4-11, accounting for nearly one-third of Mexico's industrial jobs and 45 percent of its total exports.

The Mexican government is capitalizing on the success of the maquiladoras by promoting industrial growth elsewhere in the country. What development planners desire most is that industrial firms create complete manufacturing complexes within the heart of Mexico rather than limit their investments to assembly plants that hug the U.S. border. A number of multinational corporations have recently undertaken such ventures, most notably U.S. automakers and several of their Japanese and German competitors.

To speed the opening of the rest of the country, priority is given to efforts aimed at upgrading Mexico's infrastructure to international standards, particularly its telecommunications, transportation, and electrical-power networks. Individual megaprojects to boost regional development are also being pursued, most **12** ambitiously the **dry canal** across the 150-mile (250-km) Tehuantepec isthmus to move containerized goods between the Pacific port of Salina Cruz and the Gulf port of Coatzacoalcos. This would not only be an overland trade route to rival the aging Panama Canal, but also an ultramodern road/rail corridor lined with factories that specialize in the manufacturing and assembly of goods in transit between the affluent countries of the Asian Pacific Rim and the North Atlantic Basin.

Uneven Regional Development

Despite such efforts, an economic divide is deepening between the southern and northern halves of Mexico. South of the capital, technological and social development lags far behind the rest of the country as traditional productive activities (low-output farming, mining, logging) continue to dominate. North of Mexico City, growth corridors are stirring that increasingly exhibit the landscape of the global economy in the form of new manufacturing facilities, technology training academies, and even the beginnings of a "Silicon Valley of Mexico" on the edge of Guadalajara. To capture the forces transforming northern Mexico, we focus on one of its leading urban centers, Monterrey (a second vignette of Ciudad Juárez/El Paso is included in the Supplementary Material).

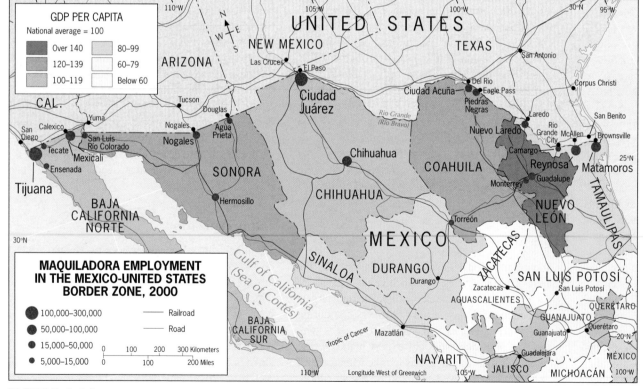

FIGURE 4-11

Burgeoning Monterrey—150 miles (250 km) inside Mexico, yet close enough to the Texas border to have benefited from all the recent development trends—is frequently singled out as a model of what a future Mexican growth center should be. Among this city's assets are a highly educated and well-paid labor force, a stable international business community, and a thriving high-technology complex of ultramodern industrial facilities that has attracted blue-chip multinational companies. In addition, a new expressway link to the Rio Grande is helping to forge an international growth corridor between Monterrey and Dallas-Fort Worth, which is becoming a primary axis of cross-border trade with Texas. For Mexico as a whole, however, there may also be a negative side to Monterrey's success. This city and its surrounding State of Nuevo León have now become so prosperous (note its standout GDP in Fig. 4-11) that there is widening talk of secession. Similar to the devolutionary movement in northern Italy's Lombardy (p. 59), affluent Nuevo León—which receives back less than 20 centavos in services for every tax peso (100 centavos) it sends to Mexico City—has increasingly complained that the federal government tries to develop the impoverished South by taking more money from the industrial North.

NAFTA and Continuing Challenges

The launching of the North American Free Trade zone in 1994 was supposed to mark an economic turning point for Mexico. Instead, NAFTA's early years were plagued by an unending string of crises that included the Chiapas rebellion, monetary devaluation, economic recession, crime waves associated with uncontrolled drug trafficking, and political scandals and transitions. Nonetheless, trade with the United States increased substantially: more than 85 percent of Mexican exports now go to the United States, which is also the source for about 75 percent of all imported goods. At the same time, Mexico has become a leading U.S. trade partner, in 1997 dislodging Japan as the United States' second-largest export market after Canada.

As Mexico completes its economic recovery, we should keep in mind that NAFTA is the first reciprocal trade agreement among high-income and upper-middle-income countries. As events since 1993 have demonstrated, there are bound to be internal as well as external problems during the transition to the new supranational economic order. Clearly, Mexico's development has been enhanced by NAFTA, although the geographic distribution of benefits to date has favored the more affluent northern States.

As Mexico strives to join the ranks of the world's advantaged states, the movements of its inhabitants indicate that the country is still a wobbly giant. About one million Mexicans, lured by the promise of jobs in manufacturing and agriculture, annually migrate to the U.S. border zone. Most migrants would prefer to keep on going north, and hundreds of thousands do manage to cross the U.S. boundary each year. Large numbers of frustrated Mexicans also attempt to leave the country illegally, but are intercepted and deported by U.S. authorities. Undoubtedly, many more get through, which has triggered heightened public support in the United States for tougher measures to curb this flow.

Mexico today is at a crossroads. Its most recent national election has installed a new regime that is expected to produce new initiatives aimed at reducing the divisiveness of the U.S. border, devolving power to the States, resolving the Chiapas conflict, and continuing the revitalization of the economy. Expectations are high throughout the country, and results should not be long in coming.

▷ THE CENTRAL AMERICAN REPUBLICS

Crowded onto the narrow segment of the Middle American land bridge between Mexico and the South American continent are the seven countries of Central America (Fig. 4-12). Territorially, they are all quite small; their population sizes range from Guatemala's 13.5 million down to Belize's 280,000. Physiographically, the land bridge consists of a highland belt flanked by coastal lowlands on both the Caribbean and Pacific sides (Fig. 4-1). These highlands are studded with volcanoes, and local areas of fertile volcanic soils are scattered throughout them. From earliest times, the region's inhabitants have been concentrated in this upland zone, where tropical temperatures are moderated by elevation and rainfall is sufficient to support a variety of crops.

Altitudinal Zonation of Environments

Continental Middle America and the western margin of South America are areas of high relief and strong environmental contrasts. Even though temperate intermontane basins and valleys are most favored, people also cluster in hot tropical lowlands as well as high plateaus just below the snow line in South America's Andes Mountains. In each of these zones, distinct local climates, soils, vegetation, crops, domestic animals, and modes of life **13** prevail. Such **altitudinal zones** (diagrammed in Fig. 4-13) are known by specific names as if they were regions with distinguishing properties—as, in reality, they are.

The lowest vertical zone, from sea level to 2500 feet **14** (about 750 m), is known as the *tierra caliente*, the "hot land" of the coastal plains and low-lying interior basins where tropical agriculture predominates. Above this zone lie the tropical highlands containing Middle and South America's largest population clusters, the **15** *tierra templada* of temperate land reaching up to about 6000 feet (1800 m). Temperatures here are cooler; prominent among the commercial crops is coffee, while corn (maize) and wheat are the staple grains. Still higher, from about 6000 feet to nearly 12,000 feet **16** (3600 m), is the *tierra fría*, the cold country of the higher Andes where hardy crops such as potatoes and barley are mainstays. Above the tree line, which marks **17** the upper limit of the *tierra fría*, lies the *tierra helada*; this fourth altitudinal zone, extending from about 12,000 to 15,000 feet (3600 to 4500 m), is so cold and barren that it can support only the grazing of sheep and other hardy livestock. The highest zone of **18** all is the *tierra nevada*, a zone of permanent snow and ice associated with the loftiest Andean peaks.

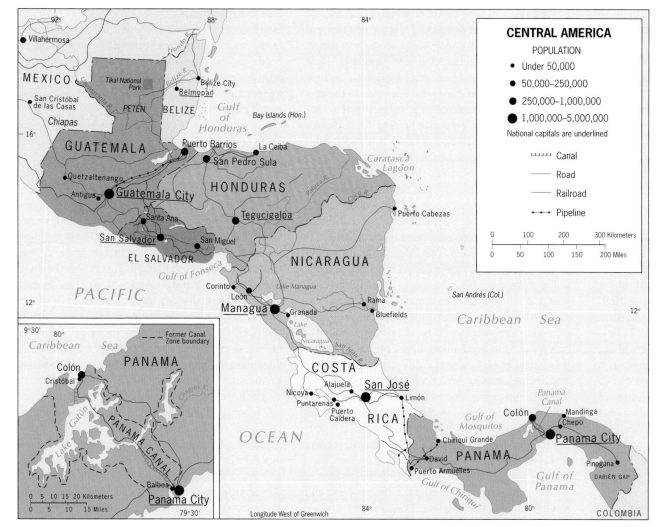

FIGURE 4-12

Population Patterns

The distribution of population within Central America, apart from its concentration in the uplands, also exhibits greater densities toward the Pacific than toward the Caribbean coastlands (Fig. I-9). El Salvador, Belize, and (to some degree) Panama are exceptions to the rule that people in continental Middle America are concentrated in the *tierra templada* zone. El Salvador, in particular, is mostly tropical

tierra caliente, and the majority of its people are crowded onto the intermontane plains lying less than 2500 feet (750 m) above sea level.

By contrast, the Caribbean coastal lowlands—hot, wet, and awash in infertile soils—support comparatively few people. In the region's most populous republic, Guatemala, the heartland also has long been in the southern highlands. Although the large majority of Costa Rica's population is clustered in the Central Valley around San José, the Pacific lowlands have

been the scene of major in-migration since banana plantations were established there. Even in Panama there is a strong Pacific orientation, with more than half of all Panamanians living in the southwestern lowlands and on adjoining mountain slopes.

Middle America's smaller republics face the same problems as the less developed parts of Mexico, and they also share many of the difficulties that confront the poorer Caribbean islands. No present or future challenge, however, is greater than Central America's overpopulation. The region's population explosion began a half-century ago, expanding from a base of about 9 million people in 1950 to 39 million in 2002. Unlike Mexico, which has markedly reduced its rate of natural increase since 1980, Central America (except for Costa Rica and Panama) is on a course that will see a doubling of today's population to 78 million by 2028. This amounts to nothing less than an onrushing demographic catastrophe in a region already unable to cope with most of its social, economic, and natural resource problems.

Emergence From a Turbulent Era

Devastating inequities, repressive governments, external interference, and the frequent unleashing of armed forces have destabilized Central America for much of its modern history. The roots of these upheavals are old and deep, and today the region continues its struggle to emerge from a period of turmoil that lasted through the 1980s into the mid-1990s.

Central America is not a large region, but because of its physiography it contains many isolated, comparatively inaccessible locales. Conflicts between Amerindian population clusters and mestizo groups are endemic to the region, and contrasts between the

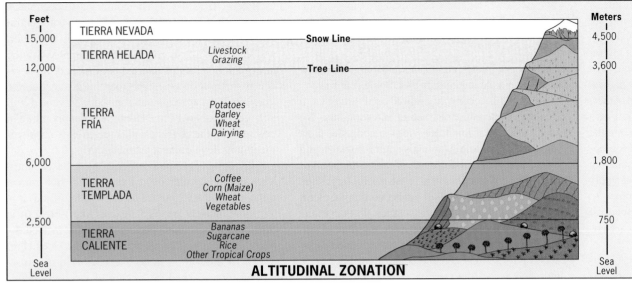

FIGURE 4-13

Most populous of the seven republics with 13.5 million inhabitants (mestizos are in the majority with 56 percent, Amerindians 44 percent), Guatemala has seen much conflict. Repressive regimes made deals with U.S. and other foreign economic interests that stimulated development, but at a high social cost. Over the past half-century, military regimes have dominated political life. The deepening split between the wretchedly poor Amerindians and the better-off mestizos, who here call themselves *ladinos*, generated a civil war that started in 1960 and has since claimed more than 200,000 lives as well as 50,000 "disappearances." An overwhelming number of the victims have been of Mayan descent; the mestizos control the government, army, and land-tenure system.

privileged and the poor are especially harsh. Dictatorial rule by local elites followed authoritarian rule by Spanish colonizers, and the latest episode of violent confrontation was simply another manifestation of this persistent polarization.

Despite the challenges that remain, battle-scarred Central America's prospects are finally brightening. The expanding opportunities of individual republics are highlighted in the country profiles that follow. An important breakthrough is also occurring at the supranational level, where a new spirit of cooperation is forging a sense of regional identity that has barely existed in the past. What began in 1993 as an exploratory effort to resuscitate intraregional trade within the framework of the 30-year-old Central American Common Market soon escalated into a series of pacts to create a more meaningful economic union. At the same time, free-trade agreements were negotiated with several nearby countries outside Central America.

Putting these new concepts into daily practice, however, requires that Central America's republics overcome some formidable obstacles. Foremost among them is the liability of a ramshackle regional infrastructure (symbolized by the decrepit Pan-American Highway), which has long hindered internal as well as

cross-border movement and interaction. Other challenges are more easily surmounted. One example is the implementation of procedures smoothing the flow of goods across international boundaries that until recently did not even have adjacent customs houses open at the same time.

Country Profiles

Guatemala, the westernmost of Central America's republics, has more land neighbors than any other. Straight-line boundaries across the tropical forest mark much of the border with Mexico, creating the box-like region of Petén between Chiapas State on the west and Belize on the east; also to the east lie Honduras and El Salvador (Fig. 4-12). This heart of the ancient Maya Empire, which remains strongly infused by Amerindian culture and tradition, has just a small window on the Caribbean but a longer Pacific coastline. Guatemala was still part of Mexico when the Mexicans threw off the Spanish yoke, and although independent from Spain after 1821, it did not become a separate republic until 1838. Mestizos, not the Amerindian majority, secured the country's independence.

Because the insurgents were able to take refuge in Guatemala's remote, rainforested plains on the Mexican border, the war was unwinnable for the ladinos based in the core area of the southern highlands. Rising international outrage over nearly 40,000 human-rights violations, perpetrated by the armed forces and their paramilitary death squads, finally forced the government to the negotiating table, and in 1996 a peace agreement was concluded. For more than a year the political violence ceased, and initial steps toward stability were taken. But in 1998 the hopeful new atmosphere was shattered by the assassination of one of the country's leading human-rights advocates, Bishop Juan José Gerardi, an event that continues to cast a deep shadow over the country.

The tragedy of Guatemala is that its economic geography has considerable potential but has long been shackled by the unending internal conflicts that have kept the income level of 80 percent of the population below the poverty line. The country's mineral wealth includes nickel in the highlands and oil in the lower-lying north. Agriculturally, soils are fertile and moisture is ample over highland areas large enough to produce a wide range of crops including excellent coffee. Ironically, this promising upland zone is now an area of out-migration for a horde of mostly indigenous, land-hungry peasants who are streaming northward into the infertile tropical rain-forests of the Petén. The push factors in this migration are the increasing pressures on traditional subsistence farmlands propelled by the realm's highest birth rate, the return of tens of thousands of Amerindians from Mexican exile following the civil war, and ineffective land reform. The pull factors are the (unjustified) beliefs that the land is rich and job opportunities are plentiful, plus a savvy understanding that the government uses the Petén as a social safety valve and looks the other way when it comes to squatter settlements as well as the enforcement of conservation regulations regarding the area's environmentally sensitive woodlands.

Belize, strictly speaking, is not a Central American republic in the same tradition as the other six. Until 1981, this country, a wedge of land between northern Guatemala, Mexico's Yucatán Peninsula, and the Caribbean, was a dependency of the United Kingdom known as British Honduras. Slightly larger than Massachusetts and with a minuscule population of only 280,000 (many of African descent), Belize has been more reminiscent of a Caribbean island than of a continental Middle American state. Today, all that is changing as the demographic complexion of Belize is being reshaped. Thousands of residents of African descent have recently emigrated (many went to the United States) and were replaced by tens of thousands of Spanish-speaking immigrants. Most of the latter are refugees from strife in nearby Guatemala, El Salvador, and Honduras, and their proportion of the Belizean population has risen from 33 to nearly 50 percent since 1980. Within the next few years the newcomers will be in the majority, Spanish will become the *lingua franca*, and Belize's cultural geography will exhibit an expansion of the Mainland at the expense of the Rimland.

The Belizean transformation extends to the economic sphere as well. No longer just an exporter of sugar and bananas, Belize is producing new commercial crops, and its seafood-processing and clothing industries have become major revenue earners. Also important is tourism, which annually lures more than 150,000 vacationers to the country's Mayan ruins, re-sorts, and newly legalized casinos; a growing speciality is ecotourism, based on the natural attractions of the country's near-pristine environment. Belize is also known as a center for *offshore banking*—a financial haven for foreign companies and individuals who want to avoid paying taxes in their home countries.

Honduras is a country on hold as it struggles to rebuild its battered infrastructure and economy. We noted earlier that hurricanes rank among the most dangerous of Middle America's natural hazards, and Honduras had the misfortune of taking the brunt of the region's worst in 1998. That tropical cyclone, named Mitch, rampaged all across northern Central America and proved to be the costliest disaster in the modern history of the Western Hemisphere. Honduras was hit the hardest, and the consequences were catastrophic as massive floods and mudslides were unleashed across the country, killing 9200 people, demolishing more than 150,000 homes, destroying 21,000 miles of roadway and 335 bridges, and rendering 2 million homeless. Also devasted was the critical agricultural sector that employed two-thirds of Honduras's labor force, accounted for nearly a third of its gross domestic product, and earned more than 70 percent of its foreign revenues.

With 6.4 million inhabitants, about 90 percent mestizo, bedeviled Honduras still has years to go even to restore what was already the third poorest economy in the Americas (after Haiti and Nicaragua). Agriculture, livestock, forestry, and limited mining formed the mainstays of the pre-1998 economy, with the familiar Central American products—bananas, coffee, shellfish, apparel—earning most of the external income. There has also been some growth of maquiladora-type light industry around San Pedro Sula near the northwestern coast, dominated by a complex of sweatshops that finish clothing designed and mostly produced elsewhere. Foreign investors, however, have been wary of risking additional funds because of the country's reputation as one of the world's most corrupt (U.S.-bound drug smuggling flourishes here). Nonetheless, Honduras has a democratically elected government, although the military wield considerable power.

Honduras, in direct contrast to Guatemala, has a lengthy Caribbean coastline and a small window on the Pacific (Fig. 4-12). The country also occupies a critical place in the political geography of Central America, flanked as it is by Nicaragua, El Salvador, and Guatemala—all continuing to grapple with the aftermath of years of internal conflict and, most recently, natural disaster. The road back to economic viability is an arduous one, but once traversed will still leave four out of five Hondurans deeply mired in poverty and the country with little overall improvement in its development prospects.

El Salvador is Central America's smallest country territorially, smaller even than Belize, but with a population about 25 times as large (6.6 million) it is the most densely peopled. Again, like Belize, it is one of only two continental republics that lack coastlines on both the Caribbean and Pacific sides (Fig. 4-12). El Salvador adjoins the Pacific in a narrow coastal plain backed by a chain of volcanic mountains, behind which lies the country's heartland. Unlike neighboring Guatemala, El Salvador has a quite homogeneous population (94 percent mestizo and just 5 percent Amerindian). Yet ethnic homogeneity has not translated into social or economic equality or even opportunity. Whereas other Central American countries were called banana republics, El Salvador was a coffee republic, and the coffee was produced on the huge landholdings of a few landowners and on the backs of a subjugated peasant labor force. The military supported this system and repeatedly suppressed violent and desperate peasant uprisings.

From 1980 to 1992, El Salvador was torn by a devastating civil war that was worsened by outside arms supplies from the United States (supporting the government) and Nicaragua (aiding the Marxist rebel forces). But ever since the negotiated end to that war, efforts have been under way to prevent a recurrence because El Salvador is having difficulty overcoming its legacy of searing inequality. The civil war did have one positive result: affluent citizens who left the country and did well in the United States and elsewhere send substantial funds back home, which now provide the largest single source of foreign revenues. This has helped stimulate such industries as apparel and footwear manufacturing, as well as food processing. But a major stumbling block to revitalization of the agricultural sector has again been land reform, and El Salvador's future still hangs in the balance.

Nicaragua is best approached by reexamining the map (Fig 4-12), which underscores the country's pivotal position in the heart of Central America. The Pacific coast follows a southeasterly direction, but the Caribbean coast is oriented north-south so that Nicaragua forms a triangle of land with its lakeside capital, Managua, located in a valley on the mountainous, earthquake-prone, Pacific side (the country's core area has always been located here). The Caribbean side, where the uplands yield to a coastal plain of rainforest, savanna, and swampland, has for centuries been home to Amerindian peoples such as the Miskito, who have been remote from the focus of national life.

Until the end of the 1970s, Nicaragua was the typical Central American republic, ruled by a dictatorial government and exploited by a wealthy land-owning minority, its export agriculture dominated by huge plantations owned by foreign corporations. It was a situation ripe for insurgency, and in 1979 leftist rebels overthrew the government; but the new regime quickly produced its own excesses, resulting in civil war for most of the 1980s. The conflict ended in 1990, and more democratic governments have since been voted into office. Nonetheless, the former rebels continue to wield influence in Nicaraguan society, and the country remains too divided to resolve most economic and political issues.

Nicaragua's economy has been a leading casualty of this turmoil, and for the past two decades it has ranked as continental Middle America's poorest. Still mired in the difficult aftermath of Hurricane Mitch, Nicaragua is struggling to rebuild much of its infrastructure. Thus it faces yet another challenge piled upon its already formidable burdens. Agricultural recovery is a leading priority: not only are commercial-farm products major foreign income earners, but if farming opportunities are not promptly restored, emigration from the countryside to the badly overcrowded towns and cities will accelerate and swamp prospects for urban economic development. Moreover, in the months before Mitch, the Nicaraguans had finally resolved property-ownership issues stemming from land redistribution during the civil war. With more than 200,000 families and at least one-third of the country's arable farmland affected, continued national stability necessitates that this vital social reform get back on track.

When "normal" conditions return, Nicaragua will still possess few comparative advantages in the competition for foreign investment capital against the region's more prosperous republics. Among the more promising development possibilities is the construction of a *dry canal* across the flatlands of southern Nicaragua, with high-speed trains ferrying freight-filled containers between Caribbean and Pacific ports that can be directly loaded onto ultramodern ships too large to fit through the isthmus's aging Panama Canal. Perhaps Middle America's land bridge can support additional interocean transit corridors, but here Nicaragua will confront competitors who are proposing such projects in Mexico (as we saw), Colombia, and even in Panama itself. Economic opportunity, however, must also be weighed against looming demographic disaster: unless Nicaragua can reduce the very high natural increase rate in its population of 5.4 million, the country's living standards will not have a chance to improve.

Costa Rica underscores what was said about Middle America's endless variety and diversity because it differs significantly from its neighbors and from the norms of Central America as well. Bordered by two volatile countries (Nicaragua to the north and Panama to the east), Costa Rica is a nation with an old democratic tradition and, in this cauldron, no standing army for the past half-century! Although the country's Hispanic imprint is similar to that found elsewhere on the Mainland, its early independence, its good fortune to lie remote from regional strife, and its leisurely pace of settlement allowed Costa Rica the luxury of concentrating on its economic development. Perhaps most important, internal political stability has prevailed over much of the past 175-odd years.

Like its neighbors, Costa Rica, is divided into environmental zones that parallel the coasts. The most densely settled is the central highland zone, lying in the cooler *tierra templada*, whose heartland is the Valle Central (Central Valley), a fertile basin that contains the country's main coffee-growing area and the leading population cluster focused on San José—the virtually slumless capital city that is the most cosmopolitan urban center between Mexico City and the primate cities of northern South America. To the east of the highlands are the hot and rainy Caribbean lowlands, a sparsely populated segment of Rimland where many plantations have been abandoned and replaced by subsistence farmers. Between 1930 and 1960, the U.S.-based United Fruit Company shifted most of the country's banana plantations from this crop-disease-ridden coastal plain to Costa Rica's third zone—the plains and gentle slopes of the Pacific coastlands. This move gave the Pacific zone a major boost in economic growth, and it is now an area of diversifying and expanding commercial agriculture.

The long-term development of Costa Rica's economy has given it the region's highest standard of living, literacy rate, and life expectancy. Agriculture continues to dominate (with bananas, coffee, seafood, and tropical fruits the leading exports), and tourism is expanding steadily as the country now attracts more than a million foreign visitors a year. But the most impressive economic gains are being generated by a single new industrial complex that the Intel Corporation opened just outside San José in 1998. Here, the world's leading maker of semiconductors has built one of its largest assembly/test facilities in which silicon chips manufactured at fabrication plants are transformed into the processors that constitute the brains of personal computers.

These advances notwithstanding, the country's veneer of development cannot mask serious problems. In terms of social structure, about one-quarter of the population is trapped in an unending cycle of poverty, and the huge gap between the poor and the affluent is constantly widening. In the environmental sphere, the price of recent progress has been **19** **tropical deforestation**: even though the rate of woodland destruction is now slowing, it is too late to avert an ecological disaster because more than 70 percent of Costa Rica's rainforest has vanished.

Politically, Costa Rica remains quite stable. Despite its proximity to the region's trouble spots, it has resisted involvement because the overwhelming majority of its peace-loving people prefer the country to maintain its neutrality as "the Switzerland of Central

America." One of the few clouds on the horizon is a longstanding boundary dispute with neighboring Nicaragua over the waters and banks of the San Juan River, a quarrel that extends westward to include the southern shore of Lake Nicaragua (Fig. 4-12). If this corridor were to change hands, it is likely that Costa Rica would become yet another entrant in the contest to build a new interocean routeway across Central America.

Panama owes its existence to the idea of a canal connecting the Atlantic and Pacific oceans to avoid the lengthy circumnavigation of South America. In the 1880s, when Panama was still an extension of neighboring Colombia, a French company tried and failed to build such a waterway here. By the turn of the twentieth century, U.S. interest in a Panama canal rose sharply, and the United States in 1903 proposed a treaty that would permit a renewed effort at construction across Colombia's Panamanian isthmus. When the Colombian Senate refused to go along, Panamanians rebelled, and the United States supported this uprising by preventing Colombian forces from intervening. The Panamanians, at the behest of the United States, declared their independence from Colombia, and the new republic immediately granted the United States rights to the Canal Zone, averaging about 10 miles (16 km) in width and just over 50 miles (80 km) in length.

Soon canal construction commenced, and this time the project succeeded as American technology and medical advances triumphed over a formidable set of obstacles. The Panama Canal (see inset map, Fig. 4-12) was opened in 1914, a symbol of U.S. power and influence in Middle America. The Canal Zone was held by the United States under a treaty that granted it "all the rights, powers, and authority" in the area "as if it were the sovereign of the territory." Such language might suggest that the United States held rights over the Canal Zone in perpetuity, but the treaty nowhere stated specifically

that Panama permanently yielded its own sovereignty in that transit corridor. In the 1970s, as the canal was transferring more than 14,000 ships per year (that number is now only slightly lower, but the cargo tonnage is up significantly) and generating hundreds of millions of dollars in tolls, Panama sought to terminate U.S. control in the Canal Zone. Delicate negotiations began. In 1977, an agreement was reached on a staged withdrawal by the United States from the territory, first from the Canal Zone and then from the Panama Canal itself (a process completed on December 31, 1999).

Panama today reflects some of the usual geographic features of the Central American republics. Its population of 3 million is more than two-thirds mestizo and also contains substantial Amerindian, white, and black minorities. Spanish is the official language, but English is also widely used. Ribbonlike and oriented east-west, Panama's topography is mountainous and hilly. Eastern Panama, especially Darien Province adjoining Colombia, is densely forested, and here is the only remaining gap in the intercontinental Pan American Highway. Most of the rural population lives in the uplands west of the canal; there, Panama produces bananas, shrimps and other seafood, sugarcane, coffee, and rice. Much of the urban population is concentrated in the vicinity of the waterway, anchored by the cities at each end of the canal, Colón and Panama City.

The Panama Canal, despite its age and inability to accommodate the largest 20 percent of the vessels in today's global merchant fleet, remains the country's focus, its lifeline, and—with a significant proportion of the world's cargoes moving through the waterway each year—its future. To help assure that future astride this crucial international trading artery, the canal's new Panamanian owners are busily pursuing opportunities they hope will transform their trading hub into a full-fledged Central American economic tiger. Their initial task, now successfully completed,

was to demonstrate to the international community that they are fully capable of operating and maintaining the canal without sharply increasing its tolls. They have also begun to attract major new foreign investments, which are being channeled into a number of maritime, railroad, and manufacturing projects.

One of the anchors of this development is Colón, the port city at the Caribbean end of the canal. Here, in 1948, a free-trade zone was opened, which has since become the world's second-largest such facility after Hong Kong. Most of the zone's activities involved importing and distributing products bound for South America. Today, however, this warehousing operation is being challenged by competitors in Brazil, Paraguay, and Chile. In order to maintain their supremacy, Colón's leaders are upgrading their goods-handling technology to accommodate containers (whose importance in global commerce was underscored in our earlier discussions of *dry canal* projects in Mexico and Nicaragua). One result is the huge new Manzanillo International Terminal, adjacent to the free-trade zone, an ultramodern port facility capable of transshipping more than 1000 containers a day.

Today, the Pacific end of the canal is an even more active hub of new development. Among its most prominent projects are cargo-related industrial parks and a tourist port to lure the thousands of passengers aboard the 300-plus cruise ships that annually traverse the waterway. As for Panama City itself, the only coastal capital in continental Middle America, an expanding cluster of downtown high-rises suggests Miami rather than San Salvador or San José. What generates this world-class skyline that towers over an urban area anchoring a country of only 3 million people? The official answer: international banking. The geographic answer suggests the power of relative location, including a growing linkage to drug-plagued Colombia, the country on which the Panamanians turned their backs exactly a century ago.

5 / South America

Rio de Janeiro, Brazil: its capital functions lost to Brasilia, its primacy to Sao Paulo — but its incomparable site and setting will always sustain this as the country's quintessential city.

This Chapter's Media Highlights include:

Photo Gallery: South America

GeoDiscoveries: Land-Use Patterns

3 Interactive Map Quizzes: Countries, Cities, Physical Features

Chapter Quiz

Concept Quiz

Virtual Field Trip: A Trip Up the Amazon and Rio Negro www.wiley.com/college/deblij

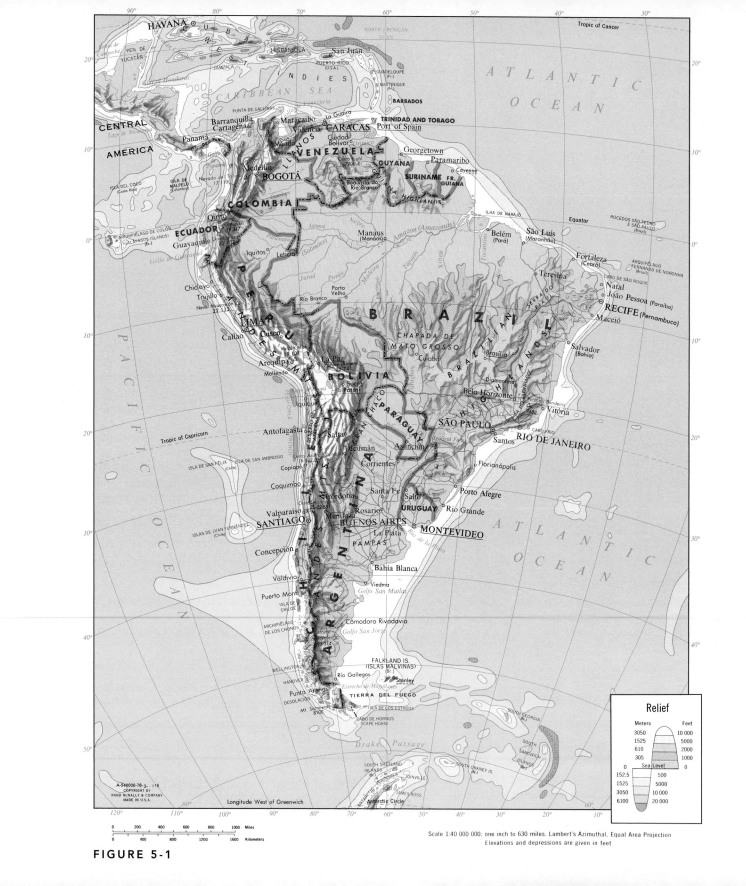

FIGURE 5-1

Scale 1:40 000 000; one inch to 630 miles. Lambert's Azimuthal, Equal Area Projection
Elevations and depressions are given in feet

chapter 5 / South America

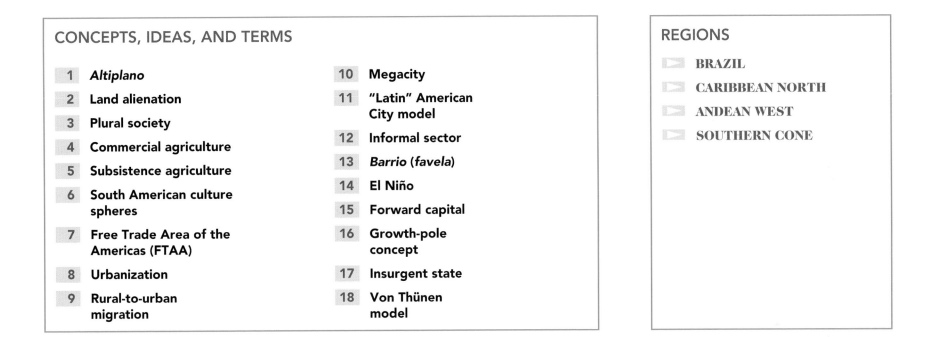

CONCEPTS, IDEAS, AND TERMS

1 *Altiplano*

2 Land alienation

3 Plural society

4 Commercial agriculture

5 Subsistence agriculture

6 South American culture spheres

7 Free Trade Area of the Americas (FTAA)

8 Urbanization

9 Rural-to-urban migration

10 Megacity

11 "Latin" American City model

12 Informal sector

13 *Barrio (favela)*

14 El Niño

15 Forward capital

16 Growth-pole concept

17 Insurgent state

18 Von Thünen model

REGIONS

- BRAZIL
- CARIBBEAN NORTH
- ANDEAN WEST
- SOUTHERN CONE

Of all the continents, South America has the most familiar shape—a giant triangle connected by mainland Middle America's tenuous land bridge to its sister continent in the north. South America also lies not only south but mostly east of its northern counterpart. Lima, the capital of Peru—one of the continent's westernmost cities—lies farther east than Miami, Florida. Thus South America juts out much more prominently into the Atlantic Ocean toward southern Europe and Africa than does North America. But lying so far eastward means that South America's western flank faces a much wider Pacific

Ocean, with the distance from Peru to Australia nearly twice that from California to Japan.

As if to reaffirm South America's northward and eastward orientation, the western margins of the continent are rimmed by one of the world's longest and highest mountain ranges, the Andes, a gigantic wall that extends unbroken from Tierra del Fuego near the southern tip of the triangle to Venezuela in the far north (Fig. 5-1). The other major physiographic feature of South America dominates its central north—the Amazon Basin; this vast humid-tropical amphitheater is drained by the mighty Amazon, which

is fed by several major tributaries. Much of the remainder of the continent can be classified as plateau, with the most important components being the Brazilian Highlands that cover most of Brazil southeast of the Amazon Basin, the Guiana Highlands located north of the lower Amazon Basin, and the cold Patagonian plateau that blankets the southern third of Argentina. Figure 5-1 also reveals two other noteworthy river basins beyond Amazonia: the Paraná-Paraguay Basin of south-central South America, and the Orinoco Basin in the far north that drains interior Colombia and Venezuela.

DEFINING THE REALM

Long characterized by regional disparities, political turmoil, and developmental inertia, South America has entered a new era of opportunity. Its major countries, heretofore accustomed to going their separate ways, are discovering the benefits of forging closer individual and multinational ties. More democratic forms of government now mark the political landscape. New transport routes are opening settlement frontiers in many once-remote parts of the continent. The perception now taking hold is that, finally, things are improving and this realm is at the threshold of a period of unprecedented economic growth.

Such optimism, however, must be tempered by the recognition that serious challenges remain to be overcome. As certain areas of South America make real progress, too many others continue to be plagued by infrastructure shortcomings, inefficiency and corruption, and endless rounds of no-gain, boom/bust cycles. Most importantly, throughout the realm, the gulf between the rich and the poor is steadily widening, with the richest 20 percent of the population controlling *70 percent* of South America's wealth while the poorest 20 percent controls only 2 percent. These numbers reveal the greatest gap between affluence and poverty to be found in any geographic realm—and show that South America is still very much a continent of stupendous contrasts.

THE HUMAN SEQUENCE

Although modern South America's largest populations are situated in the east and north, during the height of the Inca Empire the Andes Mountains contained the most densely peopled and best organized state on the continent. Although the origins of Inca civilization are still shrouded in mystery, it has become generally accepted that the Incas were descendants of ancient peoples who came to South America via the Middle American land bridge. Thus for thousands of years before the Europeans arrived in the sixteenth century, indigenous Amerindian societies had been developing in South America.

The Inca Empire

About one thousand years ago, a number of regional cultures thrived in Andean valleys and basins and at places along the Pacific coast. By AD 1300, the Incas had established themselves in the intermontane basin of Cuzco (Fig. 5-2). With their hearth consolidated, they were now ready to begin forging the greatest pre-European empire in the Americas by steadily conquering and extending their authority over the peoples of coastal Peru and other Andean basins. When the Inca civilization is compared to that of ancient Mesopotamia, Egypt, and the Mexican Aztec Empire, it quickly becomes clear that this civilization was an unusual achievement. Everywhere else, rivers and waterways provided avenues for the circulation of goods and ideas. Here, however, an empire was forged from a **1** series of elongated basins (called *altiplanos*) in the high Andes, created when mountain valleys between parallel and converging ranges filled with erosional materials from surrounding uplands. These *altiplanos* are often separated by some of the world's most rugged terrain, with high snowcapped mountains alternating with precipitous canyons.

More impressive than the Incas' military victories was their subsequent capacity to integrate the peoples and regions of the Andean domain into a stable and efficiently functioning state. The odds would

◆ Major Geographic Qualities of South America

1. South America's physiography is dominated by the Andes Mountains in the west and the Amazon Basin in the central north. Much of the remainder is plateau country.

2. Half of the realm's area and half of its population are concentrated in one country—Brazil.

3. South America's population remains concentrated along the continent's periphery. Most of the interior is sparsely peopled, but sections of it are now undergoing significant development.

4. Interconnections among the states of the realm are improving rapidly. Economic integration has become a major force, particularly in southern South America.

5. Regional economic contrasts and disparities, both in the realm as a whole and within individual countries, are strong.

6. Cultural pluralism exists in almost all of the realm's countries, and is often expressed regionally.

7. Rapid urban growth continues to mark much of the South American realm, and the urbanization level overall is today on a par with the levels in the United States and Europe.

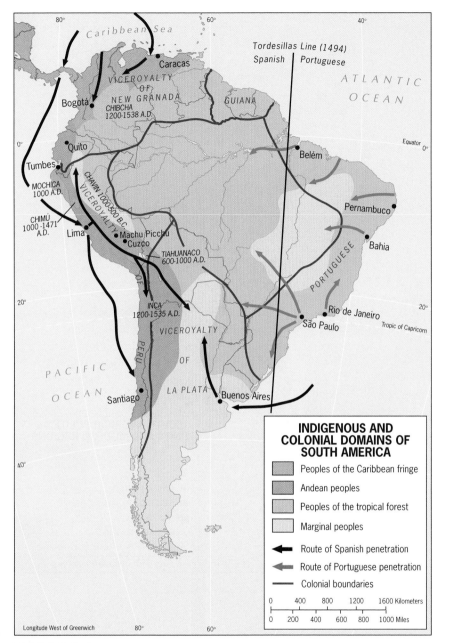

FIGURE 5-2

Map legend:

INDIGENOUS AND COLONIAL DOMAINS OF SOUTH AMERICA

- Peoples of the Caribbean fringe
- Andean peoples
- Peoples of the tropical forest
- Marginal peoples
- ← Route of Spanish penetration
- ← Route of Portuguese penetration
- — Colonial boundaries

0   400   800   1200   1600 Kilometers

0   200   400   600   800   1000 Miles

seem to have been against them because as they progressed their domain became ever more elongated, making effective control much more difficult. The Incas, however, were expert road and bridge builders, colonizers, and administrators, and in an incredibly short time they unified their new territories that stretched from Colombia southward to central Chile (brown zone, Fig. 5-2).

The Incas themselves were always in a minority in this huge state, and their position became one of a ruling elite in a rigidly class-structured society. A bureaucracy of Inca administrators strictly controlled the life of the empire's subjects, and the state was so highly centralized that a takeover at the top was enough to gain power over the entire empire—as the Spaniards quickly proved in the 1530s. The Inca Empire disintegrated abruptly under the impact of the Spanish invaders, but it left behind spectacular ruins such as those at Peru's Machu Picchu. It also bequeathed a legacy of social values that have remained a part of Amerindian life in the Andes to this day and still contribute to fundamental divisions between the Hispanic and Amerindian population in this part of South America.

The Iberian Invaders

In South America as in Middle America, the location of indigenous peoples largely determined the direction of the thrusts of European invasion. The Incas, like Mexico's Maya and Aztec peoples, had accumulated gold and silver at their headquarters, possessed productive farmlands, and constituted a ready labor force. Not long after the defeat of the Aztecs in 1521, Francisco Pizarro sailed southward along the continent's northwestern coast, learned of the existence of the Inca Empire, and withdrew to Spain to organize its overthrow. He returned to the Peruvian coast in 1531 with 183 men and two dozen horses, and the events that followed are well known. In 1533, his party rode victorious into Cuzco.

At first, the Spaniards kept the Incan imperial structure intact by permitting the crowning of an emperor who was under their control. But soon the breakdown of the old order began. The new order that gradually emerged in western South America placed the indigenous peoples in serfdom to the Spaniards. **2** Great haciendas were formed by **land alienation** (the takeover of former Amerindian lands), taxes were instituted, and a forced-labor system was introduced to maximize the profits of exploitation.

Lima, the west coast headquarters of the Spanish conquerors, soon became one of the richest cities in the world, its wealth based on the exploitation of vast Andean silver deposits. The city also served as the capital of the viceroyalty of Peru, as the Spanish authorities quickly integrated the new possession into their colonial empire (Fig. 5-2). Subsequently, when Colombia and Venezuela came under Spanish control and, later, when Spanish settlement expanded in what is now Argentina and Uruguay, two additional viceroyalties were added to the map: New Granada and La Plata.

Meanwhile, another vanguard of the Iberian invasion was penetrating the east-central part of the continent, the coastlands of present-day Brazil. This area had become a Portuguese sphere of influence because Spain and Portugal had signed a treaty in 1494 to recognize a north-south line 370 leagues west of the Cape Verde Islands as the boundary between their New World spheres of influence. This border ran approximately along the meridian of 50°W longitude, thereby cutting off a sizeable triangle of eastern South America for Portugal's exploitation (Fig. 5-2). But a brief look at the political map of South America (Fig. 5-3) shows that this treaty did not limit Portuguese colonial territory to the east of the 50th meridian. Instead, Brazil's boundaries were bent far inland to include almost the entire Amazon Basin, and the country came to be only slightly smaller in territorial size than all the other South American countries combined. This westward thrust was the work of many Brazilian elements, particularly the *Paulistas*, the settlers of São Paulo who needed Amerindian slave labor to run their plantations.

The Africans

As Figure 5-2 shows, the Spaniards initially got very much the better of the territorial partitioning of South America—not just in land quality but also in the size

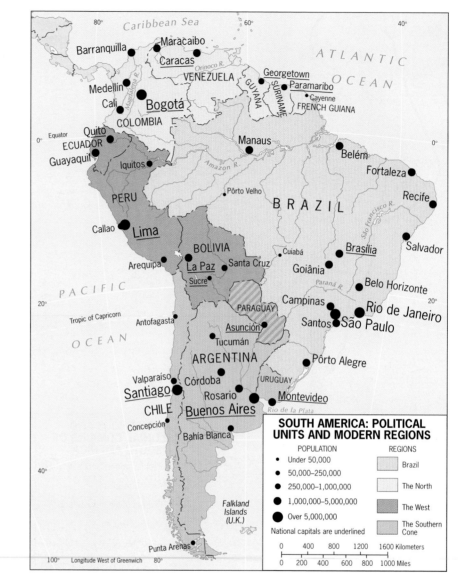

FIGURE 5-3

of the aboriginal labor force. When the Portuguese began to develop their territory, they turned to the same lucrative activity that their Spanish rivals had pursued in the Caribbean—the plantation cultivation of sugar for the European market. And they, too, found their labor force in the same source region, as millions of Africans were brought in slavery to the

tropical Brazilian coast north of Rio de Janeiro (see map, p. 219). Not surprisingly, Brazil now has South America's largest black population, which is still heavily concentrated in the country's poverty-stricken northeastern States. With blacks of "pure" and mixed ancestry today accounting for 44 percent of Brazil's population of 175 million, the Africans decidedly constitute the third major immigration of foreign peoples into South America.

Longstanding Isolation

Despite their adjacent location on the same continent, their common language and cultural heritage, and their shared national problems, the countries that arose out of South America's Spanish viceroyalties (as well as Brazil) have existed in a considerable degree of isolation from one another. Distance and physiographic barriers reinforced this separation, and the realm's major population agglomerations still adhere to the coast, mainly the eastern and northern coasts (Fig. I-9). The viceroyalties existed primarily to extract riches and fill Spanish coffers. In Iberia there was little interest in developing the American lands for their own sake. Only after those who had made Spanish and Portuguese America their home and who had a stake there rebelled against Iberian authority did things begin to change, and then very slowly. South America was saddled with the values, economic outlook, and social attitudes of eighteenth-century Iberia—not the best tradition from which to begin the task of forging modern nation-states.

Independence

Certain isolating factors had their effect even during the wars for independence. Spanish military strength was always concentrated at Lima, and those territories that lay farthest from their center of power—Argentina and Chile—were the first to establish their independence from Spain (in 1816 and 1818, respectively). In the north, Simón Bolívar led the burgeoning independence movement, and in 1824 two decisive military defeats there spelled the end of Spanish power in South America.

This joint struggle, however, did not produce unity because no fewer than nine countries emerged from the three former viceroyalties. It is not difficult to understand why this fragmentation took place. With the Andes intervening between Argentina and Chile and the Atacama Desert between Chile and Peru, overland distances seem even greater than they really were, and these obstacles to contact proved

quite effective. Hence, from their outset the new countries of South America began to grow apart, and friction and even wars have been frequent. Only within the past ten years have the countries of this realm finally begun to recognize the mutual advantages of increasing cooperation and to make lasting efforts to steer their relationships in this direction.

CULTURAL FRAGMENTATION

When we speak of the "interaction" of South American countries, it is important to keep in mind just who does the interacting. The fragmentation of colonial South America into ten individual republics, and the subsequent postures of each of these states, was the work of a small minority that constituted the landholding, upper-class elite. Thus in every country a vast majority—be they Amerindians in Peru or people of African descent in Brazil—could only watch as their European masters struggled with each other for supremacy.

3 South America, then, is a continent of **plural societies**, where Amerindians of different cultures, Europeans from Iberia and elsewhere, blacks from western tropical Africa, and Asians from India, Japan, and Indonesia cluster in adjacent areas but do not mix. The result is a cultural kaleidoscope of almost endless variety, whose internal divisions are also reflected in the realm's economic landscape. This is readily visible in the map of South America's dominant livelihood, agriculture **4** (Fig. 5-4). Here **commercial** (for-profit) and **5** **subsistence** (minimum-life-sustaining) farming exist side by side to a greater degree than anywhere else in the world, where one usually dominates the other. The geography of commercial agricultural systems (map categories 1, 2, 3, 5, 6, and 9) is heavily tied to the distribution of landholders of European background, while subsistence farming (categories 4, 7, and 8) is overwhelmingly associated with the spatial patterns of indigenous peoples as well as populations of African and Asian descent.

Certainly, calling this complex human spatial mosaic "Latin" America (as is so often the case) is not

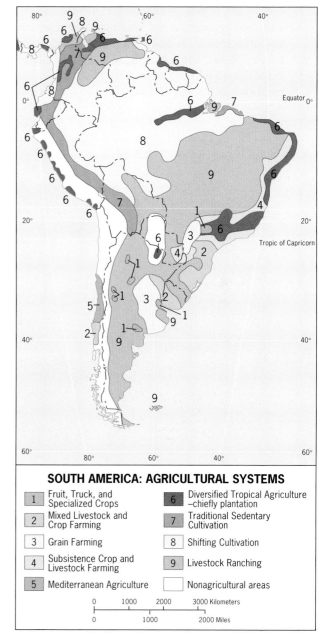

SOUTH AMERICA: AGRICULTURAL SYSTEMS

1 Fruit, Truck, and Specialized Crops
2 Mixed Livestock and Crop Farming
3 Grain Farming
4 Subsistence Crop and Livestock Farming
5 Mediterranean Agriculture
6 Diversified Tropical Agriculture –chiefly plantation
7 Traditional Sedentary Cultivation
8 Shifting Cultivation
9 Livestock Ranching
   Nonagricultural areas

0 1000 2000 3000 Kilometers
0 1000 2000 Miles

FIGURE 5-4

very useful. Is there a more meaningful approach to a regional generalization that would better represent and differentiate the continent's cultural and economic spheres? John Augelli, who also developed the Mainland-Rimland concept for Middle America, made such an attempt. His map (Fig. 5-5) shows that

6 five **South American culture spheres**—internal cultural regions—blanket the realm.

- **Tropical-plantation sphere**: Reminiscent of Middle America's Rimland, consists of five Atlantic- or Caribbean-facing coastal strips in

northern South America. Location and tropical environmental conditions favored plantation crops, especially sugar. Millions of African slave laborers brought here and still strongly influence local cultures. Failure of plantation economy reduced people to the poverty and subsistence that now dominate these five areas.

- **European-commercial sphere**: Covers most of southern, mid-latitude South America, including the core areas of Argentina, Brazil, Chile, and Uruguay. Dominated by populations of European descent with a strong Hispanic cultural imprint. Productive commercial agriculture is the major reason why this region is economically more advanced than the rest of the continent.

- **Amerind-subsistence sphere**: Elongated zone lying astride the central Andes from southern Colombia to northern Chile/northwestern Argentina. The feudal socioeconomic structure established by Spanish conquerors still prevails, with the Amerindian population forming a huge, landless peonage. Much of the population lives in high-altitude environments that are marginal for farming. The region includes some of the realm's poorest areas.

- **Mestizo-transitional sphere**: Surrounds Amerind-subsistence region and covers much of central and northern South America. Zone of mixture between Europeans and Amerindians—or Africans around coastal strips of the tropical-plantation culture sphere. Transitional economy with respect to extremes of European-commercial and Amerind-subsistence regions.

- **Undifferentiated sphere**: The remaining, least accessible areas of the realm that lie in and around the Amazon Basin and in southernmost Chile. Sparsely populated and exhibit only very limited economic development.

ECONOMIC INTEGRATION

As noted above, the separatism that has so long characterized international relations is giving way

SOUTH AMERICA: CULTURE SPHERES

- Tropical-plantation
- European-commercial
- Amerind-subsistence
- Mestizo-transitional
- Undifferentiated

FIGURE 5-5

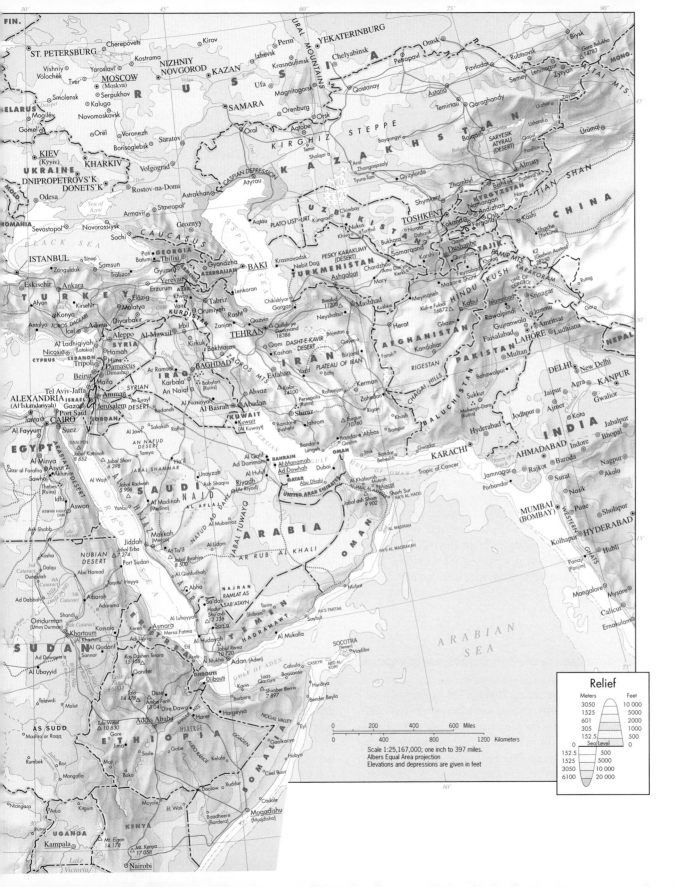

Relief

Meters		Feet
3050		10 000
1525		5000
601		2000
305		1000
152.5		500
0	Sea Level	0
152.5		500
1525		5000
3050		10 000
6100		20 000

200   400   600 Miles

400   800   1200 Kilometers

Scale 1:25,167,000; one inch to 397 miles.
Albers Equal Area projection
Elevations and depressions are given in feet

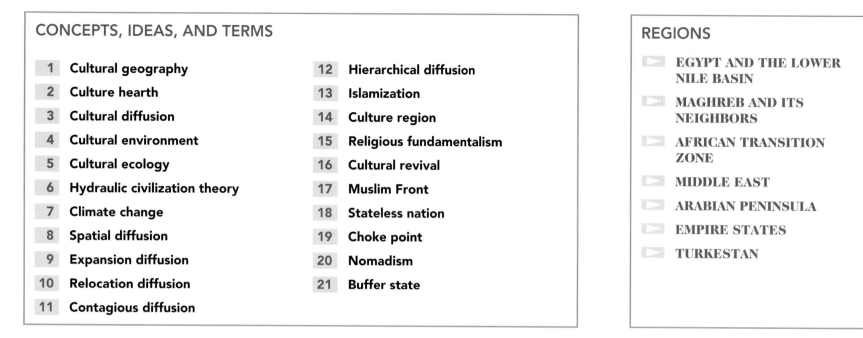

CONCEPTS, IDEAS, AND TERMS

1 Cultural geography
2 Culture hearth
3 Cultural diffusion
4 Cultural environment
5 Cultural ecology
6 Hydraulic civilization theory
7 Climate change
8 Spatial diffusion
9 Expansion diffusion
10 Relocation diffusion
11 Contagious diffusion
12 Hierarchical diffusion
13 Islamization
14 Culture region
15 Religious fundamentalism
16 Cultural revival
17 Muslim Front
18 Stateless nation
19 Choke point
20 Nomadism
21 Buffer state

REGIONS

- EGYPT AND THE LOWER NILE BASIN
- MAGHREB AND ITS NEIGHBORS
- AFRICAN TRANSITION ZONE
- MIDDLE EAST
- ARABIAN PENINSULA
- EMPIRE STATES
- TURKESTAN

From Morocco on the shores of the Atlantic to the mountains of Afghanistan, and from the Horn of Africa to the steppes of inner Asia, lies a vast geographic realm of enormous cultural complexity. It stands at the crossroads where Europe, Asia, and Africa meet, and it is part of all three (Fig. 6-1). Throughout history, its influences have radiated to these continents and to practically every other part of the world as well. This is one of humankind's primary source areas. On the Mesopotamian Plain between the Tigris and Euphrates rivers (in modern-day Iraq) and on the banks of the Egyptian Nile arose several of the world's earliest civilizations. In its soils, plants were domesticated that are now grown from the Americas to Australia. Along its paths walked prophets whose religious teachings are still followed by hundreds of millions. And at the opening of the twenty-first century, the heart of this realm is beset by some of the most bitter and dangerous conflicts on Earth.

DEFINING THE REALM

It is tempting to characterize this geographic realm in a few words and to stress one or more of its dominant features. It is, for instance, often called the "Dry GEODISCOVERIES World," containing as it does the vast Sahara as well as the Arabian Desert. But most of the realm's people live where there is water—in the Nile Delta, along the hilly Mediterranean coastal strip (the *tell*, meaning "mound" in Arabic) of northwesternmost Africa, along the Asian eastern and northeastern shores of the Mediterranean Sea, in the Tigris-Euphrates Basin, in far-flung desert oases, and along the lower mountain slopes of Iran south of the Caspian Sea and of Turkestan to the northeast. We know this world region as one where water is almost always at a premium, where peasants often struggle to make soil and moisture yield a small harvest, where nomadic peoples and their animals circulate across dust-blown flatlands, where oases are islands of sedentary farming and trade in a sea of aridity. But it also is the land of the Nile, the lifeline of Egypt, the crop-covered *tell* of northwestern Africa, the verdant coasts of western

◆ Major Geographic Qualities of North Africa/Southwest Asia

1. North Africa and Southwest Asia were the scene of several of the world's great ancient civilizations, based in its river valleys and basins.

2. From this realm's culture hearths diffused ideas, innovations, and technologies that changed the world.

3. The North Africa/Southwest Asia realm is the source of three world religions: Judaism, Christianity, and Islam.

4. Islam, the last of the major religions to arise in this realm, transformed, unified, and energized a vast domain extending from Europe to Southeast Asia and from Russia to East Africa.

5. Drought and unreliable precipitation dominate natural environments in this realm. Population clusters exist where water supply is adequate to marginal.

6. Certain countries of this realm have enormous reserves of oil and natural gas, creating great wealth for some but doing little to raise the living standards of the majority.

7. The boundaries of the North Africa/Southwest Asia realm consist of volatile transition zones in several places in Africa and Asia.

8. Conflict over water sources and supplies is a constant threat in this realm, where population growth rates are high by world standards.

9. The Middle East, as a region, lies at the heart of this realm; and Israel lies at the center of the Middle East conflict.

10. Religious, ethnic, and cultural discord frequently cause instability and strife in this realm.

Turkey, the meltwater-fed valleys of Central Asia. Compare Figure 6-1 to Figure I-8, and the dominance of *B* climates becomes evident. Also consult Figure I-9, and you will see how water-dependent this realm's clustered population is.

An "Arab World"?

North Africa/Southwest Asia is also often referred to as the Arab World. This term implies a uniformity that does not actually exist. First, the name *Arab* is applied loosely to the peoples of this area who speak Arabic and related languages, but ethnologists normally restrict it to certain occupants of the Arabian Peninsula—the Arab "source." In any case, the Turks are not Arabs, and neither are most Iranians or Israelis. Moreover, although the Arabic language prevails from Mauritania in the west across all of North Africa to the Arabian Peninsula, Syria, and Iraq in the east, it is not spoken in other parts of this realm. In Turkey, for example, Turkish is the major language,

and it has Ural-Altaic rather than Arabic's Semitic or Hamitic roots. The Iranian language belongs to the Indo-European linguistic family. Other "Arab World" languages that have separate ethnological identities are spoken by the Jews of Israel, the Tuareg people of the Sahara, the Berbers of northwestern Africa, and the peoples of the transition zone between North Africa and Subsaharan Africa to the south.

An "Islamic World"?

Yet another name given to this realm is the World of Islam. The prophet Muhammad (Mohammed) was born in Arabia in AD 571, and in the centuries after his death in 632, Islam spread into Africa, Asia, and Europe. This was the age of Arab conquest and expansion. Their armies penetrated southern Europe, their caravans crossed the deserts, and their ships plied the coasts of Asia and Africa. Along these routes they carried the Muslim (Islamic) faith, converting the ruling classes of the states of the West African savanna,

threatening the Christian stronghold in the highlands of Ethiopia, penetrating the deserts of inner Asia, and pushing into India and even the island extremities of Southeast Asia. Today, the Islamic faith extends far beyond the limits of the realm under discussion (Fig. 6-2). Nor is the World of Islam entirely Muslim. Judaism, Christianity (notably in Egypt and Lebanon), and other faiths survive in the heartland of the Islamic World. So this connotation is not satisfactory either.

"Middle East"?

Finally, this realm is frequently called the Middle East. That must sound odd to someone in, say, India, who might think of a Middle West rather than a Middle East! The name, of course, reflects the biases of its source: the "Western" world, which saw a "Near" East in Turkey, a "Middle" East in Egypt, Arabia, and Iraq, and a "Far" East in China and Japan. Still, the term has taken hold, and it can be seen and heard in everyday usage by scholars, journalists, and members

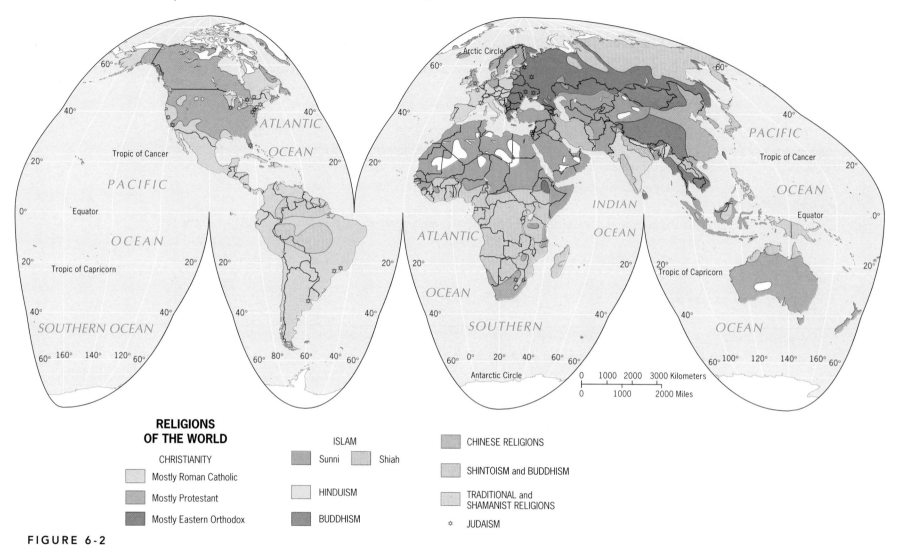

**RELIGIONS
OF THE WORLD**

CHRISTIANITY

Mostly Roman Catholic

Mostly Protestant

Mostly Eastern Orthodox

ISLAM

Sunni   Shiah

HINDUISM

BUDDHISM

CHINESE RELIGIONS

SHINTOISM and BUDDHISM

TRADITIONAL and
SHAMANIST RELIGIONS

✡   JUDAISM

FIGURE 6-2

of the United Nations. Even so, it should be applied only to one of the regions of this vast realm, not to the realm as a whole.

HEARTHS OF CULTURE

This geographic realm occupies a pivotal part of the world: here Eurasia, crucible of human cultures, meets Africa, source of humanity itself. A million years ago, the ancestors of our species walked from East Africa into North Africa and Arabia and spread from one end of Asia to the other. One hundred thousand years ago, *Homo sapiens* crossed these lands on the way to Europe, Australia, and, eventually, the Americas. Ten thousand years ago, human communities in what we now call the Middle East began to domesticate plants and animals, learned to irrigate their fields, enlarged their settlements into towns, and formed the earliest states. One thousand years ago, the heart of the realm was stirred and mobilized by the teachings of Muhammad and the Quran (Koran), and Islam was on the march from North Africa to India. Today this realm is a cauldron of religious and political activity, weakened by conflict but empowered by oil, plagued by poverty but fired by a wave of religious fundamentalism.

In the Introduction (p. 13) we discussed the concept of culture and its regional expression in the cultural landscape. **Cultural geography**, we noted, is a wide-ranging and comprehensive field that studies spatial aspects of human cultures, focusing not only on **2** cultural landscapes but also on **culture hearths**— the crucibles of civilization, the sources of ideas, innovations, and ideologies that changed regions and realms. Those ideas and innovations spread far and wide through a set of processes that we study under the **3** rubric of **cultural diffusion**. Because we understand these processes better today, we can reconstruct ancient routes by which the knowledge and achievements of culture hearths spread (that is, diffused) to other areas. Another topic of cultural geography, also relevant in the context of the North African/Southwest **4** Asian realm, is the **cultural environment** that a dominant culture creates. Human cultures exist in long-term accommodation with (and adaptation to)

their natural environments, exploiting opportunities that these environments present and coping with the extremes they can impose. The study of the relationship between human societies and natural environments has become a separate branch of cultural - **5** geography called **cultural ecology**. As we will see, the North African/Southwest Asian realm presents many opportunities to investigate cultural geography in regional settings.

Mesopotamia and the Nile

In the basins of the major rivers of this realm (the Tigris and Euphrates of modern-day Turkey, Syria, and Iraq, and the Nile of Egypt) lay two of the world's earliest culture hearths (Fig. 6-3). Mesopota-

mia, "land amidst the rivers," had fertile alluvial soils, abundant sunshine, ample water, and animals and plants that could be domesticated. Here, in the Tigris-Euphrates lowland between the Persian Gulf and the uplands of present-day Turkey, arose one of humanity's first culture hearths, a cluster of communities that grew into larger societies and, eventually, into the world's first states. (Early state development probably was going on simultaneously in East Asia's river basins as well.) Mesopotamians were innovative farmers who knew when to sow and harvest crops, water their fields, and store their surplus. Their knowledge diffused to villages near and far, and a *Fertile Crescent* evolved extending from Mesopotamia across southern Turkey into Syria and the Mediterranean coast beyond (Fig. 6-3).

Irrigation was the key to prosperity and power in Mesopotamia, and urbanization was its reward. Among many settlements in the Fertile Crescent, some thrived, grew, enlarged their hinterlands, and diversified socially and occupationally; others failed. What determined success? One theory, the **6** **hydraulic civilization theory**, holds that cities that could control irrigated farming over large hinterlands held power over others, used food as a weapon, and thrived. One such city, Babylon on the Euphrates River, endured for nearly 4000 years (from 4100 BC). A busy port, its walled and fortified center endowed with temples, towers, and palaces, Babylon for a time was the world's largest city.

Egypt's cultural evolution may have started even earlier than Mesopotamia's, and its focus lay upstream from (south of) the Nile Delta and downstream from (north of) the first of the Nile's series of rapids, or cataracts (Fig. 6-3). This part of the Nile Valley is surrounded by inhospitable desert, and unlike Mesopotamia (which lay open to all comers), the Nile provided a natural fortress here. The ancient

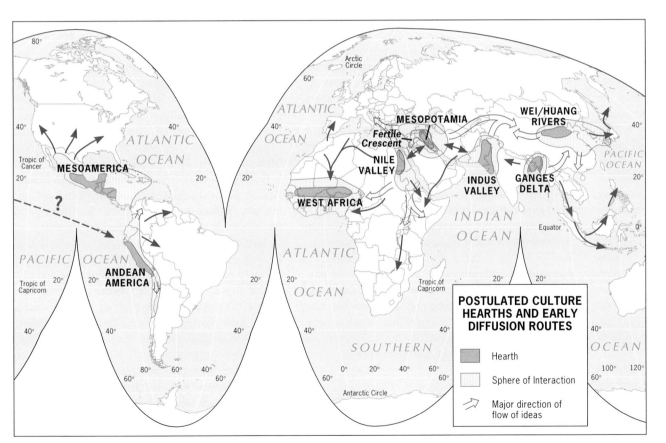

FIGURE 6-3

POSTULATED CULTURE HEARTHS AND EARLY DIFFUSION ROUTES

Hearth

Sphere of Interaction

Major direction of flow of ideas

Egyptians converted their security into progress. The Nile was their highway of trade and interaction; it also supported agriculture through irrigation. The Nile's cyclical ebb and flow was much more predictable than that of the Tigris-Euphrates river system. By the time Egypt finally fell victim to outside invaders (about 1700 BC), a full-scale urban civilization had emerged. Ancient Egypt's artist-engineers left a magnificent legacy of massive stone monuments, some of them containing treasure-filled crypts of god-kings called Pharaohs. These tombs have enabled archeologists to reconstruct the ancient history of this culture hearth.

To the east, separated from Mesopotamia by more than 1200 miles (1900 km) of mountain and desert, lies the Indus Valley (Fig. 6-3). By modern criteria, this eastern hearth lies outside the realm under discussion here; but in ancient times it had cultural and commercial ties with the Tigris-Euphrates region. Mesopotamian innovations reached the Indus region early, and eventually the cities of the Indus became power centers of a civilization that extended far into present-day northern India.

Today, the world continues to benefit from the accomplishments of the ancient Mesopotamians and Egyptians. They domesticated cereals (wheat, rye, barley), vegetables (peas, beans), fruits (grapes, apples, peaches), and many animals (horses, pigs, sheep). They also advanced the study of the calendar, mathematics, astronomy, government, engineering, metallurgy, and a host of other skills and technologies. In time, many of their innovations were adopted and then modified by other cultures in the Old World and eventually in the New World as well. Europe was the greatest beneficiary of these legacies of Mesopotamia and ancient Egypt, whose achievements constituted the foundations of "Western" civilization.

Decline and Decay

As Figure I-8 reminds us, many of the early cities of this culture realm lay in what is today desert territory. Assuming that there were no good reasons to build large settlements in the middle of deserts, we may **7** hypothesize that **climate change** sweeping over this region, not a monopoly over irrigation tech-

niques, gave certain cities in the ancient Fertile Crescent an advantage over others. Climate change, associated with shifting environmental zones after the last Pleistocene glacial retreat, may have destroyed the last of the old civilizations. Perhaps overpopulation and human destruction of the natural vegetation contributed to the process. Indeed, some cultural geographers suggest that the momentous innovations in agricultural planning and irrigation technology were not "taught" by the seasonal flooding of the rivers but were forced on the inhabitants as they tried to survive changing environmental conditions.

The scenario is not difficult to imagine. As outlying areas began to fall dry and farmlands were destroyed, people congregated in the already crowded river valleys—and made every effort to increase the productivity of the land that could still be watered. Eventually overpopulation, destruction of the watershed, and perhaps reduced rainfall in the rivers' headwater areas dealt the final blow. Towns were abandoned to the encroaching desert; irrigation canals filled with drifting sand; croplands dried up. Those who could migrated to areas that were still reputed to be productive. Others stayed, their numbers dwindling, increasingly reduced to subsistence.

As old societies disintegrated, power emerged elsewhere. First the Persians, then the Greeks, and later the Romans imposed their imperial designs on the tenuous lands and disconnected peoples of North Africa/Southwest Asia. Roman technicians converted North Africa's farmlands into irrigated plantations whose products went by the boatload to Roman Mediterranean shores. Thousands of people were carried as slaves to the cities of the new conquerors. Egypt was quickly colonized, as was the area we now call the Middle East. One region that lay distant, and therefore remote from these invasions, was the Arabian Peninsula, where no major culture hearth or large cities had emerged and where the turmoil had not affected Arab settlements and nomadic routes.

STAGE FOR ISLAM

In a remote place on the Arabian Peninsula, where the foreign invasions of the Middle East had had little ef-

fect on the Arab communities, an event occurred early in the seventh century that was to change history and affect the destinies of people in many parts of the world. In a town called Mecca (Makkah), about 45 miles (70 km) from the Red Sea coast in the Jabal Mountains, a man named Muhammad in the year AD 611 began to receive revelations from Allah (God). Muhammad (571–632) was then in his early forties and had barely 20 years to live. Convinced after some initial self-doubt that he was indeed chosen to be a prophet, Muhammad committed his life to fulfilling the divine commands he believed he had received. Arab society was in social and cultural disarray, but Muhammad forcefully taught Allah's lessons and began to transform his culture. His personal power soon attracted enemies, and in 622 he fled from Mecca to the safer haven of Medina (Al Madinah), where he continued his work. This moment, the *hejira* ("migration"), marks the starting date of the Muslim era, Year 1 on Islam's calendar. Mecca, of course, later became Islam's holiest place.

The precepts of Islam in many ways constituted a revision and embellishment of Judaic and Christian beliefs and traditions. All of these faiths have but one god, who occasionally communicates with humankind through prophets; Islam acknowledges that Moses and Jesus were such prophets but considers Muhammad to be the final and greatest prophet. What is Earthly and worldly is profane; only Allah is pure. Allah's will is absolute; Allah is omnipotent and omniscient. All humans live in a world that Allah created for their use but only to await a final judgment day.

Islam brought to the Arab World not only the unifying religious faith it had lacked but also a new set of values, a new way of life, a new individual and collective dignity. Islam dictated observance of the Five Pillars: (1) repeated expressions of the basic creed, (2) the daily prayer, (3) a month each year of daytime fasting (Ramadan), (4) the giving of alms, and (5) at least one pilgrimage in each Muslim's lifetime to Mecca. And Islam prescribed and proscribed in other spheres of life as well. It forbade alcohol, smoking, and gambling. It tolerated polygamy, although it acknowledged the virtues of monogamy. Mosques appeared in Arab settlements, not only for the (Friday) sabbath prayer, but also as social gathering places to knit communities closer together. Mecca became the spiritual center for a divided, widely dispersed people for whom a collective focus was something new.

The Arab-Islamic Empire

Muhammad provided such a powerful stimulus that Arab society was mobilized almost overnight. The prophet died in 632, but his faith and fame spread like wildfire. Arab armies carrying the banner of Islam formed, invaded, conquered, and converted wherever they went. As Figure 6-4 shows, by AD 700 Islam had

reached far into North Africa, into Transcaucasia, and into most of Southwest Asia. In the centuries that followed, it penetrated Southern and Eastern Europe, Central Asia's Turkestan, West Africa, East Africa, and South and Southeast Asia, even reaching China by AD 1000.

The spread of Islam provides a good illustration **8** of a series of processes called **spatial diffusion**, focusing on the way ideas, inventions, and cultural practices propagate through a population in space and time. In 1952, the Swedish geographer Torsten Hägerstrand published a fundamental study of spa-

tial diffusion entitled *The Propagation of Diffusion Waves*. He reported that diffusion takes place in two **9** forms: **expansion diffusion**, when propagation waves originate in a strong and durable source area and spread outward, affecting an ever larger region **10** and population; and **relocation diffusion**, in which an innovation, idea, or (for example) a virus is carried by migrants from the source to distant locations and diffuses from there. The global spread of AIDS is a case of relocation diffusion.

Both expansion diffusion and relocation diffusion include several types of processes. The spread of Islam, as Figure 6-4 shows, initially proceeded by a form of expansion **11** diffusion called **contagious diffusion** as the faith moved from village to village across the Arabian Peninsula and the Middle East. But Islam got powerful boosts when kings, chiefs, and other high officials were converted, who in turn propagated the faith downward through their bureaucracies to far-flung subjects. This form of ex- **12** pansion diffusion, called **hierarchical diffusion**, served Islam well.

But the map leaves no doubt that Islam later spread by relocation diffusion as well, notably to the Ganges Delta in South Asia, to present-day Indonesia, and to East Africa. That process continues to this day, but the heart of Islam remains in Southwest Asia. There, Islam became the cornerstone of an Arab Empire with Medina as its first capital. As the empire grew by expansion diffusion, its headquarters were relocated from Medina to Damascus (in present-day Syria) and later to Baghdad on the Tigris River (Iraq). And it prospered. In architecture, mathematics, and science, the Arabs overshadowed their European contemporaries. The Arabs established institutions of higher learning in many cities including Baghdad, Cairo, and Toledo (Spain), and their distinctive

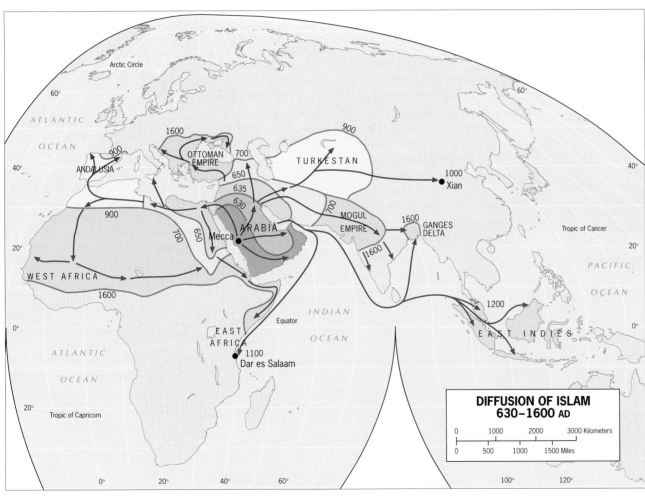

FIGURE 6-4

cultural landscapes united their vast domain. Non-Arab societies in the path of the Muslim drive were not only Islamized, but also Arabized, adopting other Arab traditions as well. Islam had spawned a culture; it still lies at the heart of that culture today.

As we noted, Islam's expansion eventually was checked in Europe, Russia, and elsewhere. But a map showing the total area under Muslim sway in Eurasia and Africa reveals the enormous dimensions **13** of the domain affected by **Islamization** at one time or another (Fig. 6-5). Islam continues to expand, now mainly by relocation diffusion. There are

Islamic communities in cities as widely scattered as Vienna, Singapore, and Cape Town, South Africa; Islam is also growing rapidly in the United States. With more than 1.3 billion adherents today, Islam is a vigorous and burgeoning cultural force around the world.

ISLAM DIVIDED

For all its vigor and success, Islam still fragmented into sects. The earliest and most consequential division arose after Muhammad's death. Who should be his legitimate successor? Some believed that only a blood relative should follow the prophet as leader of Islam. Others, a majority, felt that any devout follower of Muhammad was qualified. The first chosen successor was the father of Muhammad's wife (and thus not a blood relative). But this did not satisfy those who wanted to see a man named Ali, a cousin of Muhammad, made *caliph* (successor). When Ali's turn came, his followers, the Shi'ites, proclaimed that Muhammad finally had a legitimate successor. This offended the Sunnis, those who did not see a blood relationship as necessary for the succession. From the beginning of this disagreement, the numbers of Muslims who took the Sunni side far exceeded those who regarded themselves as Shiah (followers) of Ali. The great expansion of Islam was largely propelled by Sunnis; the Shi'ites survived as small minorities scattered throughout the realm. Today, about 85 percent of all Muslims are Sunnis.

But the Shi'ites vigorously promoted their version of the faith. In the early sixteenth century their work paid off: the royal house of Persia (modern-day Iran) made Shi'ism the only legal religion throughout its vast empire. That domain extended from Persia into lower Mesopotamia (modern Iraq), into Azerbaijan, and into western Afghanistan and Pakistan. As the map of religions (Fig. 6-2) shows, this created for Shi'ism **14** a large **culture region** and gave the faith unprecedented strength. Iran remains the bastion of Shi'ism in the realm today, and the appeal of Shi'ism continues to radiate into neighboring countries and even farther afield.

During the late twentieth century, Shi'ism gained unprecedented influ-

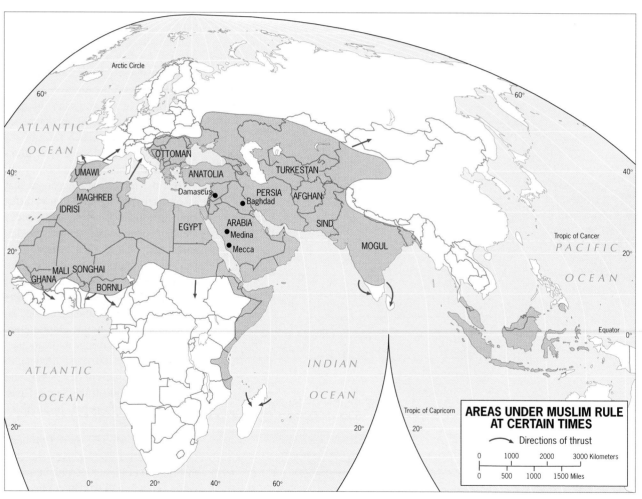

FIGURE 6-5

ence in the realm. In its heartland, Iran, a *shah* (king) tried to secularize the country and to limit the power of the *imams* (mosque officials); he provoked a revolution that cost him the throne and made Iran a Shi'ite Islamic republic. Before long, Iran was at war with neighboring, Sunni-dominated Iraq, and Shi'ite parties and communities elsewhere were invigorated by the newfound power of Shi'ism. From Arabia to Africa's northwestern corner, Sunni-ruled countries warily watched their Shi'ite minorities, newly imbued with religious fervor. Mecca, the holy place for both Sunnis and Shi'ites, became a battleground during the week of the annual pilgrimage, and for a time the (Sunni) Saudi Arabian government denied entry to Shi'ite pilgrims.

Recently, that schism has healed somewhat, but intra-Islamic sectarian differences run deep.

Smaller Islamic sects further diversify this realm's religious landscape. Some of these sects play a disproportionately large role in the societies and countries of which they are a part, as we will see in our regional discussion.

Religious Fundamentalism in the Realm

Another cause of intra-Islamic conflict lies in the **15** resurgence of **religious fundamentalism**, or as Muslims refer to it, *religious revivalism*. In the

1970s, the imams in Shi'ite Iran wanted to reverse the shah's moves toward liberalization and secularization: they wanted to (and did) recast society in traditional, revivalist Islamic molds. An *ayatollah* (leader under Allah) replaced the shah in 1979; Islamic rules and punishments were instituted. Urban women, many of whom had been considerably liberated and educated during the shah's regime, resumed more traditional Islamic roles. Vestiges of Westernization, encouraged by the shah, disappeared. Under these new conditions even the war against Iraq (1980–1990), which began as a conflict over territory, became a holy war that cost more than a million lives.

Islamic fundamentalism did not rise in Iran alone, nor was it confined to Shi'ite communities. Many Muslims—Sunnis as well as Shi'ites—in all parts of the realm disapproved of the erosion of traditional Islamic values, the corruption of society by European colonialists and later by Western modernizers, and the declining power of the faith in the secular state. As long as economic times were good, such dissatisfaction remained submerged. But when jobs were lost and incomes declined, a return to fundamental Islamic ways became more appealing.

This set Muslim against Muslim in all the regions of the realm. Revivalists fired the faith with a new militancy, challenging the status quo from Afghanistan to Algeria. The militants forced their governments to ban "blasphemous" books, to resegregate the sexes in schools, to enforce traditional dress codes, to legitimize religious-political parties, and to heed the wishes of the *mullahs* (teachers of Islamic ways). Militant Muslims proclaimed that democracy inherited from colonialists and adopted by Arab nationalists was incompatible with the rules of the Quran.

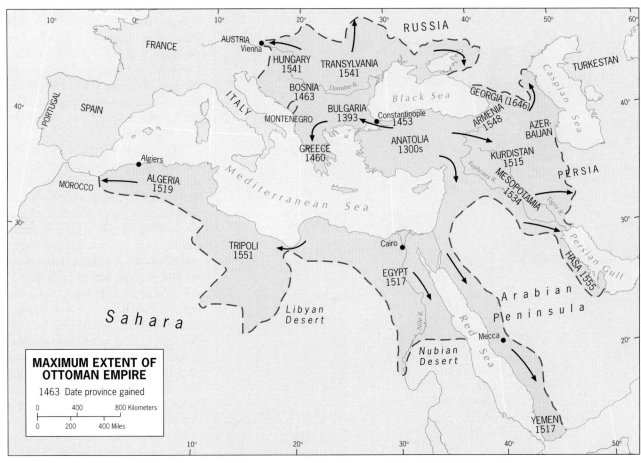

MAXIMUM EXTENT OF OTTOMAN EMPIRE

1463 Date province gained

0   400   800 Kilometers

0   200   400 Miles

FIGURE 6-6

The rift between more liberal Muslims and fundamentalists poses a major challenge for the future. Militant Muslims confront the governments of Egypt, Jordan, Tunisia, and even Turkey. These governments have reacted in various ways (in Egypt, a combination of appeasement and containment; in Tunisia and Algeria, repression; in Jordan, co-option). Other countries, such as Morocco and Saudi Arabia, clearly are vulnerable. Hanging in the balance is the future of the realm.

Fundamentalism and religious militancy are not exclusively Islamic phenomena. They also infect Christian, Judaic, Hindu, and Buddhist societies. Nowhere, however, does the fundamentalist drive exhibit the intensity and vigor it displays in the Islamic realm.

Islam in Europe

As Figure 6-5 shows, Islam's expansion carried deep into Europe. An Arab-Berber alliance, the *Moors*, invaded Spain in 711 and controlled all but northern Castile and Catalonia before the end of the eighth century. By the time the Catholic armies ousted the last of the Muslims, Islamic culture had made an indelible imprint on Iberian cultural landscapes.

It took the Europeans seven centuries to reclaim all of Iberia, but even before the last Muslims were driven out, Islamic power was making itself felt elsewhere in Europe. The Ottomans (named after their leader, Osman I), based in what is today Turkey, conquered Christian Constantinople (now Istanbul) in 1453 and pushed into Greece and Eastern Europe (Fig. 6-6). Less than a century later, Ottoman forces were on the doorstep of Vienna.

Muslim armies also overpowered much of the Middle East and part of present-day Persia as well as North Africa. Under Suleyman the Magnificent, who ruled from 1522 to 1560, the Ottoman Empire was the most powerful state in western Eurasia.

The Ottoman Empire survived for more than four centuries, but it lost territory as time went on, first to the Hungarians, then to the Russians, and later to the Greeks and Serbs until, after World War I, the European powers took over its provinces and made them colonies—colonies we now know by the names of Syria, Iraq, Lebanon, and Yemen (Fig. 6-7). As the map shows, the French and the British took large possessions; even the Italians annexed part of the Ottoman domain.

The boundary framework that the colonial powers created to delimit their holdings was not satisfactory. As Figure I-9 reminds us, this realm's population of more than 500 million is clustered, fragmented, and strung out in river valleys, coastal zones, and crowded oases. The colonial powers laid out long stretches of boundary as ruler-straight lines across uninhabited territory; they saw no need to adjust these boundaries to cultural or physical features in the landscape. Other boundaries, even some in desert zones, were poorly defined and never marked on the ground. Later, when the colonies had become independent states, such boundaries led to quarrels, even armed conflicts, among neighboring Muslim states.

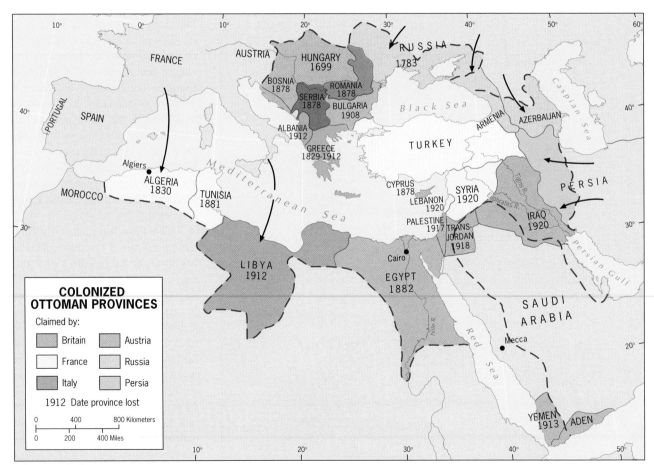

FIGURE 6-7

THE POWER AND PERIL OF OIL

Travel through the cities and towns of North Africa and Southwest Asia, talk to students, shopkeepers, taxi drivers, and migrants, and you will hear the same refrain: "Leave us in peace, let us do things our traditional way. Our problems—with each other, with the world—seem always to result from outside interference. The stronger countries of the world exploit our weaknesses and magnify our quarrels. We want to be left alone."

That wish might be closer to fulfillment were it not for two relatively recent events: the creation of the state of Israel and the discovery of some of the world's largest oil reserves. We will discuss the founding of Israel in the regional section of this chapter. Here we focus on the realm's most valuable export product: oil.

Location of Known Reserves

GEODISCOVERIES In general terms, oil (and associated natural gas) exists in this realm in three discontinuous zones (Fig. 6-8). The most productive of these zones extends from the southern and southeastern

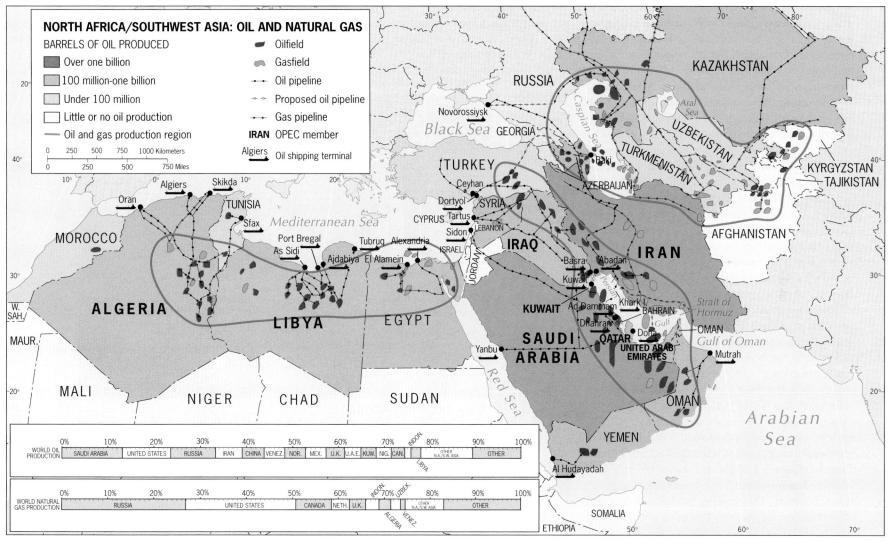

FIGURE 6-8

part of the Arabian Peninsula northwestward around the rim of the Persian Gulf, reaching into Iran and continuing northward into Iraq, Syria, and southeastern Turkey, where it peters out. The second zone lies across North Africa and extends from north-central Algeria eastward across northern Libya to Egypt's Sinai Peninsula, where it ends. The third zone begins on the margins of the realm in eastern Azerbaijan, continues eastward under the Caspian Sea into Turkmenistan and Kazakhstan, and also reaches into Uzbekistan, Tajikistan, Kyrgyzstan, and Afghanistan.

The search for oil goes on, both in this realm and in many other areas of the world. As a result, estimates of the realm's reserves are subject to change as new discoveries are announced. Current assessments suggest that more than 65 percent of the world's known oil reserves lie in the North Africa/Southwest Asia realm.

In terms of production, Saudi Arabia long has been among the world leaders, and it is the undisputed top exporter. Two countries that rival Saudi Arabia as producers, the United States and Russia, consume most of their own output. The United States, in fact, is the world's leading importer. Thus the oil production in Southwest Asia and North Africa is crucial to the rest of the world. As Figure I-10 indicates, oil wealth has elevated several of this realm's countries into the upper-middle and high-income categories. Petroleum wealth also has enmeshed these Islamic societies in world affairs, so that a threat to stability in an oil-rich country risks foreign intervention.

When the colonial powers laid down the boundaries that partitioned this realm among themselves, no one knew about the riches that lay beneath the ground. A few wells had been drilled, and production in Iran had begun as early as 1908 and in Egypt's Sinai Peninsula in 1913. But the major discoveries came later, in some cases after the colonial powers had already withdrawn. Some of the newly independent countries, such as Libya, Iraq, and Kuwait, found themselves with wealth undreamed of when the Turkish Ottoman Empire collapsed. As Figure 6-8 shows, however, others were less fortunate. A few countries had (and still have) potential. The smaller, weaker emirates and sheikdoms on the Arabian Peninsula always feared that powerful neighbors would try to annex them (Kuwait faced this prospect in 1990 when Iraq in-

vaded it). The unevenly distributed oil wealth, therefore, created another source of division and distrust among Islamic neighbors.

A Foreign Invasion

The oil-rich countries of the realm had a coveted energy source, but they lacked the equipment or skills to exploit it. These had to come from the outside world and entailed what many tradition-bound Muslims feared most: penetration by the vulgarities of Western ways. In his book *The Middle East*, geographer William Fisher reports that Saudi Arabia had two major cultural forces: Islam and Aramco, the joint Arab-American oil company. Yet while some

Western ideas and practices were diffused, Islamic societies proved resistant to foreign cultural influences. Most schools, for example, remain segregated by sex; the role of women in society varies by country but remains traditional in many countries. While women in Turkey enjoy considerable freedom and opportunity, Afghan women were compelled to adhere to the strictest Islamic orthodoxy.

In geographic terms, the impact of oil, its production and sale, in the exporting countries of this realm can be summarized as follows:

1. *High Incomes.* When oil prices on international markets were high, several countries in this realm ranked as the highest-income societies in the

FROM THE FIELD NOTES
"The impact of oil wealth is nowhere better illustrated than it is along the waterfront of the famous Creek of Dubai (adjoining the city of the same name, largest in the United Arab Emirates). We stopped on the bridge that links the two sectors of Dubai to observe the old *dhows* (most of these wooden boats now with motors as well as triangular sail) still at their centuries-old moorings, overlooked not by traditional Arab buildings but by modern skyscrapers. Oil may have transformed the local economy, but don't count the role of the dhows out just yet. They still carry trade and contraband. We went to the dock and asked what was being transported. 'Jeans and cassettes to Iran,' we were told, 'and caviar and carpets from Iran. It's a good business.' So the traditional role of the dhows goes on, even in the age of oil."

world. Even when oil prices declined, virtually all the petroleum-exporting states remained in the upper-middle-income category (Fig. I-10).

2. *Modernization.* Huge oil revenues transformed cultural landscapes throughout the realm, producing a façade of modernization in the cityscape of ports and capitals. Gleaming glass-encased office buildings towered over mosques; superhighways crossed ancient camel paths; state-of-the-art port facilities handled oil and oil-funded trade.

3. *Industrialization.* Farsighted governments among those with oil wealth, realizing that petroleum reserves will not last forever, have invested some of their income in industrial plants that will outlast the era of oil. Petrochemical industries, plastics fabrication, and desalinization plants are among these facilities.

4. *Intra-Realm Migration.* The oil wealth has attracted millions of workers from less favored parts of the realm to work in the oilfields, in the ports, and in many other, mainly menial capacities. This has brought many Shi'ites to the countries of eastern Arabia; hundreds of thousands of Palestinians also work there as laborers. Saudi Arabia, with a population of 23 million, has more than 5 million foreign workers.

5. *Inter-Realm Migration.* The willingness of workers from such countries as Pakistan, India, and Sri Lanka to work for wages even lower than those the oil industry pays has attracted a substantial flow of temporary immigrants from outside the realm. These workers serve mostly as domestics, gardeners, refuse collectors, and the like.

6. *Regional Disparities.* Oil wealth and its manifestations in the cultural landscape create strong contrasts with areas not directly affected. The ultramodern east coast of Saudi Arabia is a world apart from large areas of its interior, where it becomes a land of desert, oasis, and camel, of vast distances, slow change, and isolated settlements. To some degree, this phenomenon affects all oil-rich countries.

7. *Foreign Investment.* Governments and Arab businesspeople have invested oil-generated wealth in foreign countries. These investments have created a network of international involvement that links many of this realm's countries not only to the economies of foreign states, but also to growing Arab (and thus Islamic) communities in those states.

The map (Fig. 6-8) contains a warning that came home to the oil-rich states (not only in this realm but elsewhere as well) in the 1980s when, after a period of high oil prices and huge revenues, oil prices fell and incomes plummeted. Even the power of an 11-member cartel, OPEC (Organization of Petroleum Exporting Countries), could not recover the lost advantage. Figure 6-8 shows a system of oil and gas pipelines that strongly resembles the exploitative interior-to-coast railroad lines in a mineral-rich colony of the past. Such a pattern spells disadvantage for the exporter, whether colony or independent country. Markets, not raw-material exporters, dominate international trade. Oil brought this realm into contact with the outside world in ways unforeseen just a century ago. Oil has strengthened and empowered some of its peoples; it has dislocated and imperiled others. It has truly been a double-edged sword.

REGIONS OF THE REALM

Identifying and delimiting regions in this vast geographic realm is quite a challenge. Population clusters tend to be widely scattered in some areas, highly concentrated in others. Cultural transitions and landscapes—internal as well as peripheral—make it difficult to discern a regional framework. This, furthermore, is a highly changeable realm and always has been. Several centuries ago it extended into Eastern Europe; now it reaches into Central Asia, where an Islamic **cultural revival** (the regeneration of a long-dormant culture through internal renewal and external infusion) is under way.

The following are the regional components of this far-flung realm today (Fig. 6-9):

1. **Egypt and the Lower Nile Basin**. This region in many ways constitutes the heart of the realm as a whole. Egypt (together with Iran and Turkey) is one of the realm's three most populous countries. It is the historic focus of this part of the world and a major political and cultural force. It shares with its southern neighbor, Sudan, the waters of the lower Nile River.

2. **The Maghreb and Its Neighbors**. Western North Africa (the Maghreb) and the areas that border it also form a region, consisting of Algeria, Tunisia, and Morocco at the center and Libya, Chad, Niger, Mali, and Mauritania along the broad periphery. The last four of these countries also lie astride or adjacent to the broad transition zone where the Arab-Islamic realm of northern Africa merges into Subsaharan Africa.

3. **The African Transition Zone**. From southern Mauritania in the west to Somalia in the east, across the entire African landmass at its widest extent, the realm dominated by Islamic culture interdigitates with that of Subsaharan Africa. No sharp dividing line can be drawn here: people of African ethnic stock have adopted the Muslim faith and Arabic language and traditions. As a result, this is less a region than a broad zone of transition.

4. **The Middle East**. This region includes Israel, Jordan, Lebanon, Syria, and Iraq. In effect, it is the crescent-like zone of countries that extends from the eastern Mediterranean coast to the head of the Persian Gulf.

5. **The Arabian Peninsula**. Dominated by the large territory of Saudi Arabia, the Arabian Peninsula also includes the United Arab Emirates, Kuwait, Bahrain, Qatar, Oman, and Yemen. Here lies the source and focus of Islam, the holy city of Mecca;

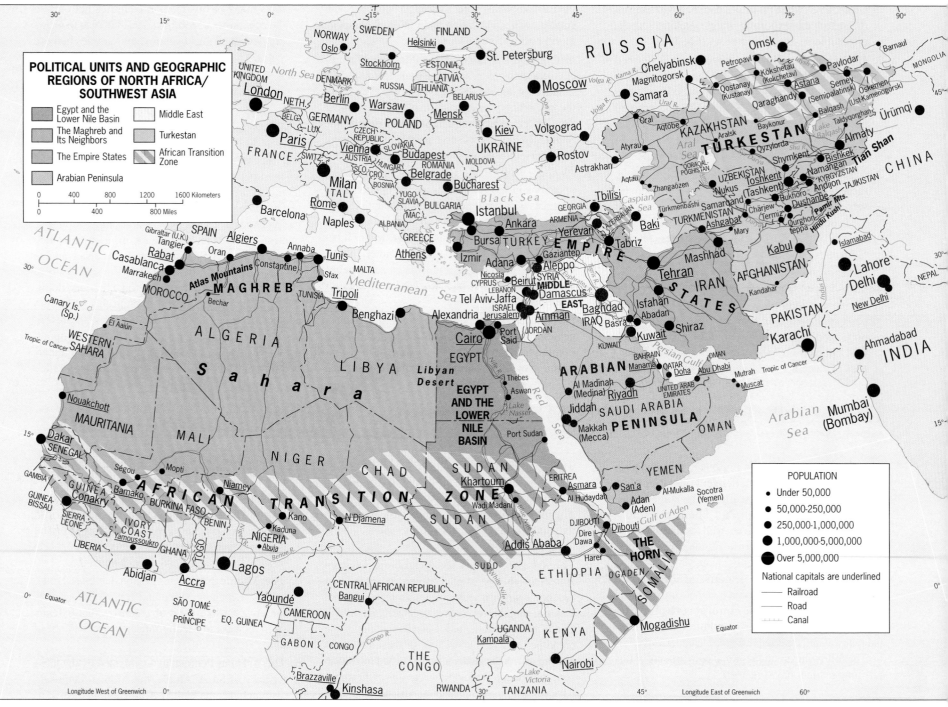

POLITICAL UNITS AND GEOGRAPHIC REGIONS OF NORTH AFRICA/ SOUTHWEST ASIA

Egypt and the Lower Nile Basin

The Maghreb and Its Neighbors

The Empire States

Arabian Peninsula

Middle East

Turkestan

African Transition Zone

0   400   800   1200   1600 Kilometers

0   400   800 Miles

POPULATION

• Under 50,000

• 50,000–250,000

● 250,000–1,000,000

● 1,000,000–5,000,000

● Over 5,000,000

National capitals are underlined

—— Railroad

—— Road

⊢⊢⊢⊢ Canal

FIGURE 6-9

here, too, lie many of the world's greatest oil deposits.

6. **The Empire States**. Two of the realm's giants, states with imperial histories and majestic cultures, dominate this region: Turkey and Iran. Turkey today is the realm's most secular state (most Turks are Sunni Muslims); Shi'ite Iran is one of its Islamic republics. To the north, Azerbaijan, former Soviet republic and once a part of Persia's empire, lies in the turbulent, Muslim-infused Transcaucasian Transition Zone. To the south, the northern part of the island of Cyprus is under Turkey's thrall.

7. **Turkestan**. Turkish influence ranged far and wide in Southwest and Central Asia, and following the Soviet collapse that influence proved to be durable and strong. In the five former Soviet republics the strength and potency of Islam vary, and the new governments deal warily (sometimes forcefully) with Islamic revivalists. Boundaries of this region, as in the African Transition Zone, do not always coincide with national borders. Kazakhstan is the most prominent case, but regional cultural influences also radiate into China and Pakistan. Afghanistan also forms part of this large, landlocked, turbulent region.

EGYPT AND THE LOWER NILE BASIN

Egypt occupies a pivotal location in the heart of a realm that extends over 6000 miles (9600 km) longitudinally and some 4000 miles (6400 km) latitudinally. At the northern end of the Nile and of the Red Sea, at the eastern end of the Mediterranean Sea, in the northeastern corner of Africa across from Turkey to the north and Saudi Arabia to the east, adjacent to Israel, to Islamic Sudan, and to militant Libya, Egypt lies in the crucible of this realm. Because it owns the Sinai Peninsula (recently lost to and regained from Israel), Egypt, alone among states on the African continent, has a foothold in Asia, a foothold that gives it a coast overlooking the strategic Gulf of

Aqaba (the northeasternmost arm of the Red Sea). Egypt also controls the Suez Canal, vital link between the Indian and Atlantic oceans and lifeline of Europe. We hardly need to further justify Egypt's designation (together with northern Sudan) as a discrete region.

Egypt's Nile is the aggregate of two great branches upstream: the White Nile, which originates in the streams that feed Lake Victoria in East Africa, and the Blue Nile, whose source lies in Lake Tana in Ethiopia's highlands. The two Niles converge at Khartoum in modern-day Sudan.

About 95 percent of Egypt's 71.1 million people live within a dozen miles (20 km) of the great river's banks or in its delta (Fig. 6-10). It has always been this way: the Nile rises and falls seasonally, watering and replenishing soils and crops on its banks.

The ancient Egyptians used *basin irrigation*, building fields with earthen ridges and trapping the floodwaters with their fertile silt, to grow their crops. That practice continued for thousands of years until, during the nineteenth century, the construction of permanent dams made it possible to irrigate Egypt's farmlands year round. These dams, with locks for navigation, controlled floods, expanded the country's cultivable area, and allowed the farmers to harvest more than one crop per year on the same field. In a single century, all of Egypt's farmland was brought under *perennial irrigation*.

The greatest of all Nile projects, the Aswan High Dam (which began operating in 1968), creates Lake Nasser, one of the world's largest artificial lakes (Fig. 6-10). As the map shows, the lake extends into Sudan, where 50,000 people had to be resettled to make way for it. The Aswan High Dam increased Egypt's irrigable land by nearly 50 percent and today provides the country with about 40 percent of its electricity. But, as is so often the case with megaprojects of this kind, the dam also produced serious problems. Snail-carried schistosomiasis and mosquito-transmitted malaria thrived in the dam's standing water, afflicting hundreds of thousands of people living nearby. By blocking most of the natural fertilizers in the Nile's annual floodwaters, the dam necessitated the widespread use of artificial fertilizers, very costly to small farmers and

damaging to the natural environment. And the now fertilizer- and pesticide-laden Nile no longer supports the fish fauna offshore, reducing the catch and depriving coastal populations of badly needed proteins. Egypt's elongated oasis along the Nile, just 3 to 15 miles (5 to 25 km) wide, broadens north of Cairo across a delta anchored in the west by the great city of Alexandria and in the east by Port Said, gateway to the Suez Canal. The delta contains extensive farmlands, but it is a troubled area today. The ever more intensive use of the Nile's water and silt upstream is depriving the delta of much needed replenishment. And the low-lying delta is geologically subsiding, raising fears of salt-water invasion from the Mediterranean Sea that would damage soils here.

Egypt's millions of subsistence farmers, the *fellaheen*, still struggle to make their living off the land, as did the peasants of the Egypt of five millennia past. Rural landscapes seem barely to have changed; ancient tools are still used, and dwellings remain rudimentary. Poverty, disease, high infant mortality rates, and low incomes prevail. The Egyptian government is embarked on grandiose plans to expand its irrigated acreage, but without modernization and greater efficiency in the existing farmlands, the future is dim.

Egypt's Regional Geography

Egypt has six subregions, mapped in Figure 6-10. Most Egyptians live and work in Lower (i.e., northern) and Middle Egypt (regions ① and ②), the country's core area anchored by Cairo and flanked by the leading port and second industrial center, Alexandria. The economy has benefited from further oil discoveries in the Sinai (region ⑥) and in the Western Desert (region ④), so that Egypt now is self-sufficient and even exports some petroleum. Cotton and textiles are the other major source of external income, but the important tourist industry has repeatedly been hurt by Islamic extremists. As the population mushrooms, the gap between food supply and demand widens, and Egypt must import grain. Since the late 1970s, Egypt has been a major recipient of U.S. foreign aid.

Egypt today is at a crossroads in more ways than one. Its planners know that reducing the high birth rate would improve the demographic situation, but revivalist Muslims object to any programs that promote family planning. Its accommodation with Israel helps ensure foreign aid but divides the people. Its government faces a fundamentalist challenge. Egypt's future, in this crucial corner of the realm, is uncertain.

Northern, Arabized Sudan

As Figure 6-10 shows, Egypt is flanked by two countries that have posed challenges to its leadership: Sudan to the south and Libya to the west. Sudan, more than twice as large as Egypt and with nearly 31 million people, lies centered on the confluence of the White Nile (from Uganda) and the Blue Nile (from Ethiopia). Here the twin capital, Khartoum-Omdurman, anchors a large agricultural area where cotton was planted during colonial times. The British administration combined northern Sudan, which was Arabized and Islamized, with a large area to the south, which was African and where many villagers had been Christianized. After the British left, the regime in Khartoum wanted to impose its Islamic rule on the south, and a bitter civil war ensued. The cost in human lives and dislocation over the better part of the past three decades is incalculable; in 2002 an estimated 4 million people remained refugees in their own land.

Port Sudan on the Red Sea is Sudan's maritime outlet, but the country's economy is symbolic of the periphery. Its per-capita income is one of the world's lowest. Trade with Saudi Arabia, Sudan's main partner, exchanges sheep and cotton for oil. But change may be on the way. The pipeline that used to transfer oil from Port Sudan to Khartoum is about to be supplemented by one whose oil will move the other way: a major reserve has been discovered east of Muglad, about 400 miles (640 km) southwest of Khartoum but still within the northern, Arabized sector of Sudan (Fig. 6-10). An underground pipeline nearly 1000 miles (1600 km) long to Port Sudan via Khartoum will bring self-sufficiency and make the country an oil exporter. Critics of the foreign petroleum companies involved in this development argue that the oil revenues will also enable Khartoum to wage more effective war on its southern peoples.

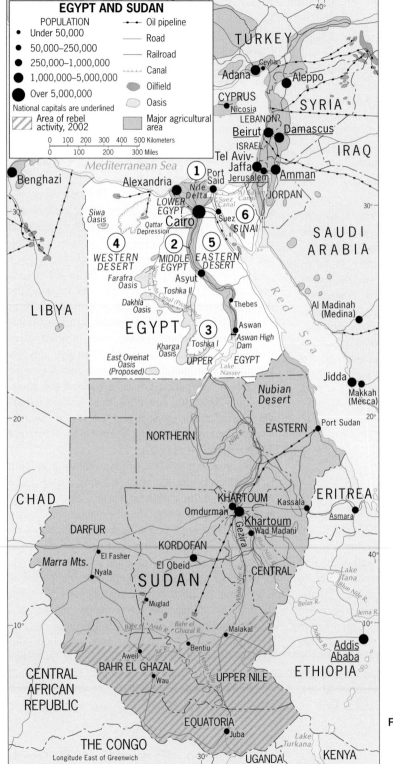

EGYPT AND SUDAN

POPULATION
- Under 50,000
- 50,000–250,000
- 250,000–1,000,000
- 1,000,000–5,000,000
- Over 5,000,000

→•→ Oil pipeline
—— Road
—— Railroad
···· Canal
◯ Oilfield
◯ Oasis

National capitals are underlined

▨ Area of rebel activity, 2002

▨ Major agricultural area

0   100 200 300 400 500 Kilometers
0   100   200   300 Miles

FIGURE 6-10

THE MAGHREB AND ITS NEIGHBORS

The countries of northwestern Africa are collectively called the *Maghreb*, but the Arab name for them is more elaborate than that: *Djezira-al-Maghreb*, or "Isle of the West," in recognition of the great Atlas Mountain range rising like a huge island from the Mediterranean Sea to the north and the sandy flatlands of the immense Sahara to the south.

The countries of the Maghreb (sometimes spelled *Maghrib*) are Morocco, last of the North African kingdoms; Algeria, a secular republic beset by the religious-political problems we noted earlier; and Tunisia, smallest and most Westernized of the three (Fig. 6-11). Libya, facing the Mediterranean between the Maghreb and Egypt, is unlike any other North African country: an oil-rich desert state whose population is almost entirely clustered in settlements along the coast.

Whereas Egypt is the gift of the Nile, the Atlas Mountains form the nucleus of the settled Maghreb. These high ranges wrest from the rising air enough orographic rainfall to sustain life in the intervening valleys, where good soils support productive farming. From the vicinity of Algiers eastward along the coast into Tunisia, annual rainfall averages more than 30 inches (75 cm), a total more than three times as high as that recorded for Alexandria in Egypt's delta. Even 150 miles (240 km) inland, the slopes of the Atlas still receive over 10 inches (25 cm) of rainfall. The effect of the topography can be read on the world map of precipitation (Fig. I-7): where the highlands of the Atlas terminate, desert conditions immediately begin.

The Atlas Mountains are structurally an extension of the Alpine system that forms the orogenic backbone of Europe, of which the Alps and Italy's Appennines are also parts. In northwestern Africa, these mountains trend southwest-northeast and begin in Morocco as the High Atlas, with elevations close to 13,000 feet (4000 m). Eastward, two major ranges dominate the landscapes of Algeria proper: the Tell Atlas to the north, facing the Mediterranean, and the Saharan Atlas to the south, overlooking the great desert. Between these two mountain chains, each consisting of several parallel ranges and foothills, lies a series of intermontane basins (analogous to South America's Andean *altiplanos* but at lower elevations), markedly drier than the northward-facing slopes of the Tell Atlas. In these valleys, the rain shadow effect of the Tell Atlas is reflected not only in the steppe-like natural vegetation but also in land-use patterns: pastoralism replaces cultivation, and stands of short grass and bushes blanket the countryside.

During the colonial era, which began in Algeria in 1830 and lasted until the early 1960s, well over a million Europeans came to settle in North Africa—most of them French, and a large majority bound for Algeria—and these immigrants soon dominated commercial life. They stimulated the renewed growth of the region's towns; Casablanca, Algiers, Oran, and Tunis became the urban foci of the colonized territories. Although the Europeans dominated trade and commerce and integrated the North African countries with France and the European Mediterranean world, they did not confine themselves to the cities and towns. They recognized the agricultural possibilities of the favored parts of the tell (the lower Tell Atlas slopes and narrow coastal plains that face the Mediterranean) and established thriving farms. Not surprisingly, agriculture here is Mediterranean. Algeria soon became known for its vineyards and wines, citrus groves, and dates; Tunisia has long been one of the world's leading exporters of olive oil; and Moroccan oranges went to many European markets.

Between desert and sea, the Maghreb states display considerable geographic diversity. Morocco, a conservative kingdom in a revolutionary region, is tradition-bound, weak economically, and has only recently found an oilfield within its borders. Its core area lies in the north, anchored by four major cities; but the Moroccans' attention is focused on the south, where the government is seeking to absorb its neighbor, Western Sahara (a former Spanish dependency with about 300,000 inhabitants, many of them immigrants from Morocco). Even if this campaign is successful, it will do little to improve the lives of most of Morocco's 30 million people. Hundreds of

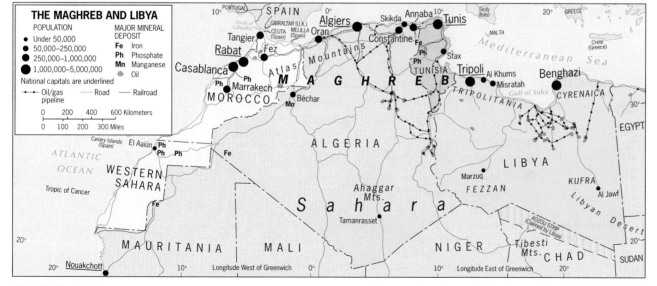

THE MAGHREB AND LIBYA

POPULATION
- Under 50,000
- 50,000–250,000
- 250,000–1,000,000
- 1,000,000–5,000,000

National capitals are underlined

Oil/gas pipeline — Road — Railroad

MAJOR MINERAL DEPOSIT
Fe Iron
Ph Phosphate
Mn Manganese
Oil

0   200   400   600 Kilometers
0   100   200   300 Miles

FIGURE 6-11

thousands have emigrated to Europe, many of them via Spain's two tiny exclaves on the Moroccan coast, Ceuta and Melilla (Fig. 6-11).

Algeria, rich in oil and natural gas, fought a bitter war of liberation against the French colonizers and has been in turmoil virtually ever since, with devastating economic and social consequences. Algiers, Algeria's primate city, is centrally situated along its Mediterranean coast and is home to 2 million of the country's 33 million citizens—but there are more Algerians in France than in the capital today. Conflict between those wanting to make Algeria an Islamic republic and the military-backed regime has cost an estimated 100,000 lives; in 2002 ethnic strife threatened in the Kabylia region east of Algiers involving Algeria's Berber minority.

As Table I-1 shows, the highest GNP per capita in the Maghreb is earned in Tunisia, smallest of the North African countries territorially but with a sizeable population of 10 million. Successive governments have severely repressed Muslim radicals, and comparative stability has allowed the relatively diversified economy, ranging from textiles and farm products to mineral production and tourism, to expand and prosper. Tunis, the historic primate city in the northeastern corner, dominates this most urbanized of the Maghreb states.

Almost rectangular in shape, Libya is a country whose four corners matter most (Fig. 6-11). What limited agricultural possibilities exist lie in the northwest in Tripolitania, centered on the capital, Tripoli, and in the northeast in Cyrenaica, where Benghazi is the urban focus. Between these two coastal clusters, which are home to 90 percent of the Libyans, lies the Gulf of Sidra, a deep Mediterranean bay. Libya has claimed this gulf as its sea, but other countries have not accepted this claim. Libya's two southern cor-

ners are the desert Fezzan, a mountainous area near the southwestern border with Algeria and Niger, and the sparsely populated Kufra oasis in the southeast. Despite its huge size and tiny population, Libya has claimed a sector (the Aozou Strip) of its southern neighbor, Chad (Fig. 6-11).

THE AFRICAN TRANSITION ZONE

As Figure 6-9 shows, the geographic realm under discussion merges into Subsaharan Africa along a wide transition zone that extends from West Africa's *Sahel* (an Arabic word for "border" or "margin") to eastern Africa's Horn. Here Islam's influence fades from north to south, as revealed by Figure 6-12.

In some of the countries in the African Transition Zone, Muslims and non-Muslims live side by side, but more often regional division creates serious problems for states and governments. As the map indicates, Nigeria's north is dominantly Muslim and its south mainly non-Muslim (Christian and animist). Cultural conflict has afflicted Nigeria repeatedly (see Chapter 7). Sudan's Islamic government's attempts to impose Islamic rule on its non-Arabized, African, Christian, and animist south has had a huge social

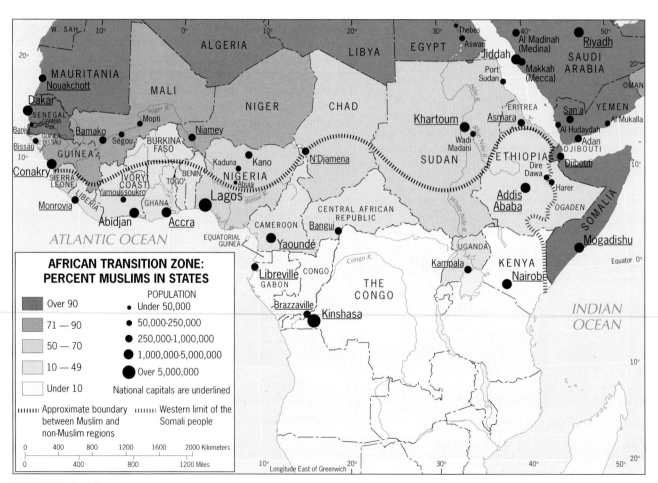

FIGURE 6-12

cost. Ethiopia's Coptic-Christian core is nearly encircled by Sudan, Eritrea, and overwhelmingly Muslim Somalia, the Somalis' traditional homeland extending across the Ogaden to Harer deep inside Ethiopian territory.

The African Transition Zone is a region in which Islam is winning converts and in which Islamic assertiveness is growing. Some observers refer **17** to this zone as Africa's **Muslim Front**, a religious frontier affecting two dozen African countries from Guinea in the west to Ethiopia in the east.

▷ THE MIDDLE EAST

As we noted earlier, the Middle East lies at the heart of this vast realm (Fig. 6-9). It is a pivotal region, not just in North Africa/Southwest Asia, but in the world. Five countries form the Middle East region (Fig. 6-13): Iraq, largest in both size and population; Syria, next in both categories; Jordan; Lebanon; and Israel.

California-sized Iraq (24.4 million) comprises nearly 60 percent of the total area of the Middle East and has 42 percent of the region's population. With major oil reserves and large areas of irrigated farmland, Iraq also is best endowed with natural resources. It is heir to the early Mesopotamian states and empires that emerged in the basin of the Tigris and Euphrates rivers, and the country is studded with archeological sites.

The regional geography of Iraq reveals the divisions within the country (Fig. 6-13). The heart of Iraq is the area around the capital, Baghdad, on the Tigris River amid the productive farmlands of the Tigris-Euphrates Plain.

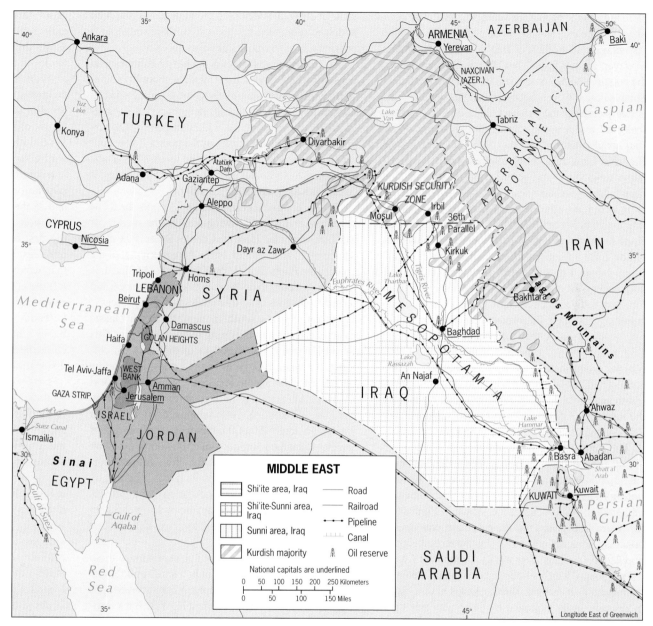

FIGURE 6-13

Here most of the people are Sunni Muslims, who dominate the core area and the country's political machine. As many as 10 million of Iraq's 24 million citizens, however, are not Sunnis but Shi'ites, concentrated in the populous south centered on Basra. Here lie some of Shiah's holiest places, but turmoil in this region, which is Iraq's outlet to the Persian Gulf and which contains major oil reserves, has caused untold misery. When Iraq invaded Kuwait and the 1991 Gulf War ensued, Baghdad's mistrust of its southern populace led to harsh subjugation and severe damage to towns, villages, and religious shrines.

In addition to the Baghdad-centered core and the Basra-anchored south, Iraq has a third subregion: the Kurdish north (Fig. 6-13). As our map shows, the Kurds form minorities in several countries. Iraq's Kurds (about 4 million out of a total of some 25 million) occupy the northern part of the country, and they, too, have felt Baghdad's wrath. Having repeatedly rebelled against their Iraqi rulers, the Kurds' loyalties were suspect during the Gulf War and so they faced severe repression. At the end of the war, the United Nations established a security zone between the 36th parallel and Iraq's northern border to allow refugees to return and to shield the Kurds from Iraqi reprisal. As the map shows, however, part of Iraq's historic Kurdish domain lies outside this security zone.

The Kurds form a nation, but they have no country to call their own; nor do they enjoy the worldwide **18** attention that peoples of some other **stateless nations** (such as the Palestinians) receive. They have inhabited the landlocked zone shown in Figure 6-13 for over 3000 years, but their domain was divided up among imperial powers. Today the largest Kurdish population (perhaps as many as 14 million) lives in Turkey; many Kurds still dream of a nation-state with Diyarbakir as its capital. But the Kurds have a history of disunity easily exploited by those who rule them. They form one of the world's largest stateless nations, ✵ hoping for a deliverance that is unlikely to come.

Syria, too, is ruled by a minority. Although Syria's population of 17.4 million is about 75 percent Sunni Muslim, the ruling elite comes from a smaller Islamic sect based there, the Alawites. Leaders of this powerful minority have retained control over the country for decades, at times by ruthless suppression of dis-

sent. In 2000, president-for-life Hafez al-Assad died and was succeeded by his son Bashar, signaling a continuation of the political status quo.

Like Lebanon and Israel, Syria has a Mediterranean coastline where crops can be raised without irrigation. Behind this densely populated coastal zone, Syria has a much larger interior than its neighbors, but its areas of productive capacity are widely dispersed. Damascus, in the southwest corner of the country, was built on an oasis and is considered to be the world's oldest continuously inhabited city. It is now the capital of Syria, with a population of 2.5 million.

The far northwest is anchored by Aleppo, the focus of cotton- and wheat-growing areas in the shadow of the Turkish border. Here the Orontes River is the chief source of irrigation water, but in the eastern part of the country the Euphrates Valley is the crucial lifeline. It is in Syria's interest to develop its eastern provinces, and recent discoveries of oil there will speed that process.

Jordan, Syria's southern neighbor, is a classic case (and victim) of changing relative location. This desert kingdom was a product of the Ottoman collapse, but when Israel was created it lost its window on the Mediterranean Sea, the (now Israeli) port of Haifa, previously in the British-administered Mandate of Palestine. Following its independence in 1946 with about 400,000 inhabitants, the creation of Israel bequeathed Jordan some 500,000 West Bank Palestinians and, later, a huge flow of refugees. Today Palestinians outnumber original residents by more than two to one in the population of 5.4 million. It may be said that the 46-year rule by King Hussein, ending in 1999, was the key centripetal force that held the country together.

With an impoverished capital, Amman, without oil reserves, and possessing only a small and remote outlet to the Gulf of Aqaba, Jordan has survived with U.S., British, and other aid. It lost its West Bank territory in the 1967 war with Israel, including its sector of Jerusalem (then the kingdom's second-largest city). No third country has a greater stake in a settlement between Israel and the Palestinians than Jordan.

The map suggests that Lebanon has significant geographic advantages in this region: a lengthy coastline on the Mediterranean Sea; a world-class capital, Beirut, on its shoreline; oil terminals on its

shores; and a major capital in its hinterland. The map at the scale of Figure 6-13 cannot reveal still another asset: the fertile, agriculturally productive Bekaa Valley in the eastern interior.

French colonialism in the post-Ottoman era created a territory which in 1930 was about equally Muslim and Christian. Beirut became known as the "Paris of the Middle East." After independence in 1946, Lebanon functioned as a democracy and did well economically as the region's leading banking and commercial center. But the Muslim sector of the population grew much faster than the Christian one, and in the late 1950s the Arabs launched their first rebellion against the established order. Following the first of several waves of influx by Palestinian refugees, Lebanon fell apart in 1975 in a civil war that wrecked Beirut, devastated the economy, and left the country at the mercy of Syrian forces which entered to stabilize the situation.

Today Lebanon has restabilized and Beirut is being rebuilt, but it will take generations to overcome the losses sustained during a quarter-century of dislocation.

Israel, the Jewish state, lies at the heart of the Arab world (Fig. 6-9). Since 1948, when Israel was created as a homeland for the Jewish people on the recommendation of a United Nations commission, the Arab-Israeli conflict has overshadowed all else in the region.

Figure 6-14 helps us understand the complex issues involved here. In 1946 the British, who had administered this area in post-Ottoman times, granted independence to what was then called "Transjordan," the kingdom east of the Jordan River. In 1948, the orange-colored area became the U.N.-sponsored state of Israel—including, of course, land that had long belonged to Arabs in this territory called Palestine.

As soon as Israel proclaimed its independence, neighboring Arab states attacked it. Israel, however, not only held its own but pushed the Arab forces back beyond its borders, gaining the green areas shown in Figure 6-14. Meanwhile, Jordanian armies crossed the Jordan River and annexed the yellow-colored area named the West Bank, including part of the city of Jerusalem. The king called his newly enlarged country Jordan.

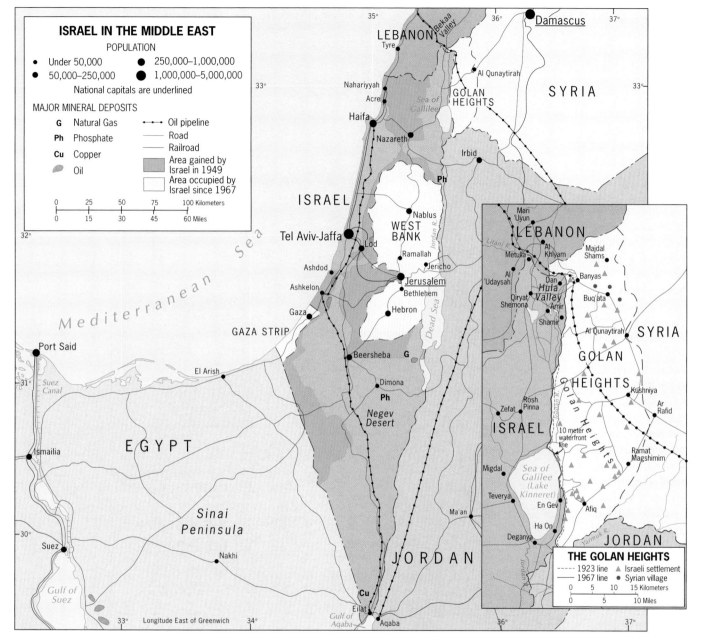

ISRAEL IN THE MIDDLE EAST

POPULATION

- Under 50,000
- 50,000–250,000
- 250,000–1,000,000
- 1,000,000–5,000,000

National capitals are underlined

MAJOR MINERAL DEPOSITS

- **G**   Natural Gas
- **Ph** Phosphate
- **Cu** Copper
- Oil

- Oil pipeline
- Road
- Railroad
- Area gained by Israel in 1949
- Area occupied by Israel since 1967

0 25 50 75 100 Kilometers
0 15 30 45 60 Miles

THE GOLAN HEIGHTS

- 1923 line
- 1967 line
- Israeli settlement
- Syrian village

0 5 10 15 Kilometers
0 5 10 Miles

FIGURE 6-14

More conflict followed. In 1967 a week-long war produced a major Israeli victory: Israel took the Golan Heights from Syria, the West Bank from Jordan, and the Gaza Strip from Egypt, and conquered the whole Sinai Peninsula all the way to the Suez Canal. In later peace agreements, Israel returned the Sinai but not the Gaza Strip.

All this strife produced a huge outflow of Palestinian Arab refugees and displaced persons. The Palestinians are a stateless nation; about 1.25 million continue to live as Israeli citizens within the borders of Israel, but nearly 2.3 million are in the West Bank and more than 1.2 million in the Gaza Strip (the Golan Heights population is comparatively insignificant). An even larger number, however, live in neighboring and nearby countries including Jordan (2.5 million), Lebanon (500,000), Syria (450,000), and Saudi Arabia (340,000); another quarter of a million reside in Iraq, Egypt, Kuwait, and Libya. Many have been assimilated into the local societies, but tens of thousands of others continue to live in refugee camps. In 2002 the scattered Palestinian population was estimated to number nearly 9.5 million.

Israel is about the size of Massachusetts and has a population of 6.4 million (including 1.25 million Arabs), but because of its location amid adversaries and its strong international links, these data do not reflect Israel's importance. Israel has built a powerful military even as its regional Arab neighbors have grown stronger; its policies arouse Arab passions; and Palestinian aspirations for a territorial state have not borne fruit. As a democracy with strong Western ties, Israel has been a recipient of massive U.S. financial aid, and U.S. foreign policy has been to seek an accommodation between Jews and Palestinians as well as between Israel and its Arab neighbors.

The geographic obstacles to such an accommodation include the following:

1. *The West Bank.* Even after its capture by Israel in 1967, the West Bank might have become a Palestinian homeland (and possibly a state), but Jewish immigration to the area made such a future difficult. In 1977 only 5000 Jews lived on the West Bank; by 2000 there were almost 200,000, making up about 10 percent of the population and cre-

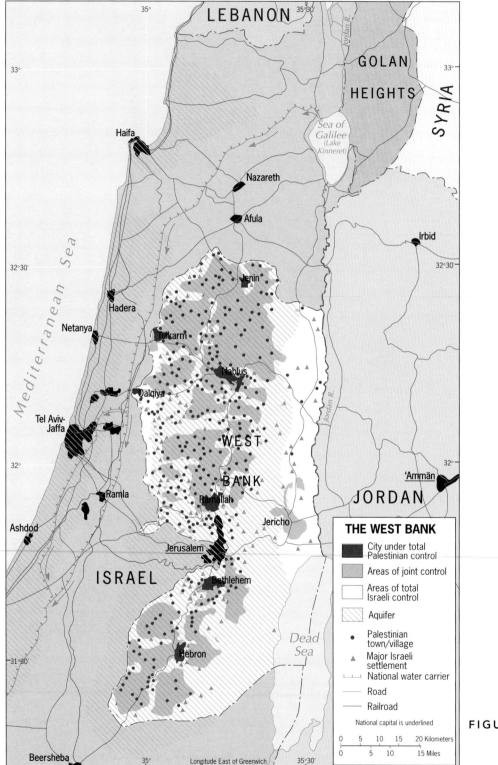

THE WEST BANK

- City under total Palestinian control
- Areas of joint control
- Areas of total Israeli control
- Aquifer
- • Palestinian town/village
- ▲ Major Israeli settlement
- National water carrier
- Road
- Railroad

National capital is underlined

| 0 | 5 | 10 | 15 | 20 Kilometers |
| 0 | 5 | 10 | 15 Miles |

FIGURE 6-15

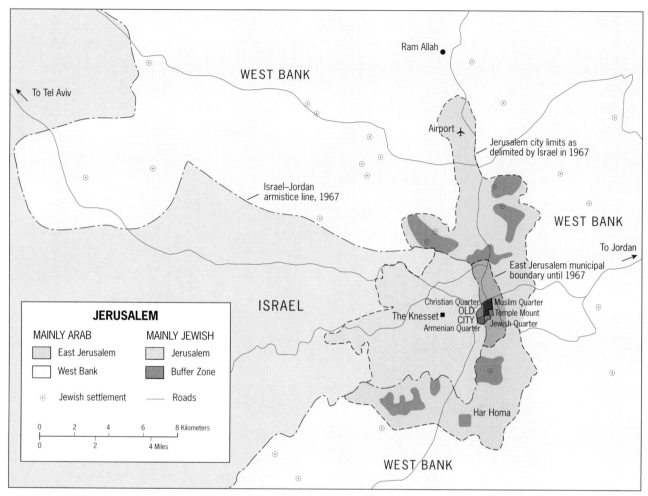

FIGURE 6-16

By the time a cease-fire was arranged, Israel held the western part of the city, and Arab forces the eastern sector. But in this eastern sector lay major Jewish historic sites, including the Western Wall. Still, in 1950 Israel declared the western sector of Jerusalem its capital, making this, in effect, a *forward capital*. Figure 6-16 shows the position of the armistice line, leaving most of the Old City in Jordanian hands. But then, in the 1967 war, Israel conquered all of the West Bank, including East Jerusalem, and in 1980 the Jewish state reaffirmed Jerusalem's status as capital, calling on all nations to move their embassies from Tel Aviv. Meanwhile, the government redrew the map of the ancient city, building Jewish settlements in a ring around East Jerusalem that would end the old distinction between Jewish west and Arab east. This enraged Palestinian leaders, who still view Jerusalem as the eventual headquarters of their hoped-for Palestinian state.

Israel lies at the center of a fast-moving geopolitical storm; the four issues raised above are only part of the overall problem (others are water rights, compensation for land expropriation, and the Palestinians' "right to return" to pre-refugee abodes). Now, in the age of nuclear, chemical, and biological weapons and longer-range missiles, Israel's search for an accommodation with its Arab neighbors is a race against time.

THE ARABIAN PENINSULA

The regional identity of the Arabian Peninsula is clear: south of Jordan and Iraq, the entire peninsula is

ating a seemingly inextricable jigsaw of Jewish and Arab settlements (Fig. 6-15).

2. *The Golan Heights.* The inset in Figure 6-14 suggests how difficult the Golan Heights issue is. The "heights" overlook a large area of northern Israel, and they flank the Jordan River and crucial Lake Kinneret (the Sea of Galilee), the water reservoir for Israel. Relations with Syria are not likely to become normal again until the Golan Heights are returned, but in democratic Israel the political climate may make ceding this territory impossible.

3. *The Gaza Strip.* Small Jewish communities continue to live under armed protection in a poverty-stricken, barren strip of land inhabited by 1.2 million Palestinians, many of whom need jobs across the border in Israel to survive. Even now, a half century later, refugee camps form a constant reminder of the consequences of Israel's creation.

4. *Jerusalem.* The United Nations intended Tel Aviv to be Israel's capital, and Jerusalem an international city. But the Arab attack and the 1948–1949 war allowed Israel to drive toward Jerusalem (see Fig. 6-14, the green wedge into the West Bank).

of the British Crown. This was unthinkable in the Portuguese colonies, where harsh, direct control was the rule. The French sought to create culturally assimilated elites that would represent French ideals in the colonies. In the Belgian Congo, however, King Leopold II, who had financed the expeditions that staked Belgium's claim in Berlin, embarked on a campaign of ruthless exploitation. His enforcers mobilized almost the entire Congolese population to gather rubber, kill elephants for their ivory, and build public works to improve export routes. For failing to meet production quotas, entire communities were massacred. Killing and maiming became routine in a colony in which horror was the only common denominator. After the impact of the slave trade, King Leopold's reign of terror was Africa's most severe demographic disaster. By the time it ended, after a growing outcry around the world, as many as 10 million Congolese had been murdered. In 1908 the Belgian government took over, and slowly its Congo began to mirror Belgium's own internal divisions: corporations, government administrators, and the Roman Catholic Church each pursued their sometimes competing interests. But no one thought to change the name of the colonial capital: it was Leopoldville until the Belgian Congo achieved independence.

12 Colonialism transformed Africa, but in its post-Berlin form it lasted less than a century. In Ghana, for example, the Ashanti (Asante) Kingdom still was fighting the British in the early years of the twentieth century; by 1957, Ghana was independent again. In a few years, much of Subsaharan Africa will have been independent for half a century, and the colonial period is becoming an interlude rather than a paramount chapter in modern African history.

FIGURE 7-7

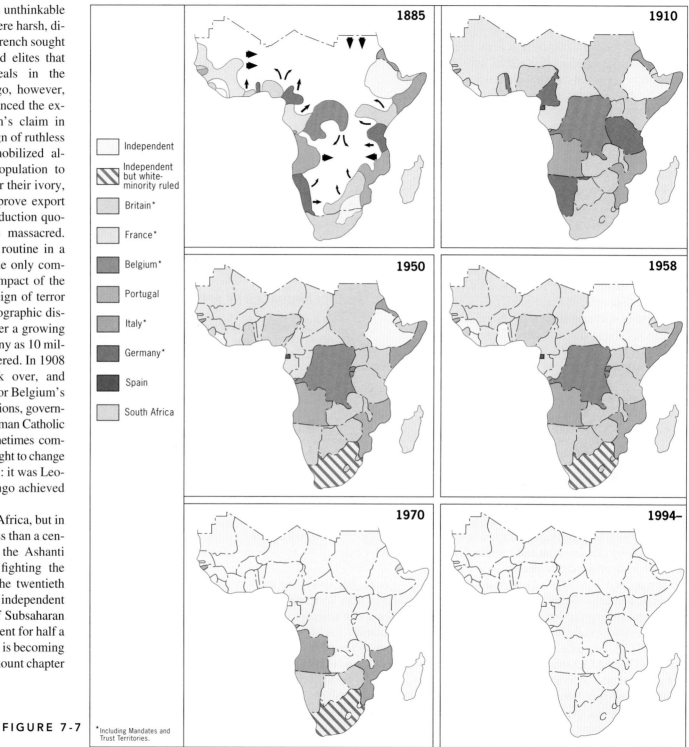

Independent

Independent but white-minority ruled

Britain*

France*

Belgium*

Portugal

Italy*

Germany*

Spain

South Africa

*Including Mandates and Trust Territories.

1885 1910 1950 1958 1970 1994–

COLONIZATION AND DECOLONIZATION, 1885–2002

CULTURAL PATTERNS

We may tend to think of Africa in terms of its prominent countries and famous cities, its development problems and political dilemmas, but Africans themselves have another perspective. The colonial period created states and capitals, introduced foreign languages to serve as the *linguae francae*, and brought railroads and roads. The colonizers stimulated labor movements to the mines they opened, and they disrupted other migrations that had been part of African life for many centuries. But they did not change the ways of life of most of the people. More than 70 percent of the realm's population still live in, and work near, Africa's hundreds of thousands of villages. They speak one of more than a thousand languages in use in the realm. The villagers' concerns are local; they focus on subsistence, health, and safety. They worry that the conflicts over regional power or political ideology will engulf them, as has happened to millions in Liberia, Sierra Leone, Ethiopia, Rwanda, The Congo, Moçambique, and Angola since the 1970s. Africa's largest peoples are major nations, such as the Yoruba of Nigeria and the Zulu of South Africa. Africa's smallest peoples number just a few thousand. As a geographic realm, Subsaharan Africa has the most complex cultural mosaic on Earth.

African Languages

Africa's linguistic geography is a key component of that cultural intricacy. Most of Subsaharan Africa's more than 1000 languages do not have a written tradition, making classification and mapping difficult. Scholars have attempted to delimit an African language map, and Figure 7-8 is a composite of their efforts. One feature is common to all language maps of Africa: the geographic realm begins approximately where the Afro-Asiatic language family (mapped in yellow in Fig. 7-8) ends, although the correlation is sharper in West Africa than to the east.

In Subsaharan Africa, the dominant language family is the Niger-Kordofanian family. It consists of two subfamilies, the tiny Kordofanian concentration in northeastern Sudan and the pervasive Niger-

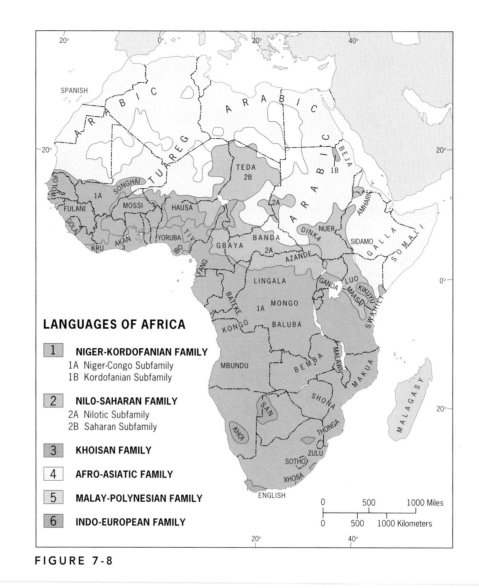

FIGURE 7-8

Congo subfamily that extends across most of the realm from West Africa to East and Southern Africa. The Bantu language forms the largest branch in this subfamily, but Niger-Congo languages in West Africa, such as Yoruba and Akan, also have millions of speakers. Another important language family is the Nilo-Saharan family, extending from Maasai in Kenya northwest to Teda in Chad. No other language families are of similar extent or importance: the Khoisan family, of ancient origins, now survives among the dwin-

dling Khoi and San peoples of the Kalahari; the small white minority in South Africa speak Indo-European languages; and Malay-Polynesian languages prevail in Madagascar, which was peopled from Southeast Asia before Africans reached it.

About 40 African languages are spoken by 1 million people or more, and a half-dozen by about 10 million or more: Hausa (50 million), Yoruba (23 million), Ibo, Swahili, Lingala, and Zulu. Although English and French have become important *linguae francae* in multilingual countries such as Nigeria and Côte d'Ivoire (where officials even insist on spelling the name of their country—Ivory Coast—in the Francophone way), African languages also serve this purpose. Hausa is a common language across the West African savanna; Swahili is widely used in East Africa. And pidgin languages, mixtures of African and European tongues, are spreading along West Africa's coast. Millions of Pidgin English (called *Wes Kos*) speakers use this medium in Nigeria and Ghana. **13 Multilingualism** can be a powerful centrifugal force in society, and African governments have tried with varying success to establish "national" alongside local languages. Nigeria, for example, made English its official language because none of its 250 languages, not even Hausa, had sufficient internal interregional use. But using a European, colonial language as an official medium invites criticism, and Nigeria remains divided on the issue. On the other hand, making a dominant local language official would invite negative reactions from ethnic minorities. Language remains a potent force in Africa's cultural life.

Religion in Africa

Africans had their own religious belief systems long before Christians and Muslims arrived to convert them. And for all of Subsaharan Africa's cultural diversity, Africans had a consistent view of their place in nature. Spiritual forces, according to African tradition, are manifest everywhere in the natural environment, not in a supreme deity that exists in some remote place. Thus gods and spirits affect people's daily lives, witnessing every move, rewarding the virtuous, and punishing (through injury or crop failure, for example) those who misbehave. Ancestral spirits can inflict misfortune on the living. They are everywhere: in the forest, rivers, mountains.

As with land tenure, the religious views of Africans clashed fundamentally with those of outsiders. Monotheistic Christianity first touched Africa in the northeast when Nubia and Axum were converted, and Ethiopia has been a Coptic Christian stronghold since the fourth century AD. But the Christian churches' real invasion did not begin until the onset of colonialism after the turn of the sixteenth century. Christianity's various denominations made inroads in different areas: Roman Catholicism in much of Equatorial Africa mainly at the behest of the Belgians, the Anglican Church in British colonies, and Presbyterians and others elsewhere. But almost everywhere, Christianity's penetration led to a blending of traditional and Christian beliefs, so that much of Subsaharan Africa is nominally, though not exclusively, Christian. Go to a church in Gabon or Uganda or Zambia, and you may hear drums instead of church bells, sing African music rather than hymns, and see African carvings alongside the usual statuary.

Islam had a different arrival and impact. Long before the colonial invasion, Islam advanced out of Arabia, across the desert, and down the east coast. Muslim clerics converted the rulers of African states and commanded them to convert their subjects. They Islamized the savanna states and penetrated into present-day northern Nigeria, Ghana, and Ivory Coast. They encircled and isolated Ethiopia's Coptic Christians and Islamized the Somali people in Africa's Horn. They established beachheads on the Kenya coast and took over Zanzibar. On the map, the African Transition Zone defines the Muslim Front (Fig. 6-12). In the field, Arabizing Islam and European Christianity competed for African minds, and Islam proved to be a far more pervasive force. From Senegal to Somalia, the population is virtually 100 percent Muslim, and Islam's rules dominate everyday life. The Sunni *mullahs* would never allow the kind of marriage between traditional and Christian beliefs seen in much of formerly colonial Africa. This fundamental contradiction between Islamic dogma and Christian accommodation creates a potential for conflict in countries where both religions have adherents.

MODERN MAP AND TRADITIONAL SOCIETY

The political map of Subsaharan Africa has 45 states but no nation-states (apart from some microstates and ministates in the islands and in the south). Centrifugal forces are powerful, and outside interventions during the Cold War, when communist and anticommunist foreigners took sides in local civil wars, worsened conflict within African states. Colonialism's economic legacy was not much better. In tropical African capitals, core areas, port cities, and transport systems were laid out to maximize profit and facilitate exploitation of minerals and soils; the colonial mosaic inhibited interregional communications except where cooperation enhanced efficiency. Colonial Zambia and Zimbabwe, for example (then called Northern and Southern Rhodesia), were landlocked and needed outlets, so railroads were built to Portuguese-owned ports. But such routes did little to create intra-African linkages. The modern map reveals the results: in West Africa you can travel from the coast into the interior of all the coastal states along railways or adequate roads. But no high-standard roadway was ever built to link these coastal neighbors to each other.

To overcome such disadvantages, African states must cooperate internationally, continentwide as well as regionally. The Organization of African Unity (OAU) was established for this purpose in 1963 and in 2001 was superseded by the African Union. In 1975 the Economic Community of West African States (ECOWAS) was founded by 15 countries to promote trade, transportation, industry, and social affairs in the region. And in the early 1990s another important step was taken when 12 countries joined in the Southern African Development Community (SADC), organized to facilitate regional commerce, intercountry transport networks, and political interaction.

Population and Urbanization

Subsaharan Africa remains the least urbanized world realm, but it is urbanizing at a fast pace. By the time you read this, the percentage of urban dwellers will have passed 30. This means that nearly 200 million people now live in cities and towns, of which many were founded and developed by the colonial powers.

African cities became centers of embryonic national core areas, and of course they served as government headquarters. This *formal sector* of the city used to be the dominant one, with government control and regulations affecting civil service, business, industry, and workers. Today, however, African cities look different. From a distance, the skyline still resembles that of a modern center. But in the streets, on the sidewalks right below the shopwindows, there are hawkers, basket weavers, jewelry sellers, garment makers, wood carvers—a second economy, most of it beyond government control. This *informal sector* now dominates many African cities. It is peopled by the rural immigrants, who also work as servants, apprentices, construction workers, and in countless other menial jobs.

Millions of urban immigrants, however, cannot find work, at least not for months or even years at a time. They live in squalid circumstances, in desperate poverty, and governments cannot assist them. As a result, the squatter rings around (and also within) many of Africa's cities are unsafe—uncomfortable, unhealthy slums without adequate shelter, water supply, or basic sanitation. Garbage-strewn (no solid-waste removal here), muddy and insect-infested during the rainy season, and stifling and smelly during the dry period, they are incubators of disease. Yet few of its residents return to their villages. Every new day brings hope.

In our regional discussion we focus on some of Subsaharan Africa's cities, all of which, to varying degrees, are stressed by the rate of population influx. Despite the plight of the urban poor and the poverty of Africa's rural areas, some of Africa's capitals remain the strongholds of privileged elites who, dominant in governments, fail to address the needs of other ethnic groups. Discriminatory policies and artificially low food prices disadvantage farmers and create even greater urban-rural disparities than the colonial period saw. But today the prospect of democracy brings hope that Africa's rural majorities will be heard and heeded in the capitals.

REGIONS OF THE REALM

On the face of it, Africa seems to be so massive, compact, and unbroken that any attempt to justify a contemporary regional breakdown is doomed to fail. No deeply penetrating bays or seas create peninsular fragments as in Europe. No major islands (other than Madagascar) provide the broad regional contrasts we see in Middle America. Nor does Africa really taper southward to the peninsular proportions of South America. And Africa is not cut by an Andean or a Himalayan mountain barrier. Given Africa's colonial fragmentation and cultural mosaic, is regionalization possible? Indeed it is.

Maps of environmental distributions, ethnic patterns, cultural landscapes, historic culture hearths, and colonial frameworks yield the four-region structure shown in Figure 7-9:

1. *West Africa* includes the countries of the western coast and Sahara margin from Senegal and Mauritania in the west to population-giant Nigeria and Niger (and part of Chad) in the east.
2. *Equatorial Africa* centers on the vast state of The Congo, and also includes Congo, Gabon, Cameroon, and the Central African Republic, a part of Chad, and southern Sudan.
3. *East Africa* also lies astride the equator, but its environments are moderated by elevation. Kenya and Tanzania are the coastal states; Uganda, Rwanda, and Burundi are landlocked. Highland Ethiopia also forms part of this region.
4. *Southern Africa* extends from the southern borders of The Congo and Tanzania to the continent's southernmost cape. Ten countries, including Angola and Zimbabwe, form part of this region, whose giant is South Africa.

The island of Madagascar, in the Indian Ocean opposite Moçambique, cannot be incorporated into either East or Southern Africa, for geographic reasons we discuss later.

▶ WEST AFRICA

West Africa occupies most of Africa's Bulge, extending south from the margins of the Sahara to the Gulf of Guinea coast and from Lake Chad west to Senegal (Fig. 7-10). Politically, the broadest definition of this region includes all those states that lie to the south of Western Sahara, Algeria, and Libya and to the west of Chad (itself sometimes included) and Cameroon. Within West Africa, a rough division is sometimes made between the large, mostly steppe and desert states that extend across the southern Sahara (Chad included) and the smaller, better-watered coastal states.

France and Britain dominated the colonial map of West Africa, and to this day the interaction among West Africa's states is limited. But West Africa's cultural vitality, historic legacies, populous cities, crowded countrysides, and bustling markets combine to create a regional imprint that is distinct and pervasive.

Nigeria

When Nigeria achieved full independence from Britain in 1960, its new government faced the task of administering a European political creation containing three major nations and nearly 250 other peoples ranging from several million to a few thousand in number.

For reasons obvious from the map, Britain's colonial imprint always was stronger in the two southern regions than in the north. Christianity became the dominant faith in the south, and southerners, especially Yoruba, took a lead role in the transition from ✳ colony to independent state. The choice of Lagos, the port of the Yoruba-dominated southwest, as the capital of a federal Nigeria (and not one of the cities in the more populous north) reflected British desires for

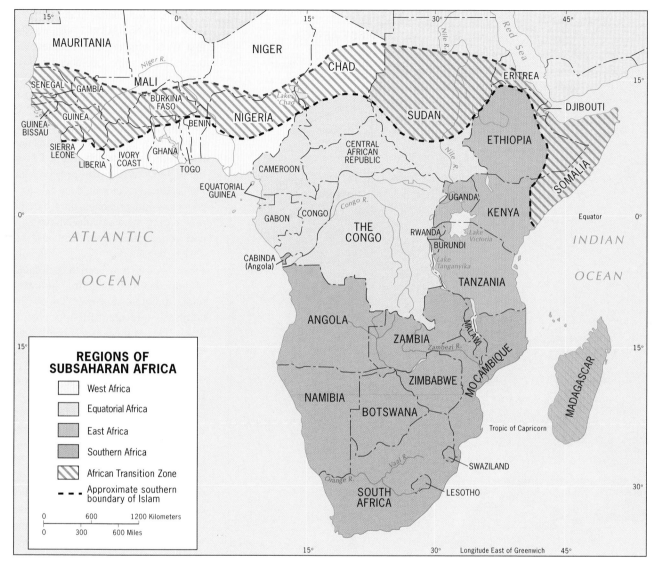

FIGURE 7-9

and State development plans. Soon, revenues from oil production dwarfed all other sources, bringing the country a brief period of prosperity and promise. But before long Nigeria's oil wealth brought more bust than boom. Misguided development plans now focused on grand, ill-founded industrial schemes and costly luxuries such as a national airline; the continuing mainstay of the vast majority of Nigerians, agriculture, fell into neglect. Worse, poor management, corruption, outright theft of oil revenues during military misrule, and excessive borrowing against future oil income led to economic disaster. The country's infrastructure collapsed. In the cities, basic services broke down. In the rural areas, clinics, schools, water supplies, and roads to markets crumbled. In the Niger Delta area, local people beneath whose land the oil was being exploited demanded a share of the revenues and reparations for ecological damage; the military regime under General Abacha responded by arresting and executing nine of their leaders. On global indices of national well-being, Nigeria sank to the lowest rungs even as its production ranked it as high as the world's tenth-largest oil producer, with the United States its chief customer.

In 1999, Nigeria's hopes were raised when, for the first time since 1983, a democratically elected president was sworn into office. But Nigeria's problems (now also including a deepening AIDS crisis) worsened when northern States, beginning with Zamfara, proclaimed Sharia law. When Kaduna State followed suit, riots between Christians and Muslims devastated the old capital city of Kaduna. There, and in ten other northern States (Fig. 7-11), the imposition of Sharia law led to the departure of thousands of Christians, intensifying the

the country's future. A three-region federation, two of which lay in the south, would ensure the primacy of the non-Islamic part of the state. But this framework did not last long. In 1967 the Ibo-dominated Eastern Region declared its independence as the Republic of Biafra, leading to a three-year civil war at a cost of 1 million lives. Since then, Nigeria's federal system has

been modified repeatedly; today there are 36 States, and the capital has been moved from Lagos to centrally-located Abuja (Fig. 7-11).

Large oilfields were discovered beneath the Niger Delta during the 1950s, when Nigeria's agricultural sector produced most of its exports (peanuts, palm oil, cocoa, cotton) and farming still had priority in national

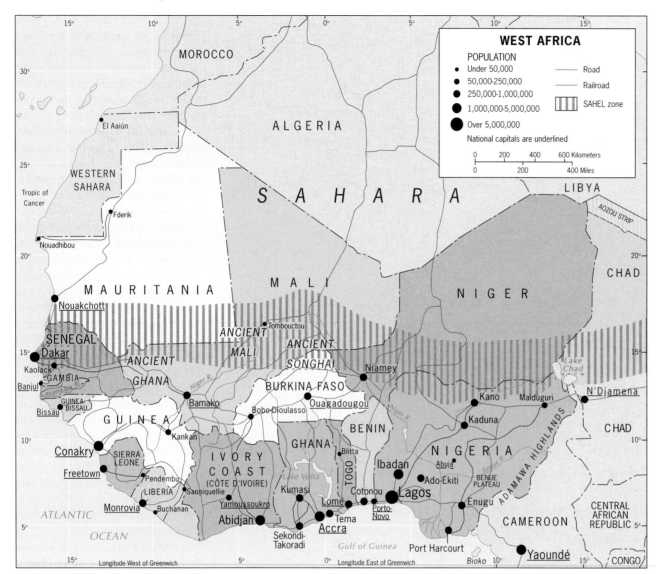

FIGURE 7-10

Among the coastal states, Ghana, once known as the Gold Coast, was the first West African state to achieve independence, with a democratic government and a sound economy based on cocoa exports. Two grandiose post-independence schemes can be seen on the map: the port of Tema, intended to serve a vast West African hinterland, and Lake Volta, which resulted from the region's largest dam project. Mismanagement contributed to Ghana's economic collapse, but in the 1990s the military regime was replaced by stable and democratic government and recovery began. Ivory Coast (officially *Côte d'Ivoire*,) translated three decades of autocratic but stable rule into economic progress, but excesses by the country's president-for-life cost it dearly. One of these excesses involved the construction of a Roman Catholic basilica to rival St. Peter's in Rome in the president's home village, Yamoussoukro, also designated the country's new capital. At the turn of the century the political succession became entangled in a north-south, Muslim-Christian schism that threatened the country's stability. Senegal on the far west coast demonstrates what stability can achieve: without oil, diamonds, or other valuable income sources and with an overwhelmingly subsistence-farming population, Senegal nevertheless managed to achieve some of the region's highest GNP levels. Over 90 percent Muslim and dominated by the

cultural fault line that threatens the cohesion of the country. Although Kaduna State temporarily repealed its decision, the Islamic revivalism now exhibited in the north raises the prospect that Nigeria, West Africa's cornerstone and one of Africa's most important states, may succumb to devolutionary forces arising from its location in the African Transition Zone. For Africa, this would be a calamity.

Coast and Interior

Nigeria is one of 17 states (counting Chad and offshore Cape Verde) that form the region of West Africa. Four of these countries, comprising a huge territory under desert and steppe environments but containing small populations, are landlocked: Mali, Burkina Faso, Niger, and Chad (Fig. 7-10).

Wolof, the ethnic group concentrated in the capital of Dakar, and with continuing close ties to France, Senegal has even been able to overcome a failed effort to unify with its English-speaking enclave, The Gambia, and a secession movement in its southern Casamance District.

Other parts of West Africa have been afflicted by civil war and horror. Liberia, founded in 1822

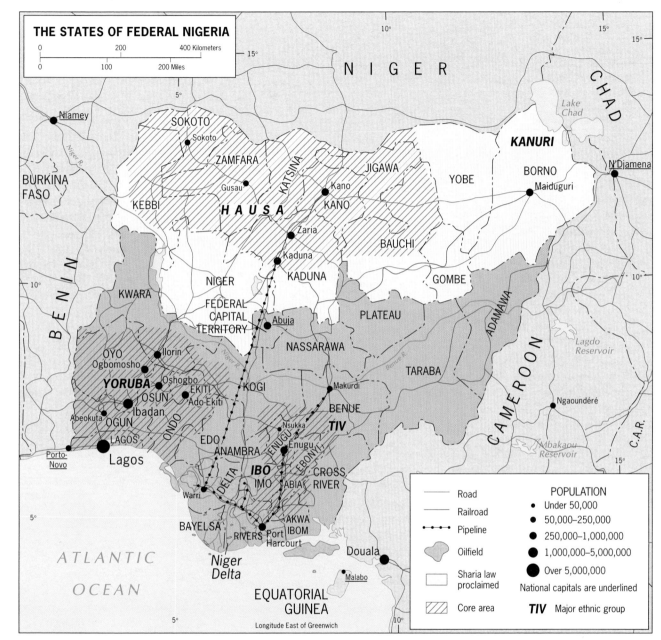

THE STATES OF FEDERAL NIGERIA

0   200   400 Kilometers

0   100   200 Miles

NIGER

CHAD

Lake Chad

KANURI

N'Djamena

SOKOTO

Sokoto

ZAMFARA

JIGAWA

YOBE

BORNO

Maiduguri

BURKINA FASO

Niamey

KEBBI

Gusau

HAUSA

KATSINA

Kano

KANO

BAUCHI

Zaria

Kaduna

KADUNA

GOMBE

BENIN

KWARA

NIGER

FEDERAL CAPITAL TERRITORY

Abuja

PLATEAU

ADAMAWA

Lagdo Reservoir

OYO   Ilorin

Ogbomosho

NASSARAWA

YORUBA

Oshogbo

OSUN   EKITI

Ado-Ekiti

KOGI

Makurdi

TARABA

CAMEROON

Ngaoundéré

C.A.R.

Ibadan

Abeokuta   OGUN   ONDO

EDO

Nsukka

BENUE

TIV

Mbakaou Reservoir

LAGOS

Porto-Novo

Lagos

DELTA

IBO

ENUGU

Enugu

EBONYI

CROSS RIVER

ANAMBRA

IMO   ABIA

Warri

BAYELSA

RIVERS   Port Harcourt

AKWA IBOM

Douala

ATLANTIC OCEAN

Niger Delta

Malabo

EQUATORIAL GUINEA

Longitude East of Greenwich

Road

Railroad

Pipeline

Oilfield

Sharia law proclaimed

Core area

POPULATION

Under 50,000

50,000–250,000

250,000–1,000,000

1,000,000–5,000,000

Over 5,000,000

National capitals are underlined

TIV   Major ethnic group

FIGURE 7-11

by freed American slaves who returned to Africa with the help of American colonization societies, was ruled by "Americo-Liberians" for six generations and sold rubber and iron ore abroad. But a military coup in 1980 ended that era, and full-scale civil war beginning in 1989 embroiled virtually every ethnic group in the country. About a quarter of a million people, one-tenth of the population, perished; hundreds of thousands of others fled as refugees, many to Sierra Leone, which also was originally founded as a haven for freed slaves, in this case by the British in 1787. This country went the all-too-familiar route from self-governing Commonwealth member to republic to one-party state to military dictatorship. In the 1990s worse was to follow: a rebel movement, funded by diamond sales, inflicted dreadful punishment on the local population. International efforts to stem the tide of violence had some effect by 2002, but untold damage had been done.

We should remember, however, that these conflicts are not representative of vast and populous West Africa. Tens of millions of farmers and herders who manage to cope with fast-changing environments in this zone between desert and ocean live remote from the news-making riots in Abidjan and the conflicts along the monsoon coast. Time-honored systems continue to serve, for example, the local village markets that drive the traditional economy. Visit the countryside, and you will find that some village markets are not open every day but, rather, every three or four days. Such a system ensures that all villages get a share in the exchange **14** network. These **periodic markets** represent one of the many traditions that endure in this region, even as the cities beckon and burst at the seams. This region's great challenges are economic survival and nation-building, constrained by a boundary framework that is as burdensome as any in Africa.

▷ EQUATORIAL AFRICA

The term *equatorial* is not just locational but also environmental. The equator bisects Africa, but only the western part of central Africa features the conditions associated with the low-elevation tropics: intense heat, high rainfall and extreme humidity, little seasonal variation, rainforest and monsoon-forest vegetation, and enormous biodiversity. To the east, beyond the Western Rift Valley, elevations rise, and cooler, more seasonal climatic regimes prevail. As a result, we recognize two regions in these lowest latitudes: Equatorial Africa to the west and East Africa to the east.

Equatorial Africa is physiographically dominated by the giant Congo Basin. The Adamawa Highlands separate this region from West Africa; rising elevations and climatic change mark its southern limits (see the *Cwa* boundary in Fig. I-8). Its political geography consists of eight states, of which The Congo (formerly Zaïre) is by far the largest in both territory and population (Fig. 7-12).

Five of the other seven states—Gabon, Cameroon, São Tomé and Príncipe, Congo, and Equatorial Guinea—all have coastlines on the Atlantic Ocean. The Central African Republic and Chad, the south of which is part of this region, are landlocked. In

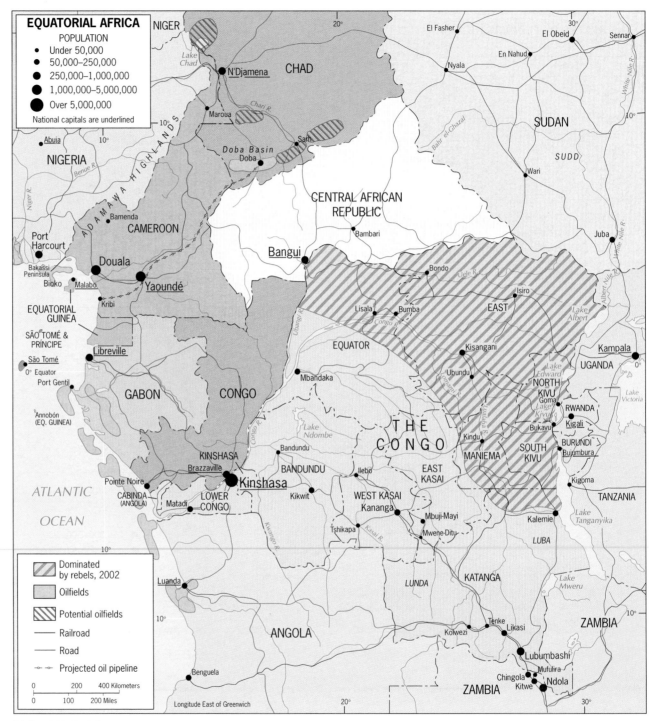

FIGURE 7-12

certain respects, the physical and human characteristics of Equatorial Africa extend even into southern Sudan. This vast and complex region is in many ways the most troubled region in the entire Subsaharan African realm.

The Congo

As the map shows, The Congo has but a tiny window (23 miles; 37 km) on the Atlantic Ocean, just enough to accommodate the mouth of the Congo River. Oceangoing ships can reach the port of Matadi, inland from which falls and rapids make it necessary to move goods by road or rail to the capital, Kinshasa. This is not the only place where the Congo River fails as a transport route. Follow it upstream on Figure 7-12, and you note that other transshipments are necessary between Kisangani and Ubundu, and at Kindu. Follow the railroad south from Kindu, and you reach another narrow corridor of The Congo's territory at the city of Lubumbashi. That vital part of The Congo contains most of its major mineral resources, including copper and cobalt.

With a territory not much smaller than the United States east of the Mississippi, a population of 55.4 million, a rich and varied mineral base, and much good agricultural land, The Congo would seem to have all the ingredients needed to lead this region and, indeed, Africa. But strong centrifugal forces, arising from its physiography and cultural geography, pull The Congo apart. The immense forested heart of the basin-shaped country creates communication barriers between east and west, north and south. Many of The Congo's productive areas lie along its periphery, separated by enormous distances. These areas tend to look across the border, to one or more of The Congo's nine neighbors, for outlets, markets, and often ethnic kinship as well.

The Congo's civil wars of the 1990s started in one such neighbor, Rwanda, and spilled over into what was then still known as Zaïre. Rwanda, Africa's most densely populated country, has for centuries been the scene of conflict between sedentary Hutu farmers and invading Tutsi pastoralists. Colonial borders and practices worsened the situation, and after independence a series of terrible crises followed. In the mid-1990s the latest of these crises generated one of the largest refugee streams ever seen in the world, and the conflict engulfed eastern (and later northern and western) Zaïre. The Zaïrian regime was toppled, rebels took control of the east (Fig. 7-12), other African countries took sides, and the renamed Congo remained in turmoil into the twenty-first century.

Across the River

To the west and north of the Congo and Ubangi rivers lie Equatorial Africa's other seven countries (Fig. 7-12). Two of these are landlocked. Chad, straddling the African Transition Zone as well as the regional boundary with West Africa, is one of Africa's most remote countries. The Central African Republic, chronically unstable and poverty-stricken, never was able to convert its agricultural potential and mineral resources (diamonds, uranium) into real progress. And one country consists of two small, densely forested volcanic islands: São Tomé and Príncipe, a ministate with a population of only 175,000 and a few exports derived from its cocoa plantations and coconut trees.

The four coastal states present a different picture. All four possess oil reserves and share the Congo Basin's equatorial forests; oil and timber, therefore, rank prominently among their exports. In Gabon, this combination has produced Equatorial Africa's only upper-middle-income economy. Of the four coastal states, Gabon also has the largest proven mineral resources, including manganese, uranium, and iron ore. Its capital, Libreville (the only coastal capital in the region), reflects all this in its high-rise downtown, bustling port, and fast-growing squatter settlements.

Cameroon, less well endowed with oil or other raw materials, has the region's strongest agricultural sector by virtue of its higher-latitude location and high-relief topography. Western Cameroon is one of the more developed parts of Equatorial Africa and includes the capital, Yaoundé, and the port of Douala.

With five neighbors, Congo could be a major transit hub for this region, especially for The Congo if it recovers from civil war. Its capital, Brazzaville, lies across the Congo River from Kinshasa and is linked to the port of Pointe Noire by road and rail. But devastating power struggles have negated Congo's geographic advantages.

As Figure 7-12 shows, Equatorial Guinea consists of a rectangle of mainland territory and the island of Bioko, where the capital of Malabo is located. A former Spanish colony that remained one of Africa's least-developed territories, Equatorial Guinea, too, has been affected by the oil business in this area. Petroleum products now dominate its exports, but, as in so many other oil-rich countries, this bounty has not significantly raised incomes for most of the people.

One other territory would seem to be a part of Equatorial Africa: Cabinda, wedged between the two Congos just to the north of the Congo River's mouth. But Cabinda is one of those colonial legacies on the African map—it belonged to the Portuguese and was administered as part of Angola. Today it is an **15 exclave** of independent Angola, and a valuable one: it contains major oil reserves.

▷ EAST AFRICA

East of the row of Great Lakes that marks the eastern border of The Congo (Lakes Albert, Edward, Kivu, and Tanganyika), central Africa takes on a different character. The land rises from the Congo Basin to the East African Plateau. Hills and valleys, fertile soils, and copious rains mark the transition in Rwanda and Burundi. Eastward the rainforest disappears and open savanna cloaks the countryside. Great volcanoes rise above a rift-valley-dissected highland. At the heart of the region lies Lake Victoria. In the north the surface rises above 10,000 feet (3300 m), and so deep are the trenches cut by faults and rivers there that the land was called, appropriately, Abyssinia (now Ethiopia).

Five countries, as well as the highland part of Ethiopia, form this East African region: Kenya, Tanzania, Uganda, Rwanda, and Burundi. Here the Bantu peoples that make up most of the population met Nilotic peoples from the north.

Kenya is neither the largest nor the most populous country in East Africa, but over the past half-century it has been the dominant state in the region.

Its skyscrapered capital at the heart of its core area, Nairobi, is the region's largest city; its port, Mombasa, is the region's busiest.

After independence, Kenya chose a capitalist path of development, aligning itself with Western interests. Without major known mineral deposits, Kenya depended on coffee and tea exports and on a tourist industry based on its magnificent national parks (Fig. 7-13). Tourism became its largest single earner of foreign exchange, and Kenya prospered, apparently proving the wisdom of its capitalist course.

But serious problems arose. Kenya during the 1980s had the highest rate of population growth in the world, and population pressure on farmlands and on the fringes of the wildlife reserves mounted. Poaching became worrisome, and tourism declined. During the late 1990s, violent weather buffeted Kenya, causing landslides and washing away large segments of the crucial Nairobi-Mombasa Highway. This was followed by a severe drought lasting several years, bringing famine to the interior. Meanwhile, government corruption siphoned off funds that should have been invested. Democratic principles were violated, and relationships with Western allies were strained. The AIDS epidemic brought another setback to a country that, in the early 1970s, had appeared headed for an economic takeoff.

Today, Kenya's prospects are uncertain. Geography, history, and politics have placed the Kikuyu (22 percent of the population of 31.6 million) in a position of power. But there are other major peoples (see Fig. 7-13) and several smaller ones. The Luhya, Luo, Kalenjin, and Kamba together constitute about 50 percent of the population, and on the territorial margins of the country there are peoples such as the Maasai, Turkana, Boran, and Galla. Creating and sustaining a political system that ensures democracy and represents the interests of these disparate peoples is Kenya's unmet challenge.

Tanzania (a name derived from Tanganyika plus Zanzibar) is the biggest and most populous East African country (37.4 million). Its total area exceeds that of the other four countries combined. Tanzania

FIGURE 7-13

has been described as a country without a core because its clusters of population and zones of productive capacity lie dispersed—mostly on its margins on the east coast (where the capital, Dar es Salaam, is located), near the shores of Lake Victoria in the northwest, near Lake Tanganyika in the far west, and near Lake Malawi in the interior south. This is in sharp contrast to Kenya, which has a well-defined core area in the Kenya Highlands (centered on Nairobi in the heart of the country). Moreover, Tanzania is a country of many peoples, none numerous enough to dominate the state. About 100 ethnic groups coexist; one-third of the population, mainly those on the coast, are Muslims.

After independence Tanzania embarked on a socialist course toward development, including a massive and disastrous farm collectivization program. The tourist industry declined sharply, and Tanzania became one of the world's poorest countries. But Tanzania did achieve remarkable political stability and a degree of democracy that none of the other East African states attained. Since 1990 the government has changed course, but the AIDS crisis, problems with the Zanzibar merger, and involvement in the troubles of neighboring countries have set Tanzania back.

Uganda contained this region's most important African state when the British colonialists arrived: the Kingdom of Buganda, peopled by the Ganda, located on the north shore of Lake Victoria (Fig. 7-13, the dark brown area). The British established their headquarters near the Ganda capital of Kampala and used the Ganda to control Uganda through indirect rule. Thus the Ganda became the dominant nation in multicultural Uganda, and when the British left, they bequeathed Uganda a complicated federal system designed to perpetuate Ganda supremacy.

The system failed, bringing to power one of Africa's most brutal dictators, Idi Amin. Uganda had a strong economy based on coffee, cotton, and other farm exports and on copper mining; its Asian minority of about 75,000 dominated local commerce. Amin ousted all the Asians, exterminated his opponents, and destroyed the economy. In addition, the AIDS epidemic struck Uganda severely, and after Amin was ousted, the country got embroiled in the political struggles of Sudan, Rwanda, and The Congo.

16 As the map shows, Uganda is a **landlocked state** and depends on Kenya for an outlet to the ocean. Its relative location adjacent to unstable Sudan, Rwanda, and The Congo constitutes a formidable challenge.

Rwanda and Burundi would seem to occupy Tanzania's northwest corner, and indeed they were part of the German colonial domain conquered before World War I. But during that war Belgian forces attacked the Germans from their Congo bases and were awarded these territories when the conflict was over in 1918. The Belgians used them as labor sources for their Katanga mines.

Rwanda (7.5 million) and Burundi (6.4 million) are physiographically part of East Africa, but their cultural geography is linked to the north and west. Here, as we noted earlier, Tutsi pastoralists from the north subjugated Hutu farmers (who had themselves made serfs of the local Twa [pygmy] population), setting up a conflict that was originally ethnic but became cultural. Certain Hutu were able to advance in the Tutsi-dominated society, becoming to some extent converted to Tutsi ways, leaving subsistence farming behind, and rising in the social hierarchy. These so-called moderate Hutu were—and are—often targeted by other Hutus, who resent their position in society. This longstanding discord, worsened by colonial policies, had repeatedly devastated both countries and, in the 1990s, spilled over into The Congo, generating the first interregional African war.

The highland zone of Ethiopia also forms part of East Africa. Addis Ababa, the historic capital, was the headquarters of a Coptic-Christian, Amharic empire that held its own against the colonial intrusion except for a brief period from 1935 to 1941, when the Italians defeated it. Indeed, the Ethiopians in their mountain fortress (Addis Ababa lies about 10,000 feet above sea level) became colonizers themselves, taking control of much of the Islamic part of the Horn.

Ethiopia's natural outlets are toward the Gulf of Aden and the Red Sea, but its government was forced to yield independence to Eritrea and the country is now effectively landlocked. Physiographically and culturally, however, it is part of East Africa, and the Amhara and Oromo peoples are neither Arabized

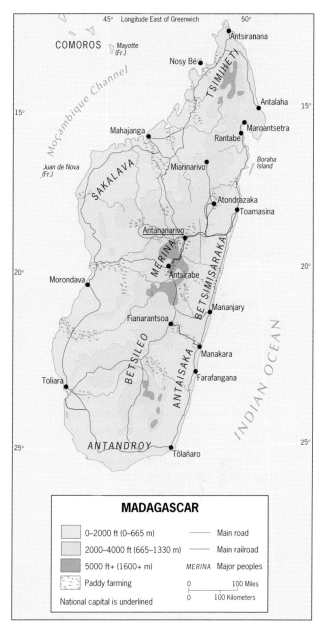

MADAGASCAR

☐ 0–2000 ft (0–665 m)	—— Main road
☐ 2000–4000 ft (665–1330 m)	—— Main railroad
■ 5000 ft+ (1600+ m)	*MERINA* Major peoples
⬚ Paddy farming	0 ——— 100 Miles
National capital is underlined	0 ——— 100 Kilometers

FIGURE 7-14

nor Muslim: they are Africans. The map shows that functional linkages between Ethiopia and East Africa remain weak, but this is likely to change in the future.

Madagascar

 Off Africa's east coast lies the world's fourth-largest island, Madagascar (Fig. 7-14). About 2000 years ago the first human settlers arrived here—not from Africa but from distant Southeast Asia. A powerful Malay kingdom of the Merina flourished in the highlands, whose language, Malagasy, became Madagascar's tongue (Fig. 7-8)

The Malay immigrants eventually brought Africans to their island, but today the Merina (4 million) and the Betsimisaraka (2 million) remain the largest of about 20 ethnic groups in the population of 16 million. After successfully resisting colonial invasion, the locals eventually yielded to France, and French became the *lingua franca* of the educated elite.

Madagascar is not part of either East Africa or Southern Africa. Its human as well as wildlife population is unique; its cultural landscapes still carry Southeast Asian imprints. Here the people eat rice, not corn, grown on terraced paddies. Rapid population growth is shrinking forests and animal refuges, the economy is weak, and poverty reigns in one of the world's most scenic outposts.

SOUTHERN AFRICA

Southern Africa, as a geographic region, consists of all the countries and

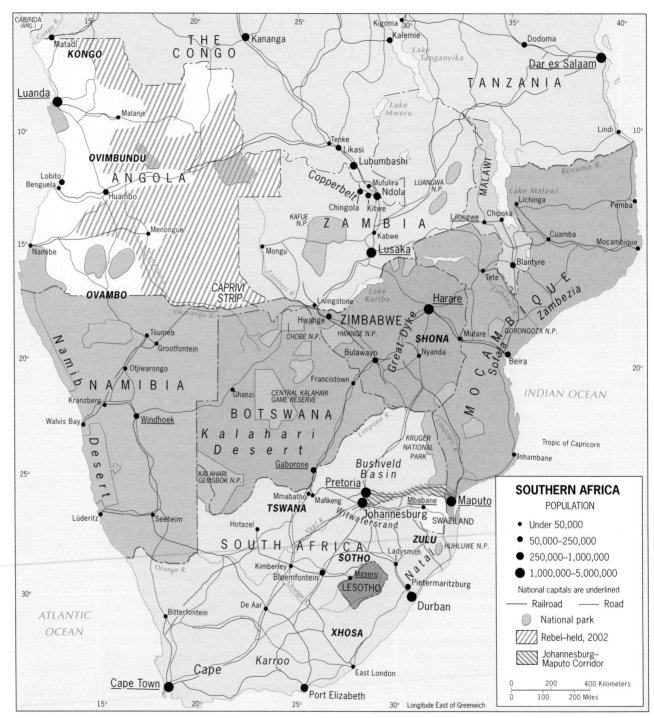

FIGURE 7-15

territories lying south of Equatorial Africa's The Congo and East Africa's Tanzania (Fig. 7-15). Thus defined, the region extends from Angola and Moçambique (on the Atlantic and Indian Ocean coasts, respectively) to South Africa and includes a half-dozen landlocked states. Also marking the northern limit of the region are Zambia and Malawi. Zambia is nearly cut in half by a long land extension from The Congo, and Malawi penetrates deeply into Moçambique. The colonial boundary framework, here as elsewhere, produced many liabilities.

Southern Africa constitutes a geographic region in both physiographic and human terms. Its northern zone marks the southern limit of the Congo Basin in a broad upland that stretches across Angola and into Zambia (the tan corridor extending eastward from the Bihe Plateau in Fig. 7-2). Lake Malawi is the southernmost of the East African rift-valley lakes; Southern Africa has none of East Africa's volcanic and earthquake activity. Most of the region is plateau country, and the Great Escarpment is much in evidence here. There are two pivotal river systems: the Zambezi (which forms the border between Zambia and Zimbabwe) and the Orange-Vaal (South African rivers that combine to demarcate southern Namibia from South Africa).

Southern Africa is the continent's richest region materially. A great zone of mineral deposits extends through the heart of the region from Zambia's Copperbelt through Zimbabwe's Great Dyke and South Africa's Bushveld Basin and Witwatersrand to the goldfields and diamond mines of the Orange Free State and northern Cape Province in the heart of South Africa. Ever since these minerals began to be exploited in colonial times, many migrant laborers have come to work in the mines.

Southern Africa's agricultural diversity matches its mineral wealth. Vineyards drape the slopes of South Africa's Cape Ranges; tea plantations hug the eastern escarpment slopes of Zimbabwe. Before civil war destroyed its economy, Angola was one of the world's leading coffee producers. South Africa's relatively high latitudes and its range of altitudes create environments for apple orchards, citrus groves, banana plantations, pineapple farms, and many other crops.

Despite this considerable wealth and potential, the countries of Southern Africa have not prospered. As Figure I-10 shows, most remain mired in the low-income category (Moçambique, with a per-capita GNP of only $210, is one of the world's poorest states); only South Africa and neighboring Botswana are in the upper-middle-income rank, the latter desert state being the realm's second most sparsely populated country. Rapid population growth, civil wars, political instability, poor management, corruption, AIDS, and environmental problems have inhibited economic growth. Nevertheless, the situation is superior to that in any of the other three regions; more countries have risen above the low-income level here than anywhere else in Subsaharan Africa. As the new millennium opened, the resurgence of South Africa gave hope that this region might finally blossom, perhaps to lead the entire realm to a higher economic and social plane.

The Northern Tier

It is helpful to visualize the states of Southern Africa in two groups: four northern and six southern countries, the region being anchored by South Africa.

In the four countries that extend across the northern tier of the region—Angola, Zambia, Malawi, and Moçambique—problems abound. Angola (13.7 million), formerly a Portuguese dependency, with its exclave of Cabinda had a thriving economy based on a wide range of mineral and agricultural exports at the time of independence in 1975. But then Angola fell victim to the Cold War, with northern peoples choosing a communist course and southerners falling under the sway of a rebel movement backed by South Africa and the United States. The results included a devastated infrastructure, idle farms, looting of diamonds, hundreds of thousands of casualties, and millions of landmines that continue to kill and maim.

On the opposite coast, the other major former Portuguese colony, Moçambique (19.9 million), fared even worse. Without Angola's mineral base and with limited commercial agriculture, Moçambiue's chief asset was its relative location. Its two major ports, Maputo and Beira, handled large volumes of exports and imports for South Africa, Zimbabwe, and Zambia. But upon independence Moçambique, too, chose a Marxist course with dire economic and dreadful political consequences. Another rebel movement supported by South Africa caused civil conflict, created famines, and generated a stream of more than a million refugees toward Malawi. Rail and port facilities lay idle, and Moçambique at one time was ranked by the United Nations as the world's poorest country. In recent years the port traffic has been somewhat revived and Moçambique and South Africa are working on a joint Maputo Development Corridor (Fig. 7-15), but it will take generations for Moçambique to recover.

Landlocked Zambia (10 million), the product of British colonialism, shares the riches of the Copperbelt with The Congo's Katanga Province. Not only have commodity prices on which Zambia depends declined severely, but Zambia's outlets—Lobito in Angola and Beira in Moçambique—and the railroads leading there were made inoperative by Cold War conflicts. Neighboring Malawi (10.8 million) has an almost totally agricultural economy, sufficiently diversified to cushion its economy against market swings. But Malawi, too, suffers from its landlocked situation.

Southern States

Six countries constitute Africa's southernmost tier of states and form a distinct subregion within Southern Africa: Zimbabwe, Namibia, Botswana, Swaziland, Lesotho, and the Republic of South Africa. As Figure 7-15 shows, four of the six are landlocked. Diamond-exporting Botswana occupies the heart of the Kalahari Desert and surrounding steppe; only 1.6 million people, most of them subsistence farmers, inhabit this Texas-sized country. In 2001, no country in Africa was more severely afflicted by the AIDS epidemic than Botswana. South Africa encircles Lesotho (2.2 million) and surrounds most of Swaziland (1.0 million), ancestral home of the Swazi nation. Botswana, Lesotho, and Swaziland depend heavily on the income that workers who labor in South African mines, factories, and fields send home.

Apart from South Africa, the region's giant, the most important southern state undoubtedly is Zimbabwe (11.5 million), landlocked but well endowed with mineral and agricultural resources. Zimbabwe (the country is named after stone ruins in its interior) is mostly an elevated plateau between the Zambezi and Limpopo rivers, with the desert to the west and the Great Escarpment to the east. Its core area is defined by the mineral-rich Great Dyke and its environs, extending southwest from the vicinity of the capital, Harare, to the country's second city, Bulawayo. Copper, asbestos, and chromium (of which Zimbabwe is one of the world's leading sources) are among its major mineral exports, but Zimbabwe is not just an ore-exporting country. Farms produce tobacco, tea, sugar, cotton, and other crops (corn is the staple).

Two nations form most of Zimbabwe's population: the Shona (71 percent) and the Ndebele (16 percent). A tiny minority of whites still owns the best farmland and sustains the agricultural economy. Mounting environmental and economic problems beginning in the 1980s led to rising social and political tensions. President Mugabe and his dominant party allowed white farms to be invaded by squatters who sometimes killed the owners; corruption rose and human rights were curbed. Meanwhile, Mugabe used Zimbabwe's armed forces to intervene in The Congo. Once-promising Zimbabwe is the tragedy of the region today.

Southern Africa's youngest independent state, Namibia (1.9 million), is a former German colony with a territorial peculiarity: the so-called Caprivi Strip linking it to the right bank of the Zambezi River (Fig. 7-15), another product of colonial partitioning. Administered by South Africa from 1919 to 1990, Namibia is named after one of the world's driest deserts. This state is about as large as Texas and Oklahoma, but only its far north receives enough moisture to permit subsistence farming, which is why most of the people live near the Angolan border. Mining in the Tsumeb area and ranching in the vast steppe country of the south form the main commercial activities. The capital, Windhoek, is centrally situated opposite Walvis Bay, the main port. German influence still lingers in what used to be called South

West Africa, as does the Afrikaner presence from South Africa's *apartheid* period. Changes are in the offing here.

South Africa

The Republic of South Africa is the giant of Southern Africa, an African country at the center of world attention, a bright ray of hope not only for Africa but for all humankind.

Long in the grip of one of the world's most notorious racial policies (*apartheid*, or "apartness," and its derivative, *separate development*), South Africa today is shedding its past and building a new future. That virtually all parties to the earlier debacle are now working cooperatively to restructure the country under a new flag, a new national anthem, and a new leadership was one of the great events of the twentieth century. Now, with a new century opened, South Africa is poised to take its long-awaited role as the economic engine for the region—and perhaps beyond.

South Africa stretches from the warm subtropics in the north to Antarctic-chilled waters in the south. With a land area in excess of 470,000 square miles (1.2 million sq km) and a heterogeneous population of 44.5 million, South Africa is the dominant state in Southern Africa. It contains the bulk of the region's minerals, most of its good farmlands, its largest cities, best ports, most productive factories, and most developed transport networks. Mineral exports from Zambia and Zim-

FROM THE FIELD NOTES

"Looking down on this enormous railroad complex, we were reminded of the fact that almost an entire continent was turned into a wellspring of raw materials carried from interior to coast and shipped to Europe and other parts of the world. This complex lies near Witbank in the eastern Rand, a huge inventory of freight trains ready to transport ores from the plateau to Durban and Maputo. But at least South Africa acquired a true transport network in the process, ensuring regional interconnections; in most African countries, railroads serve almost entirely to link resources to coastal outlets."

babwe move through South African ports. Workers from as far away as Malawi and as nearby as Lesotho work in South Africa's mines, factories, and fields.

Peoples from many sources and directions migrated to what is today South Africa. The Khoi and San are the oldest survivors, reduced now to a few tens of thousands. The Bantu came next, moving across the plateau *highveld* and along the eastern corridor between the Great Escarpment and the Indian Ocean. Then followed the Europeans, with the Dutch founding Cape Town as early as 1652. Soon the Hollanders began to bring Southeast Asians to the Cape to serve as laborers and domestics, and intermarriage began. Many of these Southeast Asian arrivals were Muslims, and the Cape became an Islamic foothold. Then the British arrived to capture the Cape from the Dutch, and when London's holdings expanded to include the Province of Natal, the arrival of tens of thousands of indentured laborers from British India made Durban a Hindu outpost.

All these foreign invaders never outnumbered the Bantu nations of South Africa, however. The Sotho on the highveld, the Xhosa in the Eastern Cape, and the Zulu in Natal could not stem the tide of foreign aggrandizement, but they were never in the kind of danger Native Americans and Australian Aborigines were. South Africa became Africa's most pluralistic and heterogeneous society, but Africans always outnumbered non-Africans by four to one or more.

Still, for more than three centuries the power lay with the white minority. The Dutch (and to a lesser extent the French) immigration spawned a new nation, the *Afrikaners* as they called themselves (or *Boers* as they were known during their conquest of the *highveld*). Often friction arose between Boer and Briton, erupting into the Boer War of 1899–1902. Although the Boers lost the war, they won the future. By the middle of the twentieth century, they outnumbered and outvoted their British compatriots and took control of South Africa's government. Now the **17** Afrikaners instituted **apartheid**, the strict segregation of South Africa's "races," and began **18** planning for the grand design of **separate development**, which involved the forced reloca-

tion of as many as 3.5 million black Africans from their dwellings to ethnically based homelands. Today the term describing what happened in South Africa is *ethnic cleansing*, and by law nearly 80 percent of South Africa was designated as white-owned.

Opposition to *apartheid* in South Africa began immediately, and many African leaders went into exile or were jailed. Among the latter was Nelson Mandela, whose influence in the banned opposition movement, the African National Congress (ANC), continued from his cell on Robben Island, South Africa's Alcatraz. Meanwhile, a combination of violent uprisings, crippling strikes, and international sanctions forced the Afrikaner government to negotiate an end to the despised system.

By the mid-1980s the Afrikaner leadership began discussions with its most famous political prisoner, and in February 1990 Mandela walked out of prison after 28 years behind bars. He and the last Afrikaner

president, F. W. de Klerk, began a process of negotiation and accommodation that led to the transfer of power from the white regime to a government elected by all the voters. In April 1994 a new era began in South Africa. Nelson Mandela, of distinguished Xhosa ancestry, had become president of an ANC-dominated government in Cape Town.

The transfer of power required the reconstruction of South Africa's administrative map. Since 1910, South Africa had had four provinces: the Cape, Natal, the Orange Free State (north of the great Orange River), and the Transvaal ("across the Vaal," the major Orange tributary). These four provinces essentially represented the European occupation of the country: the Cape and Natal were British strongholds, and the Orange Free State and Transvaal were Boer republics until their defeat in the Boer War (Fig. 7-16). The new map (Fig. 7-17) created nine provinces, leaving intact Natal (but now calling it Kwazulu-Natal) and the

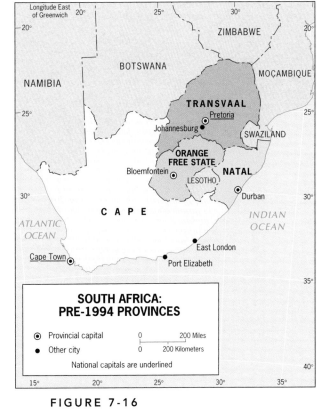

FIGURE 7-16

FIGURE 7-17

Photo Gallery   More to Explore   **20** Concepts, Ideas, and Terms   GEODISCOVERIES Geodiscoveries Module

Orange Free State. Three new provinces and part of a fourth were carved out of the Cape: the Western Cape, essentially Cape Town and its hinterland, where the electorate is mainly Coloured and white; the Eastern Cape, dominated by the Xhosa nation; the Northern Cape, by far the least populous, mainly rural province; and the Northwest, part Cape and part Transvaal, where the Tswana nation is concentrated. The remainder of the Transvaal also became three provinces: Gauteng, the heart of the country's core area centered on Johannesburg; Northern Province, home to many Afrikaners who opposed the new South Africa; and Mpumalanga, a mix of wealthy white farmers and densely peopled Swazi and other African areas.

In the momentous 1994 elections, the ANC won seven of the nine provinces. It lost in Kwazulu-Natal, where an alliance of Zulu, Asian, and white voters prevailed over the ANC, and in the Western Cape, where white and Coloured voters combined to defeat ANC candidates. Nor did the ANC gain the 70 percent majority it would have needed to govern without opposition support. What all parties in South Africa accomplished after more than four decades of oppression and mismanagement was an example to a world quick to use force to reach political goals.

South Africa passed another milestone in mid-1999, when Thabo Mbeki, Mandela's deputy, became the country's second popularly elected president. President Mbeki took office at a difficult time for the country: the AIDS epidemic was taking an ever-higher toll, neighboring Zimbabwe was in turmoil with implications for South Africa, and the economy was in trouble. South Africa still depends heavily on the export of metals and minerals, but mineworkers want better pay even as world prices sag. Foreign investment is deterred by high crime rates and a lack of confidence in the country's long-term future. Joblessness, housing needs, and land pressure can form a potent mix to destabilize a society, and the South African government faces a daunting challenge.

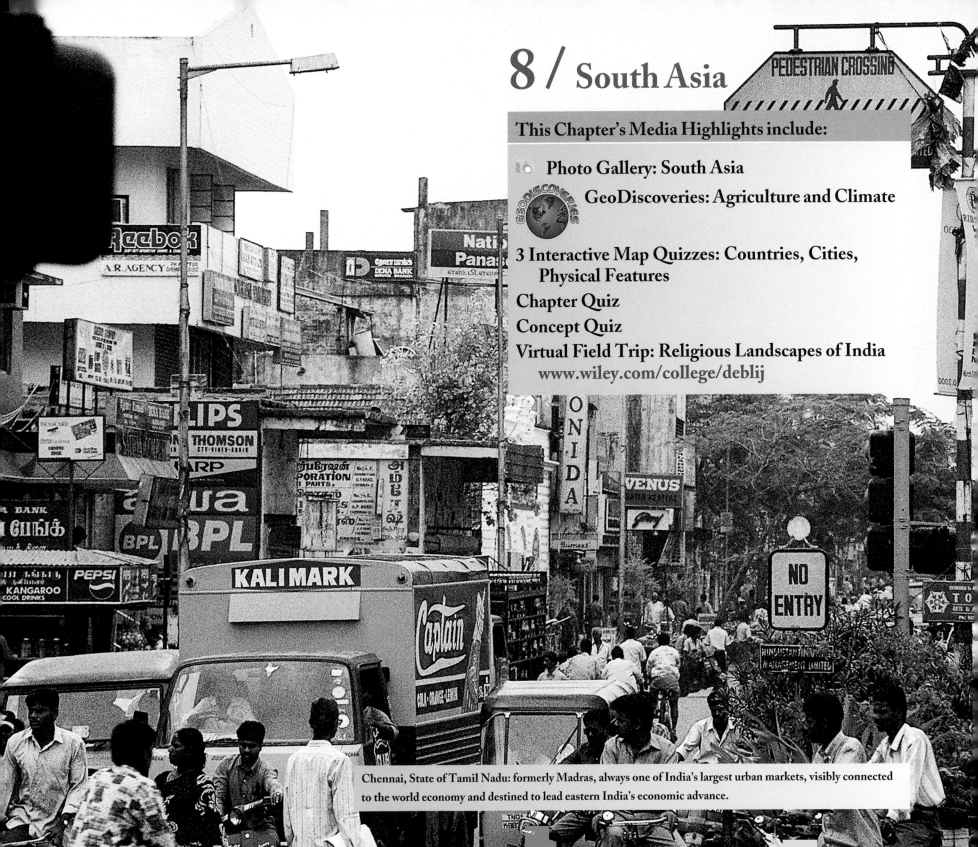

8 / South Asia

Chennai, State of Tamil Nadu: formerly Madras, always one of India's largest urban markets, visibly connected to the world economy and destined to lead eastern India's economic advance.

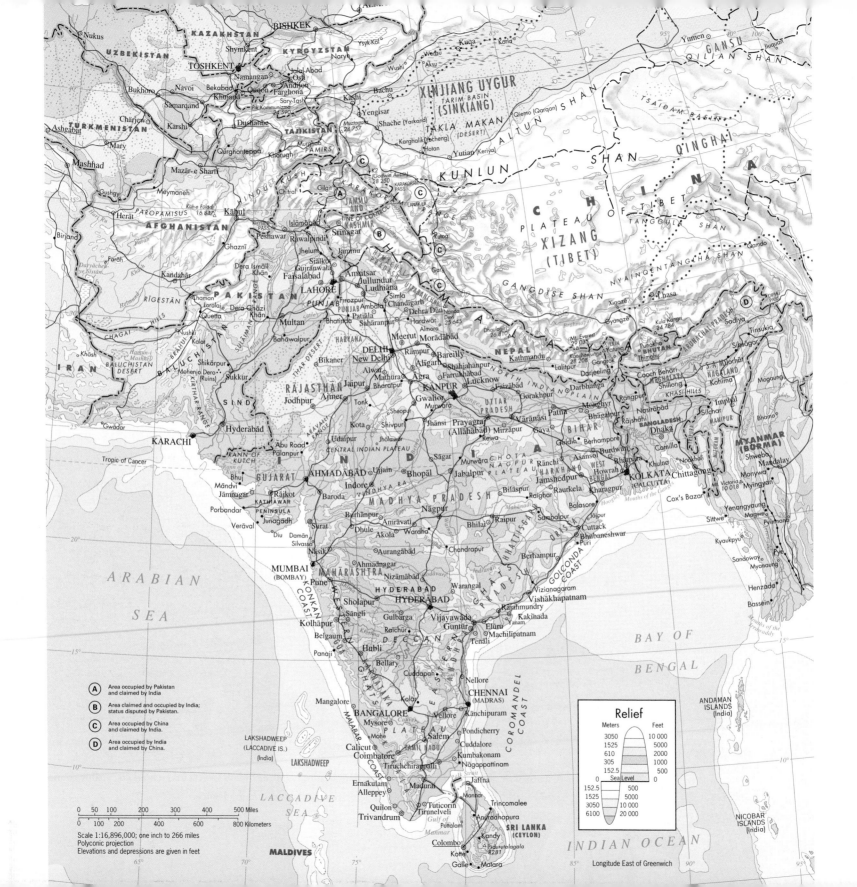

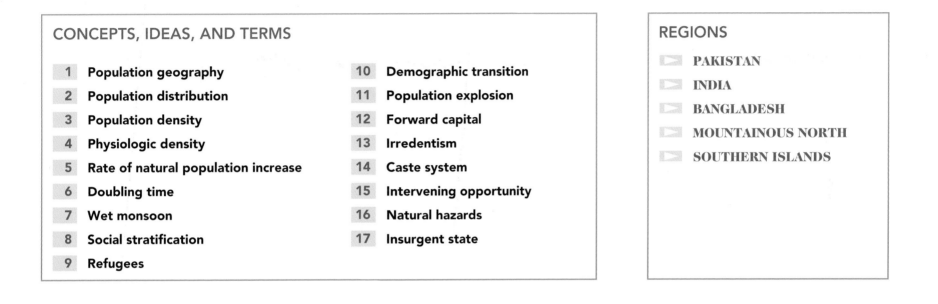

chapter 8 / South Asia

CONCEPTS, IDEAS, AND TERMS

1 Population geography
2 Population distribution
3 Population density
4 Physiologic density
5 Rate of natural population increase
6 Doubling time
7 Wet monsoon
8 Social stratification
9 Refugees
10 Demographic transition
11 Population explosion
12 Forward capital
13 Irredentism
14 Caste system
15 Intervening opportunity
16 Natural hazards
17 Insurgent state

REGIONS

PAKISTAN
INDIA
BANGLADESH
MOUNTAINOUS NORTH
SOUTHERN ISLANDS

From Iberia to Arabia and from Malaysia to Korea, Eurasia is a landmass fringed by peninsulas. The largest of all is the great triangle of India that divides the northern Indian Ocean into two seas: the Arabian Sea to the west and the Bay of Bengal to the east (Fig. 8-1). The peninsula of India forms the heart of South Asia, a vast, varied, volatile geographic realm.

DEFINING THE REALM

Mountains, deserts, and coastlines combine to make South Asia one of the world's most vividly defined physiographic realms. To the north, the Himalaya Mountains create a natural wall between South Asia and China. To the east, mountain ranges and dense forests mark the boundary between South and South-east Asia. To the west, rugged highlands and expansive deserts separate South Asia from its neighbors. Within these confines lies a geographic realm that is more densely populated than any other. If current population trends continue, it will soon be the most populous realm on Earth as well.

South Asia consists of five regions (Fig. 8-2). Its keystone is India, whose population passed the 1 billion mark in 1999. In the west lies Pakistan. South Asia's eastern flank is centered on Bangladesh. The northern region consists of the mountainous lands of Kashmir, Nepal, and Bhutan. And the southern

◆ Major Geographic Qualities of South Asia

1. South Asia is clearly defined physiographically and is bounded by mountains, deserts, and ocean; the Indian peninsula is Eurasia's largest.

2. South Asia is a poverty-afflicted realm, with low average incomes, low levels of education, poorly balanced diets, and poor overall health.

3. With only 3 percent of the world's land area but 22 percent of its population, more than half of it engaged in subsistence farming, South Asia's economic prospects are bleak.

4. Population growth rates in South Asian countries are among the highest in the world; India's population surpassed the 1 billion mark in 1999.

5. The North Indian Plain, the lower basin of the Ganges River, contains the heart of the world's second largest population cluster.

6. Despite encircling mountain barriers, invaders from ancient Greeks to later Muslims penetrated South Asia and complicated its cultural mosaic.

7. British colonialism unified South Asia under a single flag, but the empire fragmented into several countries along cultural lines after Britain's withdrawal.

8. Pakistan, South Asia's western region, lies on the flanks of two realms: largely Muslim North Africa/Southwest Asia and dominantly Hindu South Asia.

9. India is the world's largest federation and most populous democracy, but its political achievements have not been matched by enlightened economic policies.

10. Religion remains a powerful force in South Asia. Hinduism in India, Islam in Pakistan, and Buddhism in Sri Lanka all show tendencies toward fundamentalism and nationalism.

11. Active and potential boundary problems involve internal areas (notably between India and Pakistan in Kashmir) as well as external locales (between India and China in the northern mountains).

region includes the islands of Sri Lanka and the Maldives. As the map shows, India divides into several subregions.

South Asia's physiographic boundaries are formidable barriers, but they have not prevented conquerors or proselytizers from penetrating it. As a result, today this realm is a patchwork of religions, languages, traditions, and cultural landscapes. So complex is this mosaic that, remarkably, its political geography numbers just seven states (and only five on the mainland).

Among the cultural infusions was Islam. Today, Pakistan is an Islamic republic, and Islam provides the cement for the state. Pakistan's eastern border with India is a cultural divide in more ways than one: dominantly Hindu India is a secular, not a theocratic,

state. Why, then, do we include Pakistan in the South Asian rather than the Southwest Asian/North African realm? One criterion is ethnic continuity, which links Pakistan to India rather than to Afghanistan or Iran. Another is historical geography. Pakistan was part of Britain's South Asian Empire, and it originated from the partition of that domain between Muslim and Hindu majorities. Although Urdu is the official national language of Pakistan, English is the *lingua franca*, as it is in India. Furthermore, the border between India and Pakistan does not signify the eastern frontier of Islam in Asia. About 127 million of India's more than 1 billion citizens are Muslims, and in South Asia's eastern region, Bangladesh (population: 133 million) is more than 85 percent Muslim. Finally, Pakistan and India are locked in a struggle to

control a vital mountainous area in the far north, where the British withdrawal left the boundary between them unresolved. Even as Indian and Pakistani cricket teams play each other on sun-baked pitches, their armies face off in a deadly conflict—a conflict that has the potential to unleash a nuclear war.

A REALM OF POVERTY

At the opening of the twenty-first century, South Asia accounts for more than one-fifth of the world's population and two-thirds of its poorest inhabitants. Its literacy rates are among the lowest in the world. Nearly half of the people in this realm earn less than the equivalent of one U.S. dollar per day. It is estimated

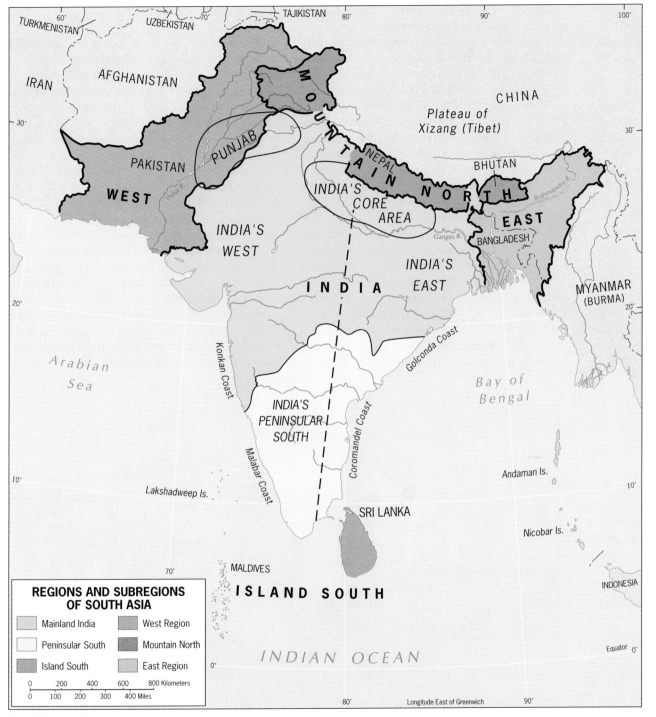

**REGIONS AND SUBREGIONS
OF SOUTH ASIA**

Mainland India   West Region

Peninsular South   Mountain North

Island South   East Region

0   200   400   600   800 Kilometers

0   100   200   300   400 Miles

FIGURE 8-2

that half the children in South Asia are malnourished and underweight, most of them girls. South Asia is often called the most deprived realm in the world.

A combination of geographic factors underlies this tragic picture. With 22 percent of the world's population but just 3 percent of its land area, South Asia lacks the natural resources to raise living standards for its hundreds of millions of subsistence farmers. Governmental policies contribute to the problem: while East and Southeast Asia forged ahead by looking outward, encouraging exports and foreign investment, and spending heavily on literacy and technical education, health care, and land reform, South Asian governments tended to adopt bureaucratic controls and state planning. Cultural traditions also play their role. Resistance to change and reluctance by the privileged to open doors of opportunity to the less advantaged inhibit economic advancement for all.

A key factor lies in the rate of population growth. This is an aspect of **1 population geography**, the spatial view of demography. Figure I-9 reveals the concentration of population in South Asia's major river valleys, underscoring the dependence of the majority on these ribbons of life. **2 Population distribution**, as this map reveals, should be compared to **3 population density**, a more specific measure. Column 4 of Table I-1 lists population density by country, but Figure I-9 proves that this average needs elaboration. Parts of India, populous as it is, are sparsely peopled; other areas are teeming. The national averages represent the *arithmetic densities* for those countries (905 in India), but they conceal regional variations.

A better measure represents the number of people in a country per unit area of agriculturally productive **4** land, and this is called its **physiologic density** (listed in column 5 of Table I-1). In India today, this index is 1615 per square mile (625 per sq km).

Critical to any assessment of South Asia's social **5** and economic problems is the realm's **rate of natural population increase**. By world standards, these rates, for the individual countries, remain high: around 2.0 percent per year. When populations grow this fast, economic advancement becomes elusive. The needs of the new arrivals absorb most of the gains. From this rate of natural increase we can calculate the **6** **doubling time** of a population. Muslim countries tend to have high growth rates and thus low doubling times. Take Pakistan, whose current population is 160 million. The latest data show its rate of natural increase as 2.8 percent and its doubling time as 25 years; at current rates, therefore, Pakistan will have 320 million people by 2027, barring major changes. India will have 2 billion people by 2039, an almost unimaginable situation. Such rates of increase can only retard, if not altogether stop, economic development.

PHYSIOGRAPHIC REGIONS OF SOUTH ASIA

Before we look into South Asia's complex and fascinating cultural geography, we need to discuss the physical stage of this populous realm. South Asia is a realm of immense physiographic variety, of snow-capped peaks and forest-clad slopes, of vast deserts and broad river basins, of high plateaus and spectacular shores. The collision of two of the Earth's great tectonic plates created the world's highest mountain ranges, their icy crests yielding meltwaters for great rivers below. The workings of the Earth's atmosphere put South Asia in the path of tropical cyclones and produce reversing seasonal windflows known as *monsoons*. This is a realm of almost infinite variety, a world unto itself.

In general terms, we can recognize three clearly defined physiographic zones in South Asia: the north-

ern mountains, the southern peninsular plateaus, and between them a belt of river lowlands.

The *northern mountains* extend from the Hindu Kush and Karakoram ranges in the northwest through the Himalayas in the center (Mount Everest, the world's tallest peak, lies in Nepal) to the ranges of Bhutan and the Indian State of Arunachal Pradesh in the east. Dry and barren in the west on the Afghanistan border, the ranges become green and tree-studded in Kashmir, forested in the lower sections of Nepal, and even more densely vegetated in Arunachal Pradesh. Transitional foothills, with many deeply eroded valleys cut by rushing meltwaters, lead to the river basins below.

The *belt of river lowlands* extends eastward from Pakistan's lower Indus Valley (the area known as Sind) through the wide plain of the Ganges Valley of India and on across the great double delta of the Ganges and Brahmaputra in Bangladesh (Fig. 8-1). In the east, this physiographic region often is called the North Indian Plain. To the west lies the lowland of the Indus River, which rises in Tibet, crosses Kashmir, and then bends southward to receive its major tributaries from the Punjab ("Land of Five Rivers").

The *southern plateaus* dominate the topography of peninsular India, as exemplified by the massive Deccan, a tableland built of basalt that poured out when India separated from Africa during the breakup of Gondwana (see Fig. 7-3). The Deccan (meaning "South") tilts to the east, so that its highest areas are in the west and the major rivers flow into the Bay of Bengal. North of the Deccan lie two other plateaus, the Central Indian Plateau to the west and the Chota-Nagpur Plateau to the east (Fig. 8-1). On the map, note the Eastern and Western Ghats ("hills") that descend from Deccan plateau elevations to the narrow coastal plains below. Here India benefits from the **7** **wet monsoon** when moist, warm oceanic air is drawn by a low-pressure system into the peninsular interior (Fig. 8-3), usually starting in June. As this air is pulled over the Western Ghats ①, it is cooled, its moisture condenses, and rains begin which may last 60 days or more. Other streams of air come from the Bay of Bengal ② and get caught up in the convection over northeastern India and Bangladesh. Seemingly

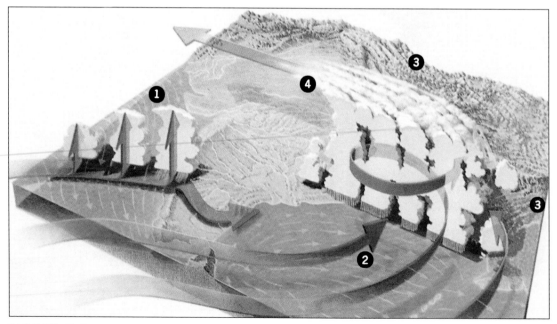

FIGURE 8-3

endless summer rains now inundate the whole North Indian Plain; the Himalayas to the north create a corridor that sends the rains westward ③ until eventually it dries out as it reaches Pakistan ④. By the end of summer the system breaks down, the wet monsoon gives way to periodic rains and, eventually, another dry season—and the anxious wait begins for next year's life-giving monsoon.

THE HUMAN SEQUENCE

Great river basins mark the physiography of South Asia; in one of these basins, that of the Indus River in present-day Pakistan, lies evidence of the realm's oldest major civilization. It existed at the same time, and interacted with, ancient Mesopotamia, and it was centered on large, well-organized cities (Fig. 8-4). From here, influences and innovations diffused into India. In fact, India's very name is believed to derive from the ancient Sanskrit word *sindhu*.

Eventually, such cities as Harappa and Mohenjo-Daro seem to have experienced the same fate as those of Mesopotamia, perhaps also because of environmental change. About 3500 years ago Aryan (Indo-European) speaking peoples migrated into the region, penetrating India and starting the process of welding the Ganges Basin's isolated tribes and villages into an organized system. Having absorbed much of the culture of the Indus Valley, they brought a new order to this region. A new belief system, *Hinduism*, arose and with it a new way of life. A complex **8** **social stratification** developed in which Brahmans, powerful priests, stood at the head of a caste system in which soldiers, merchants, artists, peasants, and all others had their place.

Hinduism was restrictive, especially for those of the lower castes, and in the sixth century BC (more than 2500 years ago) a prince born into one of northern India's kingdoms sought a better way. Prince Siddhartha, better known as Buddha, gave up his royal position to teach religious salvation through meditation, the rejection of Earthly desires, and a reverence for all forms of life. His teachings did not have a major impact on the Hindu-dominated society of his time, but what he taught was not forgotten.

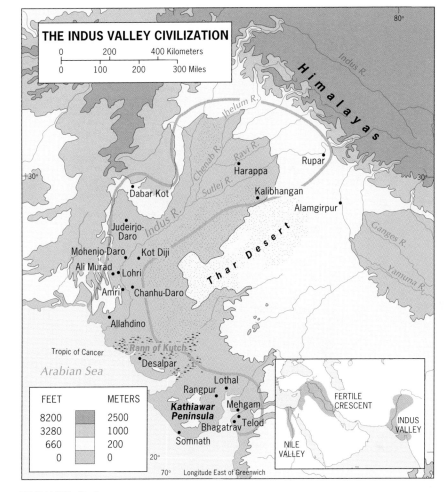

FIGURE 8-4

Centuries later, the ruler of a powerful Indian state decided to make *Buddhism* the state religion, following which the faith spread far and wide.

But first South Asia was penetrated by other foreign influences: the Persians were followed by the Greeks under Alexander the Great in the late fourth century BC. This is why, on a world map of languages, Hindi (the major domestic language of India) lies at the eastern end of the region of Indo-European languages, in the same family as Persian, Italian, and English. But the map of India's languages (Fig. 8-5) shows that southern India's languages are *not* Indo-European. Indeed, while northern India was a subregion of cultural infu-

sion and turmoil, the south lay remote and isolated. This part of the peninsula had apparently been settled long before the Indus and Ganges civilizations arose, and southern India developed into a distinctive subregion with its own cultures. The *Dravidian* languages spoken here—Telugu, Tamil, Kanarese (Kannada), and Malayalam—have long literary histories.

Aśoka's Mauryan Empire

When the Greeks withdrew from the Ganges Basin and the Hindu heartland was once again free, a powerful empire arose there—the first true empire in

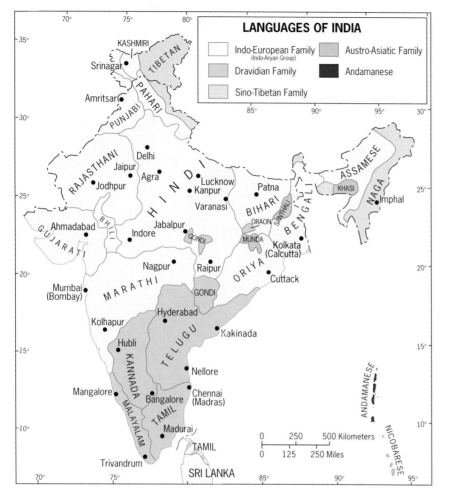

LANGUAGES OF INDIA

Indo-European Family (Indo-Aryan Group)

Dravidian Family

Sino-Tibetan Family

Austro-Asiatic Family

Andamanese

FIGURE 8-5

The Mauryan Empire represented India's greatest political and cultural achievements in its day, and when the empire collapsed, late in the second century AD, India fragmented into a patchwork of states. Once again, India lay open to infusions from the west and northwest, and across present-day Pakistan they came: Persians, Afghans, Turks, and others driven from their homelands or attracted by the lands of the Ganges.

The Power of Islam

In the late tenth century, Islam came rolling like a giant tide across the subcontinent, spreading from Persia in the west and Afghanistan in the northwest. Of course, the Indus Valley lay directly in the path of this Islamic advance, and virtually everyone was converted. Next the Muslims penetrated the Punjab, the subregion that lies astride the present Pakistan-India border, and there perhaps as many as two-thirds of the inhabitants became converts. Then Islam crossed the bottleneck where Delhi is situated and diffused east- and southeastward into the Gangetic Plain and the subregion known as Hindustan—India's evolving core area. Here Islam's proselytizers had less success, persuading perhaps one in eight Indians to become Muslims. In the meantime, Islam arrived at the Ganges Delta by boat, and present-day Bangladesh became overwhelmingly Islamic. (To the south of the Ganges heartland, however, Islam's diffusion wave lost its energy: Dravidian India never came under Muslim influence.)

Islam's vigorous, often violent, onslaught changed Indian society. As in West Africa, Islam frequently was superimposed through political control: when the rulers were converted, their subjects followed. By the early fourteenth century, a sultanate centered at Delhi controlled more of the subcontinent than even the Mauryan Empire had earlier. Later, the Islamic Mogul Empire (the similarity to the word "Mongol" is by no means a coincidence) constituted the largest political entity ever to unify the realm in precolonial times. To many Hindus of lower caste, Islam represented a welcome alternative to the rigid socioreligious hierarchy in which they were trapped at the bottom. Thus Islam was the faith of the ruling elites

the realm. This, the Mauryan Empire, extended its influence over India as far west as the Indus Valley (thus incorporating the populous Punjab) and as far east as Bengal (the double delta of the Ganges and Brahmaputra); it reached as far south as the modern city of Bangalore.

This Mauryan Empire was led by a series of capable rulers who achieved stability over a vast domain. Undoubtedly the greatest of these leaders was Aśoka, who reigned for nearly 40 years during the middle of the third century BC. Aśoka was a believer in Buddhism, and it was he who elevated this religion from obscurity to regional and ultimately global importance.

In accordance with Buddha's teachings, Aśoka reordered his government's priorities from conquest and expansion to a Buddhist-inspired search for stability and peace. He sent missionaries to the outside world to carry Buddha's teachings to distant peoples, thereby also contributing to the diffusion of Indian culture. As a result, Buddhism became permanently established as the dominant religion in Sri Lanka (formerly Ceylon), and it established footholds as far afield as Southeast Asia and Mediterranean Europe. Ironically, Buddhism thrived in these remote places even as it declined in India itself. With Aśoka's death, the faith lost its strongest supporter.

and of the disadvantaged, a powerful cultural force in the heartland of Hinduism.

Just as Islam weakened in southern Europe, so its force ultimately became spent in vast and populous India. For all the Muslims' power, they never managed to convert a majority of South Asians. They dominated the northwest corner of the realm (present-day Pakistan), where Lahore became one of Islam's greatest cities. But in all of what is today India, less than 15 percent of the population became and remained Muslim. And throughout the period of Islamic intervention, the struggle for cultural supremacy continued. Placid Hinduism and aggressive Islam did not easily coexist.

The European Intrusion

Into this turbulent complexity of religious, political, and linguistic disunity yet another element began to intrude after 1500: European powers in search of raw materials, markets, and political influence. Because the Europeans profited from the Hindu-Muslim contest, they exploited local rivalries, jealousies, and animosities. British merchants gained control over the trade with Europe in spices, cotton, and silk goods, ousting the French, Dutch, and Portuguese. The British East India Company's ships also took over the intra-Asian sea trade between India and Southeast Asia, which had long been in the hands of Arab, Indonesian, Chinese, and Indian merchants. In effect, the East India Company (EIC) became India's colonial administration.

In time, however, the EIC faced problems that made it increasingly difficult to combine commerce with administration. Eventually, a mutiny among Indian troops in the service of the EIC led to the abolition of the company. The British government took over in 1857 and maintained its rule (raj) until 1947.

Colonial Transformation

When the British took power over South Asia, they controlled a realm with considerable industrial development (notably in metal goods and textiles) and an active trade with both Southwest and Southeast Asia. The colonialists saw this as competition, and soon India was exporting raw materials and importing manufactured goods—from Europe, of course.

Local industries declined and Indian merchants lost their markets.

Unifying their realm was a tougher task for the British. In 1857 about 750,000 square miles (2 million sq km) of South Asian territory still was beyond British control, including hundreds of entities that had been guaranteed autonomy by the EIC during its administration. These "Native States," ranging in size from a few acres to Hyderabad's 80,000 square miles (200,000 sq km), were assigned British advisors, but in fact India was a near-chaotic amalgam of modern colonial and traditional feudal systems.

Colonialism did produce assets for India. The country was bequeathed one of the best transport networks of the colonial domain, especially the railroad system (although the network focused on interior-seaport linkages rather than fully interconnecting the various parts of the country). British engineers laid out irrigation canals through which millions of acres of land were brought into cultivation. Settlements that had been founded by Britain developed into major cities and bustling ports, led by Bombay (now Mumbai), Calcutta (now Kolkata), and Madras (now Chennai). These three cities are still three of India's largest urban centers, and their cityscapes bear the unmistakable imprint of colonialism. Modern industrialization, too, was brought to India by the British on a limited scale. In education, an effort was made to combine English and Indian traditions; the Westernization of India's elite was supported through the education of numerous Indians in Britain. Modern practices of medicine were also introduced. Moreover, the British administration tried to eliminate features of Indian culture that were deemed undesirable by any standards—such as the burning alive of widows on their husbands' funeral pyres, female infanticide, child marriage, and the caste system. Obviously, the task was far too great to be achieved in barely three generations of colonial rule, but independent India itself has continued these efforts where necessary.

Partition

Even before the British government decided to yield to Indian demands for independence, it was clear that British India would not survive the coming of self-rule as a single political entity. As early as the 1930s, the idea of a separate Pakistan was being promoted by Muslim activists, who circulated pamphlets arguing that British India's Muslims were a nation distinct from the Hindus and that a separate state consisting of Sind, Punjab, Baluchistan, Kashmir, and a portion of Afghanistan should be created from the British South Asian Empire in this area. The first formal demand for such partitioning was made in 1940, and, as later elections proved, the idea had almost universal support among the realm's Muslims.

As the colony moved toward independence, a political crisis developed: India's majority Congress Party would not even consider partition, and the minority Muslims refused to participate in any future unitary government. But partition would not be a simple matter. True, Muslims were in the majority in the western and eastern sectors of British India, but Islamic clusters were scattered throughout the realm (Fig. 8-6). Any new boundaries between Hindus and Muslims to create an Islamic Pakistan and a Hindu India would have to be drawn right through areas where both sides coexisted. People by the millions would be displaced.

Nor were Hindus and Muslims the only people affected by partition. The Punjab area, for example, was home to millions of Sikhs, whose leaders were fiercely anti-Muslim. But a Hindu-Muslim border based only on those two groups would leave the Sikhs in Pakistan. Even before independence day, August 15, 1947, Sikh leaders talked of revolt, and there were some riots. But no one could have foreseen the dreadful killings and mass migrations that followed the creation of the boundary and the formation of independent Pakistan and India. Just how many people felt compelled to participate in the ensuing migrations will never be known; 15 million is the most common estimate. It was human suffering on an incomprehensible scale.

9 Even so large a flow of cross-border **refugees**, however, hardly began to "purify" India of Muslims. After the initial mass exchanges, there still were tens of millions of Muslims in India (Fig. 8-6). Today, the Muslim minority in Hindu-dominated India is almost as large as the whole Islamic population of Pakistan, having more than tripled since the late 1940s. It is the

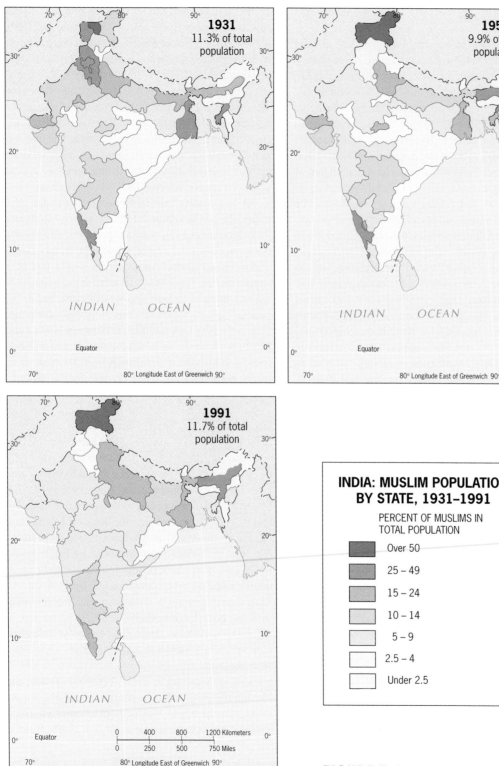

INDIA: MUSLIM POPULATION BY STATE, 1931–1991

PERCENT OF MUSLIMS IN TOTAL POPULATION

- Over 50
- 25 – 49
- 15 – 24
- 10 – 14
- 5 – 9
- 2.5 – 4
- Under 2.5

FIGURE 8-6

world's largest minority, far more than a mere remnant of the days when Islam ruled the realm. This force will play a growing role in the India of the future.

The Kashmir Issue

When the boundary between Hindu India and Muslim Pakistan was delimited in 1947, it stopped short of the northern territory of Jammu and Kashmir (Kashmir for short). In this, one of those 562 "Native States" recognized by the British administration, the Muslim population had been growing rapidly, especially in the Vale of Kashmir where the capital, Srinagar, is located (Fig. 8-7). When the time came for this State to decide whether to join Pakistan or India, about 75 percent of the population was Muslim—but the ruling elite was Hindu. Initially, the Maharajah of Kashmir tried to secure autonomy for his State, thus staying outside Pakistan as well as India. This caused a Muslim uprising against Hindu rule and a war between Pakistani and Indian armies, to be followed by two more major armed conflicts in 1965 and 1971. The result was the armistice *Line of Control* you see on Figure 8-7.

The Line of Control has been the *de facto* boundary between India and Pakistan since 1971, but it does not separate Muslim from Hindu in Kashmir. In fact, the majority of the people on the eastern (Indian) side of the line in Jammu and Kashmir are Muslims, which is why India resists a plebiscite to decide the State's future. In the late 1980s extremist Muslim groups began a long-running insurgency that escalated into sporadic combat and terrorist attacks that killed more than 10,000 people and continues to this day. But these skirmishes, costly as they are, were overshadowed by the implications of the 1998 test explosions of nuclear weapons by both India and Pakistan. This event transformed Kashmir from a problem frontier into a potentially catastrophic flashpoint for nuclear war.

SOUTH ASIA'S POPULATION DILEMMA

The population of this realm's seven countries today totals nearly 1.4 billion—more than one-fifth of all

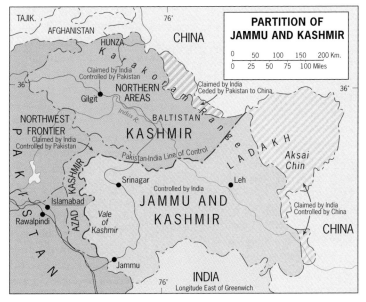

FIGURE 8-7

humankind. India alone has 1.04 billion inhabitants, second only to China among countries of the world, and is on course to overtake it. Pakistan, with a population of 160 million, and Bangladesh, with 133 million, also rank among the dozen most populous countries of the world.

Not only is South Asia populous, but its population also is growing rapidly at 1.9 percent per year, yielding a doubling time of only 36 years. Pakistan is growing even faster at 2.8 percent, doubling in 25 years. As a result, economic gains are being overtaken by growing numbers, and although the Green Revolution narrowed the food deficit, hundreds of millions of children do not get balanced meals or adequate calories.

Three-quarters of South Asia's population live in India, so we should consider this global giant in the context of its population geography and its impact on the realm as a whole. Comparing the map of world economies (Fig. I-10) with the list of population growth rates in Table I-1, we note a clear pattern: the bulk of population growth is occurring in the lower-income economies. In many of the high-income economies, population growth is small, has leveled off, or is

10 even negative. These higher-income economies have gone through the so-called **demographic transition**, a four-stage sequence that took them from high birth rates and high death rates in preindustrial times to very low birth rates and very low death rates today **11** (Fig. 8-8). Stages 2 and 3 in this model constitute the **population explosion**, a hallmark of the twentieth century: death rates in the industrializing and urbanizing countries dropped, but birth rates took longer to decline. In 1900, the world's population was about 1.5 billion; by 2000 it had surpassed 6 billion.

When the British ruled India during the nineteenth century, the country still was in the first stage, with high birth rates and high death rates; the high death rates were caused not only by a high incidence of infant and child mortality but also by famines and epidemics. As Figure 8-8 indicates, the population during Stage 1 does not grow or decline much, but it is not stable. Famines and disease outbreaks kept erasing the gains made during better times. But then India entered the second stage. Birth rates remained high, but death rates declined because medical services improved (soap came into widespread use), food distribution networks became more effective,

farm production expanded, and urbanization developed. In the 1920s, India's population still was growing at a rate of only 1.04 percent, but by the 1970s, that rate had shot up to 2.22 percent per year (Fig. 8-9). Note that India gained 28 million people during the 1920s but a staggering 135 million during the 1970s.

Has India entered the third stage, when the death rate begins to level off and birth rates decline substantially, narrowing the gap and slowing the annual increase? The rate of increase suggests it: from 2.22 percent during the 1970s, it dropped to 2.11 percent in the 1980s and to a projected 1.88 percent during the 1990s (Fig. 8-9). But India has another problem. During its population explosion, its numbers grew so large that even a declining rate of natural increase continues to add ever greater numbers to its total. In Figure 8-9, we see that while the decadal rate of increase dropped from 2.22 to 2.11 between 1970 and 1990, the millions added grew from 135 in the 1970s to 161 during the 1980s. The rate of natural increase for the 1990s is estimated to have been 1.88 percent, adding a still larger number—175 million, taking the total past 1 billion during 1999. If India has indeed entered the third stage of the demographic transition, it will not feel its effects for some time.

Some population geographers theorize that all countries' populations will eventually stabilize at some level, just as Europe's did. Certain governments, notably China's, have instituted regulations to limit family size, but this policy is more easily implemented by dictatorships than democracies. And even if such stabilization were just one doubling time away in India, the country still would have an astronomical total of more than 2 billion residents.

Geography of Demography

Statistics for a country as large as India tend to lose their usefulness unless they are put in geographic context. In its demographic as in so many of its other aspects, there is not just one India but several regionally different and distinct Indias. Figure 8-10 takes population growth down to the State level and

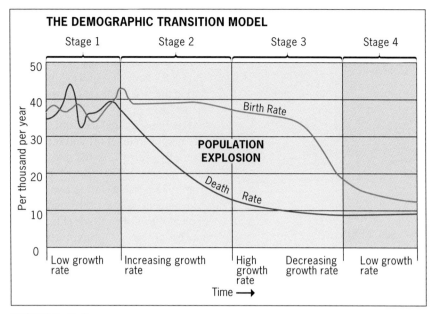

THE DEMOGRAPHIC TRANSITION MODEL

FIGURE 8-8

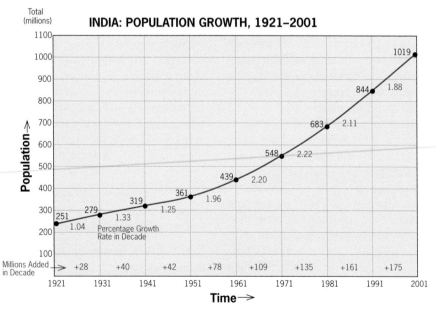

INDIA: POPULATION GROWTH, 1921–2001

FIGURE 8-9

provides a comparison between the census periods of 1971–1981 and 1981–1991.

During the period from 1971 to 1981, the highest growth rates were recorded in the States of the northeast and northwest, and only three States had growth rates below 2 percent. But between 1981 and 1991, no fewer than eight States, including Goa, had growth rates below 2 percent. Comparing these two maps tells us that little has changed in India's heartland, where the populous States of Uttar Pradesh and Bihar show slight decreases but West Bengal and Madhya Pradesh display equally slight increases. The most important reductions in rates of natural increase are recorded in the northwest, west, and south. In the tip of the peninsula, Kerala and neighboring Tamil Nadu have growth rates comparable to that of the region's leading country, Sri Lanka.

One reason for these contrasts lies in India's federal system: individual States pursued population-control policies to reduce population growth, including mass sterilizations. But another, as we note in more detail in the regional section, relates to spatial differences in India's development. As in the world at large, India's economically better-off States have lower rates of population growth.

India is a Hindu-dominated, officially secular country in which population policies can be debated and implemented. Pakistan, on the other hand, is a strictly Islamic state in which no such options exist. Population-control policies are regarded as incompatible with Islamic tenets, and Pakistan remains one of the fastest-growing nations in the world. As we noted in Chapter 6, however, Islam—like other faiths—has its regional variations. Dominantly Islamic Bangladesh is less fundamentalist in its adherence to the faith, which is a major reason why birth rates are 1.2 percent lower on South Asia's eastern flank than they are in the west.

Among world realms, only Subsaharan Africa suffers more severely from poverty and social dislocation than South Asia—but, as Table I-1 indicates, South Asian conditions are only marginally better than Africa's, and South Asia has more than twice as many people as Subsaharan Africa. We turn next to South Asia's regional components.

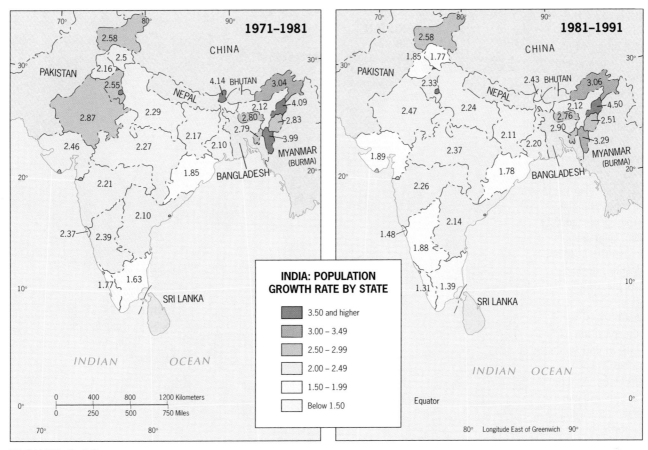

FIGURE 8-10

REGIONS OF THE REALM

PAKISTAN: ON SOUTH ASIA'S WESTERN FLANK

If India is the dominant entity in South Asia, why focus first on Pakistan? There are several reasons, both historic and geographic. Here lay South Asia's earliest urban civilizations, whose innovations radiated into the great peninsula. Here, too, lies South Asia's Muslim frontier, contiguous to the great Islamic realm to the west and irrevocably linked to the enormous Muslim minority to its east. Pakistan's

cultural landscapes bear witness to its transitional location. Teeming, disorderly Karachi is the typical South Asian city; as in India, the largest urban center lies on the coast. Historic, architecturally Islamic Lahore is reminiscent of the scholarly centers of Muslim Southwest Asia. In Pakistan's eastern borderland, the postcolonial boundary divides a Punjab that stretches beyond the horizon on both sides of the line, a contiguous cultural landscape of villages, wheatfields, and irrigation ditches. In the northwest, Pakistan resembles Afghanistan in its huge migrant

populations and its mountainous frontier. And in the far north, Pakistan and India are locked in a deathly conflict over Jammu and Kashmir. The western flank is South Asia's most critical region.

If, as is so often said, Egypt is the gift of the Nile, then Pakistan is the gift of the Indus. The Indus River and its principal tributary, the Sutlej, nourish the ribbons of life that form the heart of this populous country (Fig. 8-11). Territorially, Pakistan is not large by Asian standards; its area is about the same as that of Texas plus Louisiana. But

Pakistan's population of 159.2 million makes it one of the world's ten most populous states. Among Muslim countries (officially, it is known as the Islamic Republic of Pakistan) only Southeast Asia's Indonesia is larger, but Indonesia's Islam is much less pervasive than Pakistan's.

Upon independence in 1947, Pakistan consisted not only of the territory we know as Pakistan today, which was known as West Pakistan, but also of "East Pakistan"—present-day Bangladesh. That union was based on their shared adherence to Islam, but it did not last long. A political crisis in 1971 led to conflict and East Pakistan's declaration of independence as the People's Republic of Bangladesh.

When Pakistan became a sovereign state following the partition of British India in 1947, its capital was Karachi on the south coast, near the western end of the Indus Delta. As the map shows, however, the present capital is Islamabad, near the larger city of Rawalpindi in the north, not far from Kashmir. By moving the capital from the "safe" coast to the embattled interior and by placing it on the doorstep of contested territory, Pakistan announced its intent to stake a claim to its northern frontiers. And by naming the city Islamabad, Pakistan proclaimed its Muslim foundation, here in the face of the Hindu challenge. This politico-geographical use of a national capital can be assertive, and Islamabad exemplifies the principle **12** of the **forward capital**.

At independence, Pakistan had a bounded national territory, a capital, a cultural core, and a population—but it had few centripetal forces to bind state and nation, especially while East Pakistan was still a part of it. The disparate regions of Pakistan shared the Islamic faith and an aversion for Hindu India, but

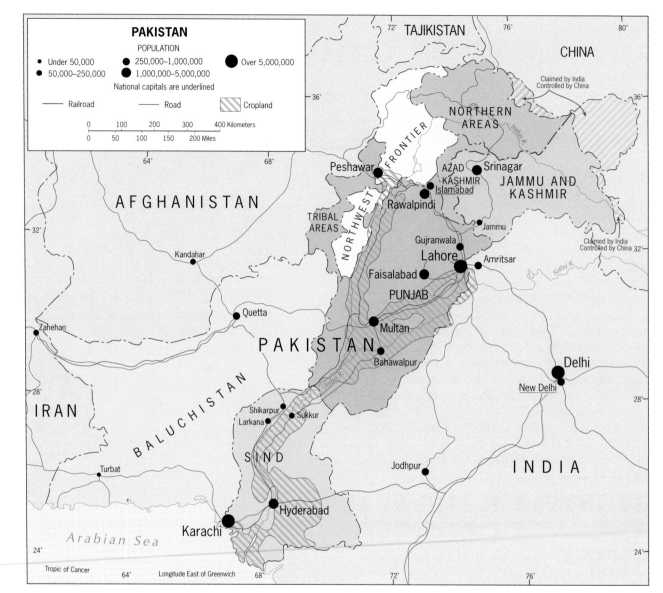

FIGURE 8-11

little else. Karachi and the coastal south, the desert of Baluchistan, the city of Lahore and the Punjab, the rugged northwest along Afghanistan's border, and the mountainous far north are worlds apart, and a Pakistani nationalism to match that of India at independence did not exist. Successive Pakistani governments, civilian as well as military, turned to Islam to provide the common bond that history and geography had denied the nation. In the process, Pakistan became one of the world's most theocratic states; its common law, based on the English model, was gradually transformed into a Quranic (Koranic) system with Islamic Sharia courts and associated punishments.

But even Islam itself is not unified in restive Pakistan. About 80 percent of the people are Sunni Muslims, and the Shia minority numbers about 16 percent. Sunni fanatics intermittently attack Shi'ites, leading to retaliation and creating grounds for subsequent revenge.

Despite the Islamization of Pakistan's plural society, it remains a strongly regionalized country in which Urdu is the official language and English is still the *lingua franca* of the elite. Yet several other major languages prevail in diverse parts, and lifeways vary from nomadism in Baluchistan to irrigation farming in the Punjab to pastoralism in the northern highlands.

The Provinces

As Figure 8-11 shows, Pakistan is administratively divided into four provinces: Punjab, Sind, Northwest Frontier, and Baluchistan. Pakistan was founded as a federal state (although it has been ruled by authoritarian and military governments plagued by corruption and inefficiency), and the relations between these provinces are important to the cohesion of the country. These relations have been difficult.

In large part this is because the *Punjab* is disproportionately dominant in Pakistan: it is the country's core area, home to about 60 percent of the entire population, contains the capital, Islamabad, the primate city (Lahore, 2000 years old, a great Muslim center with magnificent architecture), and leads the nation in almost every economic category. To the south, *Sind* lies centered on the chaotic port city of Karachi, and here the rice and wheat fields in the lower Indus Basin form the breadbasket of Pakistan. Commercially, cotton is king, supporting major textile industries in the province's cities and towns. To the west of Sind lies desert *Baluchistan*, land of ancient caravan routes still operating and home to a sizeable Shi'ite minority, reflecting its proximity to Iran. Recent oil and gas discoveries may transform this remote province in the near future. The *Northwest Frontier*, a mountainous territory of high relief, has taken the brunt of Afghanistan's disorder by becoming a haven for several million refugees, many of whom arrived via The Khyber Pass and

moved to the environs of Peshawar. There, these Afghan Pushtuns formed a nation within a nation, urged on at times by their kinspeople still in Afghanistan to demand more autonomy, a practice we know **13** as **irredentism**.

Figure 8-11 shows a part of the Northwest Frontier Province designated as "Tribal Areas." This mountainous zone on Afghanistan's border is assigned to the isolated indigenous peoples who inhabit valleys and basins remote from the rest of the country. And to the northeast, this province borders the "Northern Areas," the official name for that part of Jammu and Kashmir presently in Pakistani hands (note the Line of Control marked in red, the *de facto* boundary with India). The Northwest Frontier Province is a frontier indeed.

Prospects

For all its size and growing regional influence, Pakistan remains a low-income country with a troubled economy, huge indebtedness, worrisome social indicators (see Table I-1), and unstable government. Subsistence farming still occupies most of the people; urbanization is only at the 33 percent level; life expectancy is below 60; and about 60 percent of the population over the age of 15 is illiterate.

And yet Pakistan has made significant progress during its more than half-century of independence. Irrigated acreage has expanded enormously; self-sufficiency in staple grains (wheat) has been achieved, and some rice is even exported. The cotton industry flourishes, and despite a very limited mineral resource base, manufacturing has progressed as well, symbolized by a steel mill at Port Qasim near Karachi. And a new energy age may be dawning in Baluchistan.

But Pakistan's administrative house is not in order, and the country needs large infusions of external assistance to stay afloat. Thus it is vulnerable to outside pressure and interference: when Pakistan tested a nuclear weapon, sanctions were imposed. When an army general ousted a corrupt and incompetent civilian government, donors balked. Still, Pakistan, a weak, disorganized, and divided country two generations ago, has become a major military force and a nuclear power.

In part this was due to the Cold War, when India tilted toward Moscow while Pakistan was favored by the West as a bulwark against Soviet encroachment on Afghanistan. But today Pakistan finds itself encircled by adversity and risk. United by Islam but little else, facing domestic and international strife from Karachi to Kashmir, Pakistan is vulnerable to threats from all sides. Its support of the United States after 9/11 risks the wider radicalization of Pakistani society. Its struggle over Jammu and Kashmir risks nuclear war with a far more powerful neighbor. Even its relations with its near-neighbors in Turkestan are at risk because Turkestan fears Pakistan's power projected via Afghanistan.

No longer merely a newly decolonized, economically disadvantaged country trying to survive, Pakistan has taken a crucial place in the political geography of two neighboring realms in turbulent transition.

▷ INDIA: FIFTY-PLUS YEARS OF FEDERATION

Nearly three-quarters of the great land triangle of South Asia is occupied by a single country—India, the world's most populous democracy and, in terms of human numbers, the world's largest federation. Consider this: India has nearly as many inhabitants as live in all the countries of Subsaharan Africa *plus* North Africa/Southwest Asia combined—75 of them.

That India has endured as a unified country is a politico-geographical miracle. India is a cultural mosaic of immense ethnic, religious, linguistic, and economic diversity and contrast; it is a state of many nations. The period of British colonialism gave India the underpinnings of unity: a single capital, an interregional transport network, a *lingua franca*, a civil service. Upon independence in 1947, India adopted a federal system of government, giving regions and peoples some autonomy and identity, and allowing others to aspire to such status. Unlike Africa, where federal systems failed and where military dictatorships replaced them, India remained essentially democratic and retained a federal framework in which States have considerable local authority.

This political, democratic success has been achieved despite the presence of powerful centrifugal forces in this vast, culturally diverse country. Relations between the Hindu majority and the enormous Muslim minority, better in some States than in others, have at times threatened to destabilize the entire federation. Local rebellions, demands by some minorities for their own States, frontier wars, even involvement in a foreign but nearby civil war (in Sri Lanka) have buffeted the system—which has bent but not collapsed. India has succeeded where others in the postcolonial world have failed.

This success has not been matched in the field of economics, however. After more than 50 years of independence India remains a very poor country, and not all of this can be blamed on the colonial period or on population growth, although overpopulation remains a strong impediment to improvement of living standards. Much of it results from poor and inconsistent economic planning, too much state ownership of inefficient industries, excessive government control over economic activities, bureaucratic suppression of initiative, corruption, and restraints against foreign investment. As we shall presently see, a few bright spots in some of India's States contrast sharply to the overwhelming poverty of hundreds of millions.

States and Peoples

The map of India's political geography shows a federation of 28 States, 6 Union Territories (UTs), and 1 Na-

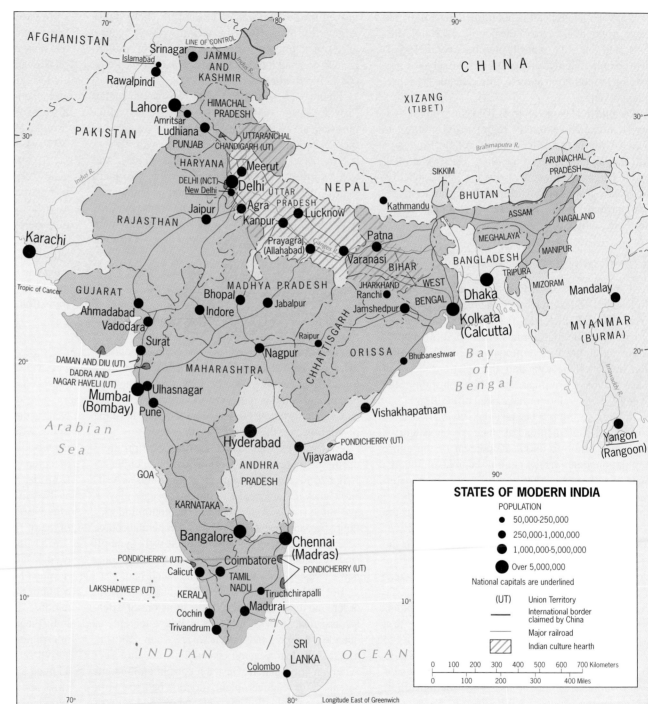

FIGURE 8-12

tional Capital Territory (NCT) (Fig. 8-12). The federal government retains direct authority over the UTs, all of which are small in both territory and population. The NCT, however, includes Delhi and the capital, New Delhi, and has over 13 million inhabitants.

The political spatial organization shown in Figure 8-12 is mainly the product of India's restructuring following independence from Britain. Its State boundaries reflect the broad outlines of the country's cultural mosaic: as far as possible the system recognizes languages, religions, and cultural traditions. Indians speak 14 major and numerous minor languages, and while Hindi is the official language (and English is the *lingua franca*), it is by no means universal. The map is the product of endless compromise—endless because demands for modifications of it continue to this day; as recently as late 2000, the federal government authorized the creation of three new States. In the northeast lie very small States established to protect the local traditions of small populations; minority groups in the larger States ask why they should not receive similar recognition.

With only 28 States for a national population of 1.039 billion, several of India's States contain more people than many countries of the world. As Figure 8-12 shows, the (territorially) largest States lie in the heart of the country and on the great southward-pointing peninsula. Uttar Pradesh (just over 160 million) and Bihar (about 77 million) constitute much of the Ganges River Basin and are the core area of modern India. Maharashtra (almost 98 million), anchored by the great coastal city of Bombay (renamed Mumbai in 1996), also has a population larger than that of most countries. West Bengal, the State that adjoins Bangladesh, has more than 84 million residents, 13.5 million of whom live in its urban focus, Calcutta (renamed Kolkata in 2000).

These are staggering numbers, and they do not decline much toward the south. Southern India consists of four States linked by a discrete history and by their distinct Dravidian languages. Facing the Bay of Bengal are Andhra Pradesh (82 million) and Tamil Nadu (64 million), both part of the hinterland of the megacity of Madras (renamed Chennai in 1997) and located on the coast near their joint border. Facing the Arabian Sea are Karnataka (54 mil-

lion) and Kerala (33 million). Kerala, often at odds with the federal government in New Delhi, has long had the highest literacy rate in India and one of the lowest rates of population growth owing to strong local government and strictly enforced policies. "It's a matter of geography," explained a teacher in the Kerala city of Cochin. "We are here about as far away as you can get from the capital, and we make our own rules."

As Figure 8-12 shows, India's smaller States lie mainly in the northeast, on the far side of Bangladesh, and in the northwest, toward Jammu and Kashmir. North of Delhi, India is flanked by China and Pakistan, and physical as well as cultural landscapes change from the flatlands of the Ganges to the hills and mountains of spurs of the Himalayas. In the State of Himachal Pradesh, forests cover the hillslopes and relief reduces living space; only 6 million people live here, many in small, comparatively isolated clusters. Before independence and political consolidation, the colonial government called this area the "Hill States."

But the map becomes even more complex in the distant northeast, beyond the narrow corridor between Bhutan and Bangladesh. The dominant State here is Assam, famed for its tea plantations and important because its oil and gas production amounts to more than 40 percent of India's total.

In the Brahmaputra Valley, Assam resembles the India of the Ganges. But in almost all directions from Assam, things change. To the north, in sparsely populated Arunachal Pradesh (1.1 million), we are in the Himalayan offshoots again. To the east, in Nagaland (1.9 million), Manipur (2.3 million), and Mizoram (0.9 million), lie the forested and terraced hillslopes that separate India from Myanmar (Burma). This is an area of numerous ethnic groups (more than a dozen in Nagaland alone) and of frequent rebellion against Delhi's government. And to the south, the States of Meghalaya (2.4 million) and Tripura (3.6 million), hilly and still wooded, border the teeming floodplains of Bangladesh. Here in the country's northeast, where peoples are always restive and where population growth is still soaring, India faces one of its strongest regional challenges.

India's Changing Map

After independence, the Indian government began phasing out the privileged "Princely States" which the British had protected during the colonial period. Next, the government reorganized the country on the basis of its major regional languages (see Fig. 8-5). Hindi, spoken by more than one-third of the population, was designated the country's official language, but 13 other major languages also were given national status by the Indian constitution, including the four Dravidian languages of the south. English, it was anticipated, would become India's common language, its *lingua franca* at government, administrative, and business levels. Indeed, English not only remained the language of national administration but also became the chief medium of commerce in growing urban India. English was the key to better jobs, financial success, and personal advancement, and the language constituted a common ground in higher education.

The newly devised framework based on the major regional languages, however, proved to be unsatisfactory to many communities in India. In the first place, many more languages are in use than the 14 that had been officially recognized. Demands for additional States soon arose. As early as 1960, the State of Bombay was divided into two language-based States, Gujarat and Maharashtra.

This devolutionary pressure has continued throughout India's existence as an independent country. In 2000, three new States were recognized: Jharkhand, carved from southern Bihar State on behalf of 18 poverty-stricken districts there; Chhattisgarh, where tribal peoples had been agitating since the 1930s for separation from the State of Madhya Pradesh; and Uttaranchal, which split from India's most populous, Ganges Basin, core-area State of Uttar Pradesh on the basis of its highland character and lifeways (Fig. 8-12).

For many years India has faced quite a different set of cultural-geographic problems in its northeast, where numerous ethnic groups occupy their own niches in a varied, forest-clad topography. The Naga, a cluster of peoples whose domain had been incorporated into Assam State, rebelled soon after India's independence. A protracted war brought federal troops into the area; after a truce and lengthy negotiations,

Nagaland was proclaimed a State in 1961. This led the way for other politico-geographical changes in India's problematic northeastern wing.

The Sikhs

A further dilemma involves India's Sikh population. The Sikhs (the word means "disciples") adhere to a religion that was created about five centuries ago to unite warring Hindus and Muslims into a single faith. This faith's principles rejected negative aspects of Hinduism and Islam, and it gained millions of followers in the Punjab and adjacent areas. During the colonial period, many Sikhs supported British administration of India, and by doing so they won the respect and trust of the British, who employed tens of thousands of Sikhs as soldiers and policemen. By 1947, there was a large Sikh middle class in the Punjab. When independence came, many left their rural homes and moved to the cities to enter urban professions. Today, they still exert a strong influence over Indian affairs, far in excess of the less than 2 percent of the population (about 19 million) they constitute.

After independence, the Sikhs demanded that the original Indian State of Panjab (Punjab) be divided into a Sikh-dominated northwest and a Hindu-majority southeast. The government agreed, so that Punjab as now constituted (Fig. 8-12) is India's Sikh stronghold, whereas neighboring Haryana State is mainly Hindu.

The Muslims

These ethnic, cultural, and regional problems are but a sample of the stresses on India's federal framework. There is no Muslim State in India, but India has about 127 million Muslims within its borders—the largest cultural minority in the world. As Figure 8-6 shows, the percentage of Muslims is highest in remote Jammu and Kashmir, but it also is substantial in such widely dispersed States as Kerala, Assam, and Uttar Pradesh. Moreover, the Muslim population today (approximately 12 percent) constitutes a larger percentage than it did after partition (9.9 percent). This Islamic minority also ranks among the most rapidly growing sectors of India's population, and is strongly urbanized as well—nearly one-third of the population of India's largest city, Mumbai (Bombay), is Muslim.

Centrifugal Forces: From India to Hindustan?

In Chapter 1 we introduced the concept of centrifugal and centripetal forces affecting the fabric of the state. No country in the world exhibits greater cultural diversity than India, and variety in India comes on a scale unmatched anywhere else on Earth. Such diversity spells strong centrifugal forces, although, as we will see, India also has powerful consolidating bonds.

Among the centrifugal forces, Hinduism's stratification of society into castes remains pervasive. Under Hindu dogma, *castes* are fixed layers in society whose ranks are based on ancestries, family ties, **14** and occupations. The **caste system** may have its origins in the early social divisions into priests and warriors, merchants and farmers, craftspeople and servants; it may also have a racial basis, for the Sanskrit term for caste is color. Over the centuries, its complexity grew until India had thousands of castes, some with a few hundred members, others containing millions. Thus, in city as well as in village, communities were segregated according to caste, ranging from the highest (priests, princes) to the lowest (the untouchables). The term *untouchable* has such negative connotations that some scholars object to its use. Alternatives include *dalits* (oppressed), the common term in Maharashtra State but coming into general use; *harijans* (children of God), which was Gandhi's designation, still widely used in the State of Bihar; and *Scheduled Castes*, the official government label.

A person was born into a caste based on his or her actions in a previous existence. Hence, it would not be appropriate to counter such ordained caste assignments by permitting movement (or even contact) from a lower caste to a higher one. Persons of a particular caste could perform only certain jobs, wear only certain clothes, worship only in prescribed ways at particular places. They or their children could not eat, play, or even walk with people of a higher social status. The untouchables occupying the lowest tier were the most debased, wretched members of this rigidly structured social system. Although the British ended the worst excesses of the caste system, and postcolonial Indian leaders—including Mohandas (Mahatma) Gandhi (the great spiritual leader who sparked the independence movement) and Jawaharlal Nehru (the first prime minister)—worked to modify it, a few decades cannot erase centuries of class consciousness. In traditional India, caste provided stability and continuity; in modernizing India, it constitutes an often painful and difficult legacy.

Today we can discern a geography of caste—a degree of spatial variation in its severity. Cultural geographers estimate that about 15 percent of all Indians are of lower caste, about 40 percent of backward caste (one important rank above the lower caste), and some 18 percent of upper caste, at the top of which are the Brahmans, men in the priesthood. (The caste system does not extend to the Muslims, Sikhs, and other non-Hindus in India, which is why these percentages do not total 100.) The colonial government and successive Indian governments have tried to help the lowest castes. This effort has had more effect in the urban than in the rural areas of India. In the isolated villages of the countryside, the untouchables often are made to sit on the floor of their classroom (if they go to school at all); they are not allowed to draw water from the village well because they might pollute it; and they must take off their shoes, if they wear any, when they pass higher-caste houses. But in the cities, untouchables have reserved for them places in the schools, a fixed percentage of State and federal government jobs, and a quota of seats in national and State legislatures. Mohandas Gandhi, who took a special interest in the fate of the untouchables (harijans) in Indian society, accomplished much of this reform.

The caste system remains a powerful centrifugal force, not only because it fragments society but also because efforts to weaken it often result in further division. Gandhi himself was killed, only a few months after independence, by a Hindu fanatic who opposed his work for the least fortunate in Indian society. Today, India is being swept by a wave of Hindu fundamentalism that is caused, at least in part, by continuing efforts to help the poorest. Higher castes see themselves as disadvantaged, and they take refuge in a "return" to fundamental Hindu values.

The radicalization of Hinduism and the infusion of Hindu nationalism into India's politics loom today as twin threats to the country's unity. Hindu nationalist political parties are polarizing the electorate at the State as well as federal level; the recent spate of name changes on India's map is one manifestation of this. If Indian politics were to fragment along religious lines, the miracle of Indian unity would be at risk.

Centripetal Forces

In the face of all these divisive forces, what bonds have kept India unified for so long? Without question, the dominant binding force in India is the cultural strength of Hinduism, its sacred writings, holy rivers, and influence over Indian life. For most Indians, Hinduism is a way of life as much as it is a faith, and its diffusion over virtually the entire country (regardless of the Muslim, Sikh, and Christian minorities) brings with it a national coherence that constitutes a powerful antidote to regional divisiveness. Over the long term, however, the key ingredients of this Hinduism have been its gentility and introspection, radical outbursts notwithstanding. Now the specter of Hindu fanaticism threatens this vital bond.

Another centripetal force lies in India's democratic institutions. In a country as culturally diverse and as populous as India, reliance on democratic institutions has been a birthright ever since independence, and democracy's survival—raucous, sometimes corrupt, always free—has been a crucial unifier.

Furthermore, communications are better in much of India than in many other countries in the global periphery, and the continuous circulation of people, ideas, and goods helps bind the disparate state together. Before independence, opposition to British rule was a shared philosophy, a strong centripetal force. After independence, the preservation of the union was a common objective, and national planning made this possible.

India's capacity for accommodating major changes and its flexibility in the face of regional and local demands are also a centripetal force. Boundaries have been shifted; internal political entities have been created, relocated, or otherwise modified; and secessionist demands have been handled with a mixture of federal power and cooperative negotiation. Indians in South Asia have accomplished what Europeans in Yugoslavia could not, and India's history of success is itself a centripetal force.

Finally, no discussion of India's binding forces would be complete without mentioning the country's strong leadership. Gandhi, Nehru, and their successors did much to unify India by the strength of their compelling personalities. For many years, leadership was a family affair: Nehru's daughter, Indira Gandhi, twice took decisive control (in 1966 and 1980) after weak governments, and her son, Rajiv Gandhi (who, in 1991, like his mother seven years earlier, also was assassinated), was prime minister in the late 1980s. Since his death, the crucial question of India's leadership has hung in the balance.

Urbanization

India is famous for its great and teeming cities, but India is not yet an urbanized society. Only 28 percent of the population lived in cities and towns in 2002—but in terms of sheer numbers, that 28 percent amounts to over 290 million people, more than the entire population of the United States.

And India's rate of urbanization is on the upswing. People by the hundreds of thousands are arriving in the already-overcrowded cities, swelling urban India by about 5 percent annually, almost three times as fast as the overall population growth. Not only do the cities attract as they do everywhere; many villagers are driven off the land by the desperate conditions in the countryside. As villagers manage to establish themselves in Mumbai or Kolkata or Chennai, they help their relatives and friends to join them in squatter settlements that often are populated by newcomers from the same area, bringing their language and customs with them and cushioning the stress of the move.

As a result, India's cities display staggering social contrasts. Squatter shacks without any amenities at all crowd against the walls of modern high-rise apartments and condominiums. Hundreds of thousands of homeless roam the streets and sleep in parks, under bridges, on sidewalks. As crowding intensifies, social stresses multiply. Disorder never seems far from the surface; sporadic rioting, often attributable to rootless urban youths unable to find employment, has become commonplace in India's cities.

India's modern urbanization has its roots in the colonial period, when the British selected Calcutta (Kolkata), Bombay (Mumbai), and Madras (Chennai) as regional trading centers and fortified ports. Madras was fortified as early as 1640; Bombay (1664) had the situational advantage of being the closest of all Indian ports to Britain; and Calcutta (1690) lay on the margin of India's largest population cluster and had the most productive hinterland, to which the Ganges Delta's countless channels connected it. This natural transport network made Calcutta an ideal colonial headquarters, but the population of Bengal was often rebellious. In 1912 the British moved their colonial government from Calcutta to the safer interior city of New Delhi, built adjacent to the old Mogul capital of Delhi.

Figure 8-12 displays the distribution of major urban centers in India. Except for Delhi-New Delhi, the largest cities have coastal locations: Kolkata (Calcutta) dominates the east, Mumbai (Bombay) the west, and Chennai (Madras) the south. But urbanization also has expanded in the interior, notably in the core area. The surface interconnections among India's cities remain inadequate (notably the road network), but an Indian urban system is emerging.

Economic Geography

If India has faced problems in its great effort to achieve political stability and national cohesion, these problems are more than matched by the difficulties that lie in the way of economic growth and development. The large-scale factories and power-driven machinery of the colonial powers wiped out a good part of India's indigenous industrial base. Indian trade routes were taken over. European innovations in health and medicine sent the rate of population growth soaring, without introducing solutions for the many problems this spawned. Surface communications improved and food distribution systems became more efficient, but local and regional food shortages occurred (and still do) as droughts frequently caused crop failures. Today, nearly half of India's one-billion-plus people live in abject poverty, and the prospects of reducing that high level of human misery anytime soon are not encouraging. (Yet even with its modest annual per-capita GNP [U.S. $440], the sheer size of India's population has created a big overall economy—the world's sixth largest according to the latest rankings.)

Agriculture

GEO*DISCOVERIES* India's underdevelopment is nowhere more apparent than in its agriculture. Traditional farming methods persist, and yields per acre and per worker remain low for virtually every crop grown under this low-technology system. Moreover, the transportation inefficiencies of the traditional farming system hamper the movement of agricultural commodities. In 2001, less than 50 percent of India's 600,000 villages were accessible by motorable road, and today animal-drawn carts still outnumber motor vehicles nationwide.

As the total population grows, the amount of cultivated land per person declines. Today, this physiologic density is 1615 per square mile (625 per sq km). However, this is nowhere near as high as the physiologic density in neighboring Bangladesh, where the figure is more than twice as great (3623 and 1399, respectively). But India's farming is so inefficient that this comparison is deceptive. Fully two-thirds of India's huge working population depends directly on the

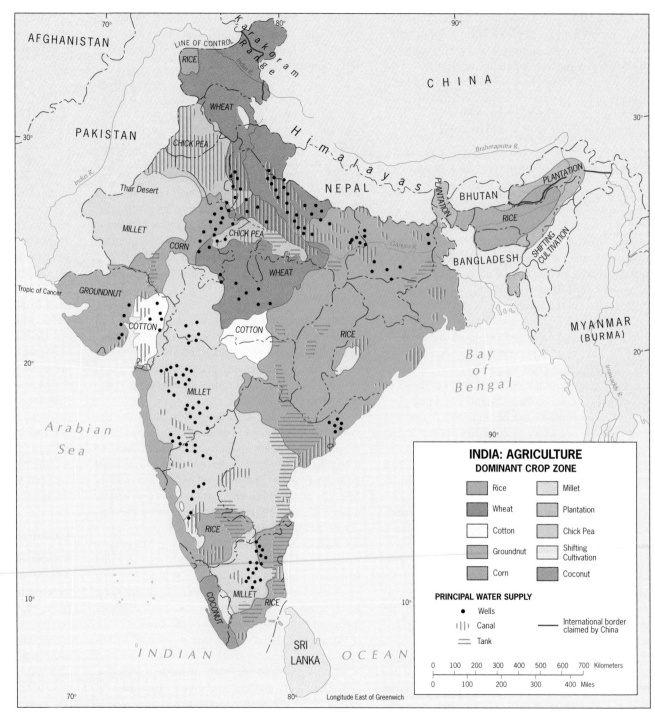

FIGURE 8-13

land for its livelihood, but the great majority of Indian farmers are poor and cannot improve their soils, equipment, or yields. Those areas in which India has substantially modernized its agriculture (as in Punjab's wheat zone) remain islands in a sea of agrarian stagnation.

This stagnation has persisted in large measure because India, after independence, failed to implement a much-needed nationwide land reform program. Roughly one-quarter of India's entire cultivated area is still owned by less than 5 percent of the country's farming families, and little land redistribution was taking place. Perhaps half of all rural families own either as little as an acre or no land at all. Independent India inherited inequities from the British colonial period, but the individual States of the federation would have had to cooperate in any national land reform program. As always, the large landowners retained considerable political influence, so the program never got off the ground.

To make matters worse, much of India's farmland is badly fragmented as a result of local rules of inheritance, thereby inhibiting cooperative farming, mechanization, shared irrigation, and other opportunities for progress. Not surprisingly, land consolidation efforts have had only limited success except in the States of Punjab, Haryana, and parts of Uttar Pradesh, where modernization has gone farthest. Official agricultural development policy, at the federal and State levels, has also contributed to India's agricultural malaise and the uneven distribution of progress. Unclear priorities, poor coordination, inadequate information dissemination, and other failures have been 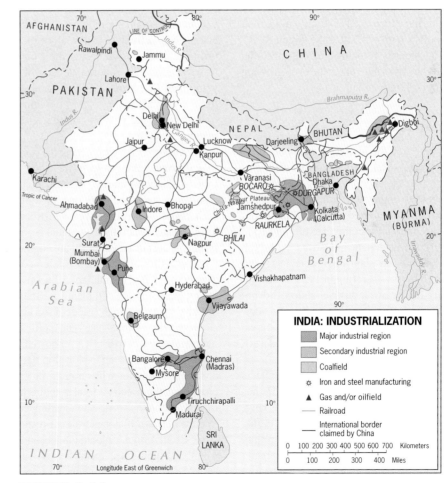 reflected in the country's disappointing output.

It is instructive to compare Figure 8-13, showing the distribution of crop regions and water supply systems in India, with Figure I-7, which shows mean annual precipitation in India and the world. In the comparatively dry northwest, notably in the Punjab and neighboring areas of the upper Ganges, wheat is the leading cereal crop. Here, India has made major gains in annual production through the introduction of high-yielding grain varieties developed under the banner of the Green Revolution, the international research program that played a key

role in overcoming the food crises of the 1960s. These "miracle crops" led to the expansion of cultivated areas, the construction of new irrigation systems, and the more intensive use of fertilizer (a mixed blessing, for fertilizers tend to be expensive and the "miracle" crops are more heavily dependent on them).

Toward the moister east, and especially in the wet-monsoon-drenched areas (Fig. 8-13), rice becomes the dominant staple. About one-fourth of India's total farmland lies under rice cultivation, most of it in the States of Assam, West Bengal, Bihar, Orissa,

and eastern Uttar Pradesh and along the Malabar-Konkan coastal strip facing the Arabian Sea. These areas receive over 40 inches (100 cm) of rainfall annually, and irrigation supplements precipitation where necessary.

India devotes more land to rice cultivation than any other country, but yields per acre remain among the world's lowest—despite the introduction of "miracle rice." Nevertheless, the gap between demand and supply has narrowed, and in the late 1980s India actually exported some grain to Africa as part of a worldwide effort to help refugees there. The

FIGURE 8-14

situation remains precarious, however. As the population map (Fig. I-9) shows, there is a considerable degree of geographic covariation between India's rice-producing zones and its most densely populated areas. India is just one poor-harvest year away from another food crisis.

Subsistence remains the fate of tens of millions of Indian villagers who cannot afford fertilizers, cannot cultivate the new and more productive strains of rice or wheat, and cannot escape the cycle of poverty. Perhaps as many as 175 million of these people do not even own a plot of land and must live as tenants, always uncertain of their fate. This is the enduring reality against which optimistic predictions of improved nutrition in India must be weighed. True, rice and wheat yields have increased at slightly more than the rate of population growth since the Green Revolution. But food security remains elusive, and India continues to face the risks inherent in the ever-growing needs of its burgeoning population.

Industrialization

In 1947 India inherited the mere rudiments of an industrial framework. After more than a century of British control over the economy, only 2 percent of India's workers were engaged in industry, and manufacturing and mining combined produced only about 6 percent of the national income. Textile and food processing were the dominant industries. Although India's first iron-making plant opened in 1911 and the first steel mill began operating in 1921, the initial major stimulus for heavy industrialization did not come until after the outbreak of World War II. Manufacturing was concentrated in the largest cities: Kolkata (Calcutta) led, Mumbai (Bombay) was next, and Chennai (Madras) ranked third.

The geography of manufacturing still reflects those beginnings, and industrialization in India has proceeded slowly, even after independence (Fig. 8-14). Kolkata now anchors India's eastern industrial region—the Bihar-Bengal District—where jute manufactures dominate, but cotton, engineering, and chemical industries also operate. On the nearby Chota-Nagpur Plateau to the west, coal-mining and iron and steel manufacturing have developed.

On the opposite side of the subcontinent, two industrial areas dominate the western manufacturing region: one is centered on Mumbai and the other on Ahmadabad. This dual region, lying in Maharashtra and Gujarat States, specializes in cotton and chemicals, with some engineering and food processing. Cotton textiles have long been an industrial mainstay in India, and this was one of the few industries to benefit from the nineteenth-century economic order the British imposed. With the local cotton harvest, the availability of cheap yarn, abundant and inexpensive labor, and the power supply from the Western Ghats' hydroelectric stations, the industry thrived.

The southern industrial region consists chiefly of a set of linear, city-linking corridors focused on Chennai, specializing in textile production and light engineering activities. Today, all of India's manufacturing regions are increasing their output of ready-to-wear garments—another legacy of the early development of cotton textiles. Clothing has become India's second-leading export by value; the production of gems and jewelry, another growing specialization, ranks first.

An important development in the south is occurring in and around Bangalore in the State of Karnataka, India's "Silicon Plateau." Several hundred software companies are based here, one-third of them foreign with names such as IBM, Texas Instruments, and Motorola. What attracts them, and makes them profitable, is the low cost of India's software engineers, who earn about one-fifth of what their foreign colleagues earn. The emigration of technicians is becoming a problem, but replacements are still plentiful. Bangalore is proof of India's potential in the modern world.

Despite some imbalances and inefficiencies, India's industrial resource base is well endowed. Limited high-quality coal deposits are exploited in the Chota-Nagpur area. In combination with large lower-grade coalfields elsewhere, the country's total output is high enough to rank it among the world's ten leading coal producers. With no known major petroleum reserves (some oil comes from Assam, Gujarat, Punjab, and offshore from Mumbai), India must spend heavily on fuel imports every year. Major investments have been made in hydroelectric plants, especially multipurpose dams that provide electricity, enhance irrigation, and facilitate

flood control. India's iron ore deposits in Bihar (northwest of Kolkata) and Karnataka (in the heart of the Deccan) may rank among the largest in the world. Jamshedpur, located west of Kolkata in the eastern industrial region, has become India's steel-making and metals-fabrication center. Yet India still exports iron ore as a raw material to the higher-income industrialized countries, mainly Japan. For low-income, revenue-needy countries, entrenched practices are difficult to break.

India East and West

The most commonly cited, and most clearly evident, regional division of India is between north and south. The north is India's heartland, the south its Dravidian appendage; the north speaks Hindi as its *lingua franca*, the south prefers English over Hindi; the north is bustling and testy, the south seems slower and less agitated.

But there is another, as yet less obvious, but potentially more significant divide across India. In Figure 8-14, draw a line from Lucknow, on the Ganges River, south to Madurai, near the southern tip of the peninsula. To the west of this line, India is showing signs of economic progress, the kind of economic activity that has brought Pacific Rim countries such as Thailand and Indonesia a new life. To the east, India has more in common with less promising countries also facing the Bay of Bengal: Bangladesh and Myanmar (Burma).

As with other regional divides, there are exceptions to our east-west delineation. Indeed, our map seems to suggest that much of India's industrial strength lies in the east. But what the map cannot reveal is the profitability of those industries. True, the east is rich in iron and coal, but the heavy industries built by the state in the 1950s are now outdated, uncompetitive, and in decline. The hinterland of Kolkata now contains India's Rustbelt. The government keeps many industries going but at a high cost. Old industries, such as carpetmaking and cottonweaving, continue to use child labor to remain viable. The State of Bihar represents the stagnation that afflicts much of India east of our line: by several measures it ranks among the poorest of the 28 States.

Compare this to western India. The State of Maharashtra, the hinterland of Mumbai, leads India in many categories, and Mumbai leads Maharashtra. Many smaller, private industries have emerged here, manufacturing goods ranging from umbrellas to satellite dishes and from toys to textiles. Across the Arabian Sea lie the oil-rich economies of the Arabian Peninsula. Hundreds of thousands of workers from western India have found jobs there, sending money back to families from Punjab to Kerala. More importantly, many have used their foreign incomes to establish service industries back home. Outward-looking western India, in contrast to the inward-looking east, has begun to establish other ties to the outside world. Satellite links have enabled Bangalore to become the center of a growing software-producing complex reaching world markets. The beaches of Goa, the small State immediately to the south of Maharashtra, appeal to the tourist markets **15** of Europe. This is, in fact, a classic case of **intervening opportunity** because resorts have sprung up along Goa's coast, and European tourists who once went to the more distant Maldives and Seychelles are coming to Goa. Maharashtra's economic success also has spilled over into Gujarat to the north, and even landlocked Rajasthan (the next State to the north) is experiencing the beginnings of what, by Indian standards, is a boom.

The boom has created political problems, however. Not only is Maharashtra State a rising economic power; it also is the base of a strong Hindu nationalist political movement whose leaders object to foreign intrusions and have blocked major development projects and other enterprises. They halted a huge industrial scheme about halfway through and closed a fast-food operation that they deemed incompatible with local culture. Such clashes between foreign interests and domestic traditions are not unique to India, but the rising tide of Hindu fundamentalism has uncertain prospects and unsettles investors to whom India remains a high-risk calculation.

Nevertheless, India's east-west divide shows a growing contrast that puts the west far ahead. The hope is that Maharashtra's success will spread northward and southward along the Arabian Sea coast and will ultimately diffuse eastward as well. But for this to happen, India will have to bring its population spiral under control.

BANGLADESH: PERSISTENT POVERTY

On the map of South Asia, Bangladesh looks like another State of India: the country occupies the area of the double delta of India's great Ganges and Brahmaputra rivers, and India almost completely surrounds it on its landward side (Fig. 8-15). But Bangladesh is an independent country, born in 1971 after its brief war for independence against Pakistan, with a territory about the size of Wisconsin. Today it is one of the poorest and least developed countries on Earth, with a population of 133 million that is growing at an annual rate of 1.9 percent.

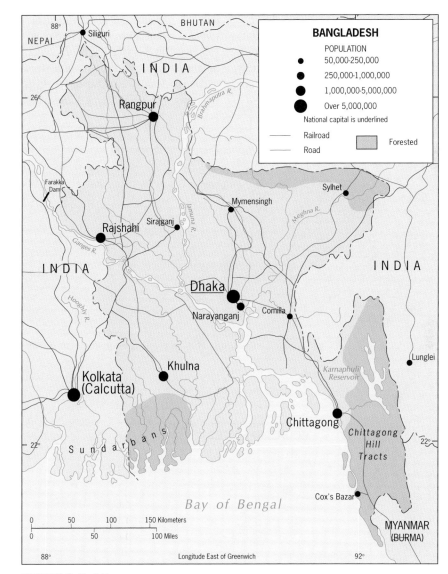

FIGURE 8-15

Not only is Bangladesh a poor country; it also is **16** highly susceptible to damage from **natural hazards**. During the twentieth century, eight of the ten deadliest natural disasters in the entire world struck this single country. In 1991, a cyclone (as hurricanes are called in this part of the world) killed over 150,000 people.

The reasons for Bangladesh's vulnerability can be deduced from Figures 8-15 and 8-1. Southern Bangladesh is the deltaic plain of the Ganges-Brahmaputra river system, combining fertile alluvial soils that attract farmers with the low elevations that endanger them when water rises. The shape of the Bay of Bengal forms a funnel that sends cyclones and their storm surges barreling into the delta coast. Without money to build seawalls, floodgates, eleva-

ted shelters in sufficient numbers, or adequate escape routes, hundreds of thousands of people are at continuous risk, with deadly consequences.

Bangladesh remains a nation of subsistence farmers; urbanization is at only 20 percent, and Dhaka, the capital, and the southeastern port of Chittagong are the only urban centers of consequence. Moreover, Bangladesh has one of the highest physiologic densities in the world (3623 people per square mile/1399 per sq km), and only higher-yielding varieties of rice and the introduction of wheat in the crop rotation (where climate allows) have improved diets and food security. But diets remain poorly balanced, and overall, the people's nutrition is unsatisfactory.

Bangladesh is a dominantly Muslim society. Its relations with neighboring India have at times been

strained over water resources (India's control over the Ganges that is Bangladesh's lifeline), cross-border migration (10 percent of the population is Hindu), and transit between parts of India across Bangladesh's north (refer to Figure 8-14 to see the reason). All the disadvantages of the global periphery afflict this populous, powerless country where survival is the leading industry and all else is luxury.

THE MOUNTAINOUS NORTH

As Figure 8-1 shows, a tier of landlocked countries and territories lies across the mountainous zone that walls

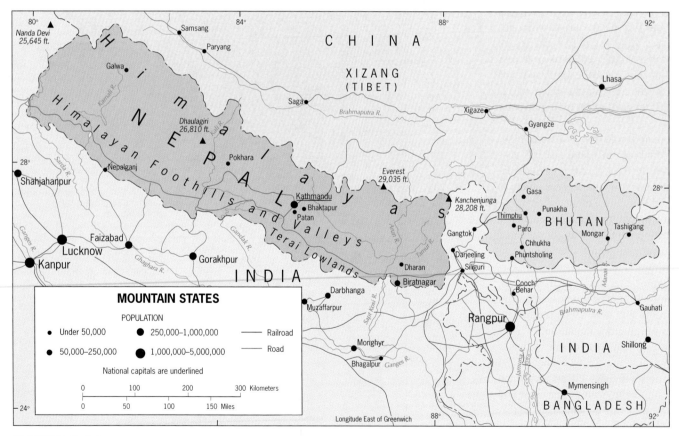

FIGURE 8-16

India off from China. One of them, Kashmir, is in a state of near-war. Another, Sikkim, was absorbed by India in 1975 and made into one of its federal States. But Nepal and Bhutan retain their independence.

Nepal, northeast of India's Hindu core, has a population of more than 25 million and is the size of Illinois. It has three geographic zones (Fig. 8-16): a southern, subtropical, fertile lowland called the Terai; a central belt of Himalayan foothills with swiftly flowing streams and deep valleys; and the spectacular high Himalayas themselves (topped by Mount Everest) in the north. The capital, Kathmandu, lies in the east-central part of the country in an open valley of the central hill zone.

Nepal is materially poor but culturally rich. The Nepalese are a people of many sources, including India, Tibet, and interior Asia. About 90 percent are Hindu, and Hinduism is the country's official religion; but Nepal's Hinduism is a unique blend of Hindu and Buddhist ideals. Thousands of temples and pagodas ranging from the simple to the ornate grace the cultural landscape, especially in the valley of Kathmandu, the country's core area. Although over a dozen languages are spoken, 90 percent of the people also speak Nepali, a language related to Indian Hindi.

As the data in Table I-1 suggest, Nepal is a troubled country suffering from severe underdevelopment; it also faces strong centrifugal social and political forces. Environmental degradation, crowded farmlands and soil erosion, and deforestation scar the countryside. The Himalayan peaks form a world-renowned tourist attraction, but tourist spending in Nepal remains relatively modest. Nepal's GNP is the lowest in all of South Asia.

Nepal's political geography also is troubled. The end of absolute monarchy in 1991 and the advent of democracy did not end the country's regional divisions: the southern Terai with its tropical lowlands is a world apart from the hills of central Nepal, and the peoples of the west have origins and traditions different from those in the east. Survival as a coherent state is Nepal's great challenge.

Mountainous Bhutan, wedged between India and China's Tibet, is the only other buffer between Asia's giants. In landlocked, fortress-like Bhutan, time seems to have stood still. Bhutan is officially a constitutional monarchy, but its king rules the country with virtually absolute power; economic subsistence and political allegiance are the norms of life for most of the population of just under one million. Thimphu, the capital, has about 50,000 inhabitants. The symbols of Buddhism, the state religion, dominate its cultural landscape. Social tensions arise from the large but diminishing Nepalese minority, most of whom are Hindus and some of whom have been persecuted by the dominant Bhutia.

Forestry, hydroelectric power, and tourism all have potential here, and Bhutan has considerable mineral resources. But isolation and inaccessibility preserve traditional ways of life in this mountainous buffer state.

▷ THE SOUTHERN ISLANDS

As Figure 8-1 shows, South Asia's continental landmass is flanked by several sets of islands: Sri Lanka off the southern tip of India, the Maldives in the Indian Ocean, and the Andaman Islands (belonging to India) marking the eastern edge of the Bay of Bengal.

The Maldives consists of more than a thousand tiny islands whose combined area is just 115 square miles (less than 300 sq km) and whose highest elevation is barely over 6 feet (2 m) above sea level. Its population of under 300,000 from Dravidian and Sri Lankan sources is now 100 percent Muslim, one-quarter of which is concentrated on the capital island named Maale. The Maldives might be unremarkable, except that, as Table I-1 shows, this country has by far the realm's highest GNP. The locals have translated their palm-studded, beach-fringed islands into a tourist mecca that attracts tens of thousands of mainly European visitors annually.

Sri Lanka: South Asian Tragedy

Sri Lanka (known as Ceylon before 1962), the compact, pear-shaped island located just 22 miles (35 km) across the Palk Strait from India, became independent from Britain in 1948 (Fig. 8-17). There were good reasons to create a separate sovereignty for Sri Lanka. This is neither a Hindu nor a Muslim country: the majority of its nearly 20 million people—about 70 percent—are Buddhists. Furthermore, unlike India or Pakistan, Sri Lanka is a plantation country, and commercial agriculture still is the mainstay of the agricultural economy.

The great majority of Sri Lanka's people are descended from migrants who came to this island from northwest India beginning about 2500 years ago. Those migrants introduced the advanced culture of their source area, building towns and irrigation systems and bringing Buddhism. Today, their descendants, known as the Sinhalese, speak a language (Sinhala) that belongs to the Indo-European language family of northern India.

The Dravidians who lived on the mainland, just across the Palk Strait, came later and in far smaller numbers—until the British colonialists intervened. During the nineteenth century the British brought hundreds of thousands of Tamils to work on their tea plantations, and soon a small minority became a substantial segment of Ceylonese society. The Tamils brought their Dravidian tongue to the island and introduced their Hindu faith. At the time of independence, they constituted more than 15 percent of the population; today they total about 18 percent.

When Ceylon became independent, it was one of the great hopes of the postcolonial world. The country had a sound economy and a democratic government, and it was renowned for its tropical beauty. Its reputation soared when a massive campaign succeeded in eradicating malaria and when family-planning campaigns reduced population growth when the rest of the realm was experiencing a population explosion. Rivers from the cool, forested interior highlands fed the paddies that provided ample rice; crops from the moist southwest paid the bills, and the capital, Colombo, grew to reflect the optimism that prevailed.

In the midst of this glowing scenario, the seeds of disaster were already being sown. Sri Lanka's Tamil minority soon began proclaiming its sense of exclusion, demanding better treatment from the Sinhalese majority. Although the government recognized Tamil as a "national language" in 1978, sporadic violence marked the Tamil campaign, and

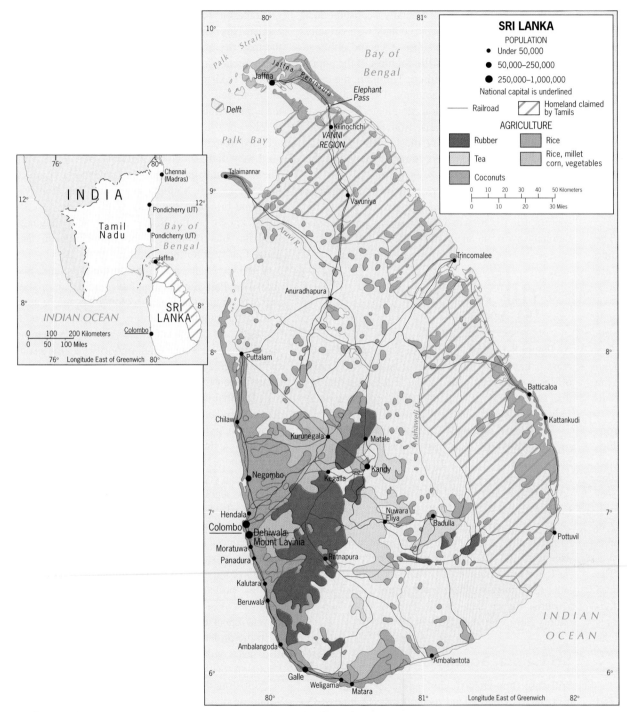

SRI LANKA

POPULATION
- • Under 50,000
- • 50,000–250,000
- ● 250,000–1,000,000

National capital is underlined

— Railroad   ⬚ Homeland claimed by Tamils

AGRICULTURE

- ▨ Rubber
- ▨ Tea
- ▨ Coconuts
- ▨ Rice
- ▨ Rice, millet corn, vegetables

0   10   20   30   40   50 Kilometers

0   10   20   30 Miles

FIGURE 8-17

in 1983 full-scale civil war began. Now many in the Tamil community demanded a separate Tamil state to encompass the north and east of the country (see Fig. 8-17), and a rebel army called the Tamil Tigers confronted Sri Lanka's national forces.

The sequence of events fits the model of **17** the evolution of the **insurgent state** discussed in Chapter 5. In Sri Lanka, the *equilibrium* stage was reached in the early 1990s, when the Tamil Tigers claimed the Jaffna Peninsula and set up their headquarters there. The *counteroffensive* stage is now in progress, but the Tamil forces strike at will (they destroyed Colombo's international airport in July 2001) and a negotiated settlement appears inevitable. Even if the Tamils do not succeed in securing the independent state they call Eelam, they will likely force the government to make some territorial concessions.

Once-promising Sri Lanka is paying dearly for its politicians' failures. The cost of the civil war is enormous. The economic consequences are incalculable, with the tourist industry having been reduced to a fraction of what it could be. And South Asia has lost a beacon of opportunity and progress.

9 / East Asia

This Chapter's Media Highlights include:

Photo Gallery: East Asia

GeoDiscoveries: Enterprise Zones and Traditional Economies

3 Interactive Map Quizzes: Countries, Cities, Physical Features

Chapter Quiz

Concept Quiz

Virtual Field Trip: Following China's Silk Route www.wiley.com/college/deblij

Shanghai, China: Pudong's futuristic skyline symbolizes China's new era.

Scale 1:16 000 000; one inch to 250 miles. Polyconic Projection
Elevations and depressions are given in feet

FIGURE 9-1

chapter 9 / East Asia

Chinese Provinces,
Autonomous Regions (AR),
Special Administrative Regions (SAR),
and Municipalities (M)

Conventional Form		Pinyin Form
Anhwei	-	Anhui
Chekiang	-	Zhenjiang
Chungking	-	Chongqing
Fükien	-	Fujian
Heilungkiang	-	Heilongjiang
Honan	-	Henan
Hong Kong (SAR)	-	Xianggang
Hopeh	-	Hebei
Hunan	-	Hunan

Ⓐ Area occupied by Pakistan and claimed by India.

Ⓑ Area claimed and occupied by India; status disputed by Pakistan.

Ⓒ Area occupied by China and claimed by India.

Ⓓ Area occupied by India and claimed by China.

Longitude East of Greenwich

0   50   100   200   300   400   500 Miles

0   100   200   400   600   800 Kilometers

Cities and Towns

0 to 50,000

50,000 to 500,000

CONCEPTS, IDEAS, AND TERMS

1 Pacific Rim
2 Confucianism
3 Sinicization
4 Extraterritoriality
5 Special Administrative Region (SAR)
6 Economic restructuring
7 Core area
8 Geography of development
9 Overseas Chinese
10 Special Economic Zone (SEZ)
11 Regional state
12 Economic tiger
13 Buffer state
14 Jakota Triangle
15 Modernization
16 Relative location
17 Areal functional organization
18 Regional complementarity
19 State capitalism

REGIONS

- CHINA PROPER
- XIZANG (TIBET)
- XINJIANG
- MONGOLIA
- JAKOTA TRIANGLE (JAPAN-KOREA-TAIWAN)

East Asia is a geographic realm like no other. At its heart lies the world's most populous country. On its periphery lies one of the globe's most powerful national economies. Along its coastline, on its peninsulas, and on its islands an economic boom has transformed cities and countrysides. Its interior contains the world's highest mountains and vast deserts. It is a storehouse of raw materials. The basins of its great rivers produce food that can sustain more than a billion people.

DEFINING THE REALM

The East Asian geographic realm consists of six political entities: China, Mongolia, North Korea, South Korea, Japan, and Taiwan. Note that we refer here to "political entities" rather than "states." In changing East Asia, the distinction is significant. Taiwan, which its government officially calls the Republic of China, functions as a state but is regarded by mainland China (the People's Republic of China) as a temporarily wayward province. North Korea is not a full member of the United Nations, and the division of the Korean Peninsula may be temporary.

As defined here, East Asia lies between the vast expanses of Russia to the north and the populous countries of South and Southeast Asia to the south. This geographic realm extends from the deserts of Central Asia to the Pacific islands of Japan and Taiwan. Environmental diversity is one of its hallmarks.

East Asia also is the hub of the evolving regional 1 phenomenon called the **Pacific Rim**. From Japan to Taiwan and from South Korea to Hong Kong (Xianggang), the Pacific frontage of East Asia is being transformed. Skyscrapers tower over old cities whose traditional housing is being swept away. Enormous industrial complexes disgorge products that flood world markets. Millions of people are on the move, abandoning farms and villages for assembly lines and sweatshops. The process started in Japan, and soon encompassed Taiwan, South Korea, Hong Kong, and Singapore. And when China's political climate changed, almost all of coastal East Asia was swept up in one of the greatest regional transformations in history.

NATURAL ENVIRONMENTS

Figure 9-1 dramatically illustrates the complex physical geography of the East Asian realm. In the southwest lie ice-covered mountains and plateaus, the

◆ Major Geographic Qualities of East Asia

1. East Asia is encircled by snowcapped mountains, vast deserts, cold climates, and Pacific waters.

2. East Asia was one of the world's earliest culture hearths, and China is one of the world's oldest continuous civilizations.

3. East Asia is the world's most populous geographic realm, but its population remains strongly concentrated in its eastern regions.

4. China, the world's largest nation-state demographically, is the current rendition of an empire that has expanded and contracted, fragmented and unified many times during its long existence.

5. China today remains a mainly rural society, and its vast eastern river basins feed hundreds of millions in a historic pattern that still continues.

6. China's sparsely peopled western regions are strategically important to the state, but they lie exposed to minority pressures and Islamic influences.

7. Along China's Pacific frontage an economic transformation is taking place, affecting all the coastal provinces and creating an emerging Pacific Rim region.

8. Increasing regional disparities and fast-changing cultural landscapes are straining East Asian societies.

9. Japan, the economic giant of the East Asian realm, has a history of colonial expansion and wartime conduct that still affects international relations here.

10. East Asia may witness the rise of the world's next superpower as China's economic and military strength and influence grow— and if China avoids the devolutionary forces that fractured the Soviet Union.

11. The political geography of East Asia contains a number of flashpoints that can generate conflict, including Taiwan, North Korea, and several island groups in the realm's seas.

Earth's crust in this region crumpled up like the folds of an accordion. A gigantic collision of tectonic plates is creating this landscape as the Indian Plate pushes northward into the underbelly of the Eurasian Plate (Fig. I-4). The result is some of the world's most spectacular scenery, but snow, ice, and cold are not the only dangers to human life here. Earthquakes and tremors occur almost continuously, causing landslides and avalanches. As the map shows, the high mountains and plateaus widen from a relatively narrow belt in the Karakoram to form Xizang's (Tibet's) vast plateau, flanked by the Himalayas to the south. Then, east of Tibet, the mountain ranges converge again and bend southward into Southeast Asia, where they lose their high relief.

As Figure I-9 shows, this Asian interior is one of the world's most sparsely populated areas, but it is nevertheless critical to the lives of hundreds of millions of people. In these high mountains, fed by the melting ice and snow, rise the great rivers that flow eastward across China and southward across Southeast and South Asia. Throughout the Holocene, these rivers have been eroding the uplands and depositing their sediments in the lowlands, in effect creating the alluvium-filled basins that now sustain huge populations. Fertile alluvial soils and adequate growing seasons, combined with ample water and millions of hands to sow the wheat and plant the rice, have allowed the emergence of one of the great population concentrations on Earth.

Physiography, therefore, has much to do with East Asia's population distribution, but even the more habitable and agriculturally productive east has its limitations. The northeast suffers from severe continentality, with long and bitterly cold winters.

High relief encircles the river basins north of the Yellow Sea and dominates much of the northern part of the Korean Peninsula, creating strong local environmental contrasts. South Korea, as Figure I-8 shows, experiences relatively moderate conditions, comparable to those of the U.S. Southeast; North Korea has a harsh continental climate like that of North Dakota.

Before we focus on East Asia's human geography, it is useful to look at a map of this realm's complex physical stage (Fig. 9-2). From the high-relief interior come three major river systems that have played crucial roles in the human drama. In the north, the Huang He (Yellow River) arises deep in the high mountains, crosses the Ordos Desert and the Loess Plateau, and deposits its fertile sediments in the vast North China Plain, where East Asia's earliest states emerged. In the center, the

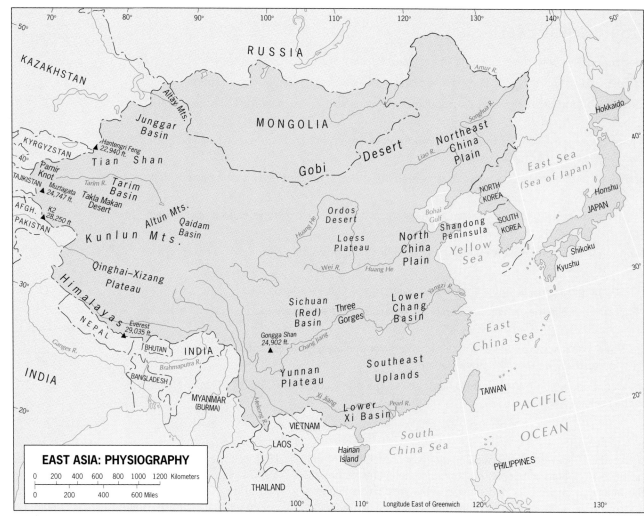

FIGURE 9-2

rocks and produced the windblown, fertile deposit known as *loess*, and where water is available it can sustain a dense agricultural population. To the south, deep in the interior, lies the Sichuan (Red) Basin, crossed by the Chang Jiang. This basin has supported human communities for a long time, and, as we noted in the Introduction, you can actually see its current population cluster on the world population map (Fig. I-9). The Sichuan Basin, encircled as it is by mountains, is one of the world's most clearly defined physiographic regions, and the concentration of its approximately 120 million inhabitants reflects that definition.

Still farther to the south lies the Yunnan Plateau, source of the tributaries that feed the Xi River. Much of southeastern China has comparatively high relief; it is hilly and in places mountainous. This high relief has helped limit contacts between China and Southeast Asia.

East Asia's Pacific margin is a jumble of peninsulas and islands. The Korean Peninsula looks like a near-bridge from Asia to Japan, and indeed it has served as such in the past. The Liaodong and Shandong peninsulas protrude into the Yellow Sea, which continues to silt up from the sediments of the Huang and Liao rivers. Off the mainland lie the islands that have played such a crucial role in the modern human geography of Asia and, indeed, the world: Japan, Taiwan, and Hainan. Japan's environmental range is expressed by cold northern Hokkaido and warm southern Kyushu, but Japan's core area lies on its main island, Honshu. As Figure 9-1 shows, myriad smaller islands flank the mainland and dot the East and South China Seas. As we will discover, some of these smaller islands have major significance in the human geography of this realm.

Chang Jiang (Long River), called the Yangzi downstream, crosses the Sichuan Basin and the Three Gorges, where a huge dam project is under way, and waters extensive ricefields in the Lower Chang Basin. And in the south, the Xi Jiang (West River) originates on the Yunnan Plateau and becomes the Pearl River in its lowest course. Its estuary, flanked by several of China's largest urban-industrial complexes, has become one of the hubs of the evolving Pacific Rim.

Further scrutiny of Figure 9-2 indicates that a fourth river system plays a role in China: the Liao River in the northeast and its basin, the Northeast China Plain. As the map suggests, however, the Liao is not comparable to the great rivers to its south, its course being shorter and its basin, in this higher-latitude area, much smaller.

In the interior, note the Loess Plateau located south of the Ordos Desert, where the Huang He (Yellow River) makes its giant loop. Glacial action pulverized

HISTORICAL GEOGRAPHY

Hominid and human histories in East Asia are lengthy and complex. Many archeological sites in this realm have yielded evidence of *Homo erectus*, including perhaps the most famous of all: Peking Man, found in a cave not far from Beijing in the 1920s. Current anthropological theory holds that *Homo sapiens* arrived in East Asia between 60,000 and 40,000 years ago and eliminated the hominids in short order. A minority of anthropologists theorize that modern humanity developed from four stocks in widely separated parts of the world, one of which was East Asia. Thus today's East Asians would trace their ancestry to Peking Man and beyond.

Early Cultural Geography

Whatever the outcome of the debate over human origins in East Asia, it is clear that humans have inhabited the plains and river basins, foothills, and islands of this realm for a very long time. Hunting sustained both the hominids and the early human communities; fishing drew them to the coasts and onto the islands. The first crossing into Japan may have occurred as long as 10,000 to 12,000 years ago, possibly much earlier, when the Jomon people, a Caucasoid population of uncertain geographic origins, entered the islands; their modern descendants, the Ainu, spread throughout the archipelago. Today, only about 20,000 persons living in northernmost Hokkaido trace their ancestry to Ainu sources.

About 2300 years ago the Yayoi people, rice farmers who had settled in Korea, appear to have crossed by boat to Kyushu, Japan's southernmost island, from where they advanced northward. The Ainu, who subsisted by fishing, trapping, and hunting, were driven back, but gene-pool studies show that much mixing of the groups took place; they also show that the Yayoi invasion was followed by other incursions from the Asian mainland. By then, powerful dynastic states had already arisen in what is today China, and early Chinese culture traits thus found their way into Japan through the process we know as *relocation diffusion*.

On the Asian mainland, plant and animal domestication had begun as early as anywhere on Earth. We in the Western world take it for granted that these momentous processes began in what we now call the Middle East and diffused from the Fertile Crescent to other parts of Eurasia and the rest of the world. But the taming of animals and the selective farming of plants may have begun as early, or earlier, here in East Asia. As in Southwest Asia, the fertile alluvial soils of the great river basins and the ebb and flow of stream water created an environment of opportunity, and millet and rice were being harvested between 7000 and 8000 years ago.

Even during this Neolithic period of increasingly sophisticated stone tools, East Asia was a mosaic of regional cultures. Their differences are revealed by the tools they made and the decorations on their bowls, pots, and other utensils. An especially important discovery of two 8000-year-old pots in the form of a silkworm cocoon, from China's Hebei Province, suggests a very ancient origin for one of the region's leading historic industries.

As noted earlier, plant and animal domestication produced surpluses and food storage, enabling population growth and requiring wider regional organization. Here as elsewhere during the Neolithic, settlements expanded, human communities grew more complex, and power became concentrated in a small group, an *elite*.

This process of *state formation* is known to have occurred in only a half dozen regions of the world, and China was one of these. But evidence about China's earliest states has long been scarce. Today, however, archeologists are focusing on the lower Yi-Luo River Valley in the western part of Henan Province, where the first documented Chinese dynasty, the Xia Dynasty (2200–1770 BC) existed. The capital of this ancient state, Erlitou, has been found, and archeologists now refer to the Xia Dynasty as the Erlitou culture. Secondary centers are being discovered, and Erlitou tools and implements in a wider area prove that the Xia Dynasty represents a substantial state.

All early states were ruled by elites, but China's political history is chronicled in *dynasties* because here the succession of rulers came from the same line of descent, sometimes enduring for centuries. In the transfer of power, family ties counted for more than anything else. Dynasties were overthrown, but the victors did not change this system. Dynastic rule lasted into the twentieth century.

The Xia Dynasty may have been the earliest Chinese state, but it lay in the area where, later, more powerful dynastic states arose: the North China Plain. Here the tenets of what was to become Chinese society were implanted early and proved to be extremely durable. From this culture hearth ideas, innovations, and practices diffused far and wide. From agriculture to architecture, poetry to porcelain making, influences radiated southward into Southeast Asia, westward into interior Asia, and eastward into Korea and Japan. In the North China Plain lay the origins of what was to become the Middle Kingdom, which its citizens considered the center of the world.

Dynastic China

If the Xia period was indeed China's first dynasty, dynastic rule in this part of East Asia lasted 4000 years, ending only in 1911 when the last emperor of the Qing (Manchu) Dynasty, a six-year-old boy, was forced to abdicate the Chinese throne. As Figure 9-3 shows, the Chinese sphere expanded quite rapidly, reaching its greatest extent during that last dynasty (there also were times when China experienced competing dynasties, internal division, and temporary losses). The lower Huang He (Yellow River) and the Wei basins were the original heartland of China during the *Shang* Dynasty (ca. 1766–ca. 1080 BC), but the following dynasty, the *Zhou* Dynasty (ca. 1027 BC–221 BC) was crucial. Taoism arose, Buddhism arrived, and Confucius (Kongfuzi in the modern Pinyin spelling) produced what was to become China's guiding philosophy for more than 2000 years.

The *Han* Dynasty (206 BC–AD 220), however, was in many ways China's formative period: enormous territorial expansion accompanied a flowering of Chinese culture. The Han Empire had about the same territorial dimensions as the Roman Empire, which existed about the same time; Xian, the "Rome of China," was one of the world's greatest

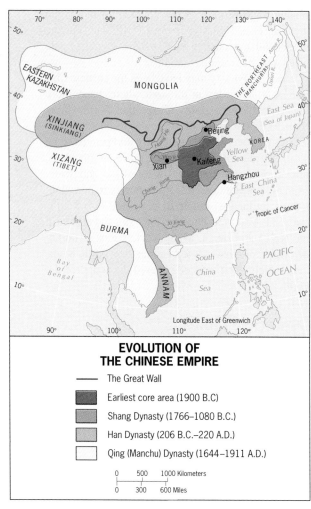

**EVOLUTION OF
THE CHINESE EMPIRE**

——— The Great Wall

Earliest core area (1900 B.C)

Shang Dynasty (1766–1080 B.C.)

Han Dynasty (206 B.C.–220 A.D.)

Qing (Manchu) Dynasty (1644–1911 A.D.)

0   500   1000 Kilometers

0   300   600 Miles

FIGURE 9-3

cities. This was the time of the Silk Route, of penetrations into Central and Southeast Asia, of authoritarian government and disciplined armies. To this day, the people in China refer to themselves as the *People of Han.*

After the Han Dynasty, China went into one of its periods of division and decline, but revival began with the brief *Sui* Dynasty (581–618) and reached a golden age during the *Tang* Dynasty (618–907), when Xian was China's cultural capital and the largest city in the world, the Silk Route was loaded with trade, Arab and Persian seafarers were arriving in Chinese ports, and Chinese influences penetrated Korea and Japan. But the glorious Tang Dynasty, too, was followed by a period of political instability until the Northern and Southern *Song* Dynasties (960–1279) brought a momentous development: the invasion and takeover by the Mongols and Kublai Khan. This was the prelude to the Mongol-dominated *Yuan* Dynasty (1264–1368), when the northern city, Beijing, grew into a major metropolis. The Mongols ruled, but instead of converting the Chinese to Mongol ways it was the Mongol rulers who adopted Chinese culture.

Eventually, Mongol power waned, and now another indigenous Chinese dynasty assumed control: the *Ming* Dynasty (1368–1644). During this era, China made major advances in science and technology; great oceangoing ships sailed in huge fleets into the Indian Ocean and reached East Africa long before the Europeans in their much smaller vessels did. Farming expanded, and with it population—until climate change caused famines and havoc, the fleets

were burned, and the Ming rulers turned isolationist and ineffective.

This created an opportunity for another group of outsiders to establish what turned out to be China's final dynasty: the Manchu, a people with Tatar links living in present-day Northeast China. The *Manchu (Qing)* Dynasty (1644–1911) began when this minority of about 1 million seized control of the nation of several hundred million by taking power in Beijing. They retained the Ming system of administration and kept Ming officials in office, and began a campaign of territorial expansion that created the largest China-centered empire ever (Fig. 9-3), including Mongolia, much of Turkestan, Xizang (Tibet), Myanmar (Burma), Indochina, Korea, and Taiwan. But the Qing rulers had the misfortune of being in charge when the European powers and Japan arrived in force; revolution and collapse of the dynasty followed.

As Figure 9-3 suggests, the growth and expansion of dynastic China was the dominant and formative process in this realm. Many of the territorial acquisitions made during the Qing Dynasty form part of the modern Chinese state today; others are historic justifications for potential claims on "lost" areas such as the Russian Far East and Mongolia. Large Chinese minorities now reside in Southeast Asian countries, an emigration that began during the Qing Dynasty and changed the social and political landscape of colonial as well as modern Southeast Asia. Only Japan, while influenced by Chinese tenets, escaped incorporation into China's East Asian sphere throughout its history.

REGIONS OF THE REALM

East Asia presents us with an opportunity to illustrate the changeable nature of regional geography. Our regional delimitation is based on current circumstances, and it predicts ways the framework may change. It is anything but static. At the beginning of the twenty-first century, we can identify five geographic regions in the East Asian realm (Fig. 9-4). These are:

1. *China Proper.* Almost any map of China's human geography—population distribution, urban centers, surface communications, agriculture, industry—emphasizes the strong concentration of Chinese activity in the country's eastern sector. This is the "real" China, where its great cities, populous farmlands, and historic sources are located. Long ago, scholars called this *China Proper,* and it is a good

regional designation. But China is a large and complex country, and a number of subregions are nested within China Proper. Some of these, such as the North China Plain and the Sichuan Basin, are old and well-established geographic units. One in particular is new: China's Pacific Rim, still growing, yet poorly defined, and shown as "formative" in Figure 9-4.

2. *Xizang* (*Tibet*). The high mountains and plateaus of Xizang, ruled by China but still widely known by its older name of Tibet, form a stark contrast to teeming China Proper. Here, next to one of the world's largest and most populous regions, lies one of the emptiest and, in terms of inhabited space, smallest regions.

3. *Xinjiang*. The vast desert basins and encircling mountains of Xinjiang form a third East Asian region. Again, physical as well as human geographic criteria come into play: here China meets Islamic Central Asia.

4. *Mongolia*. The desert state of Mongolia forms East Asia's fourth region. Like Tibet, landlocked Mongolia, vast but sparsely peopled, stands in stark contrast to populous China Proper.

5. *Jakota Triangle*. East Asia's fifth region is defined by its economic geography. *Ja*pan, South *Ko*rea, and *Ta*iwan (the name "Jakota" derives from the first two letters of each) were transformed by Pacific Rim economic developments during the second half of the twentieth century. Its recent emergence foreshadows further changes in the decades ahead as Korean unification becomes a possibility and contrasts between the Jakota Triangle and Pacific Rim China diminish.

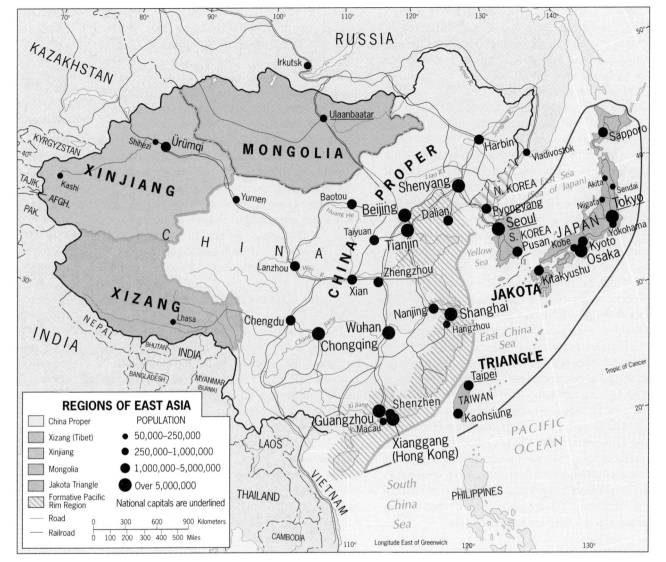

FIGURE 9-4

CHINA PROPER

When we in the Western world chronicle the rise of civilization, we tend to focus on the historical geography of Southwest Asia, the Mediterranean, and Western Europe. Ancient Greece and Rome were the crucibles of culture; Mediterranean and Atlantic wa-

ters were the avenues of its diffusion. China lay remote, so we believe, barely connected to this Western realm of achievement and progress. When an Italian adventurer named Marco Polo visited China during the thirteenth century and described the marvels he saw there, his work did little to change European minds. Europe was and would always be the center of civilization.

The Chinese, naturally, take a different view. Events on the western edge of the great Eurasian landmass were deemed irrelevant to theirs, the most advanced and refined culture on Earth. Roman emperors were rumored to be powerful, and Rome was a great city, but nothing could match the omnipotence of China's rulers. Certainly the Chinese city of Xian far eclipsed Rome as a center of sophistication.

Photo Gallery   More to Explore   **20**   Concepts, Ideas, and Terms   **GEODISCOVERIES**   Geodiscoveries Module

Chinese civilization existed long before ancient Greece and Rome emerged, and it was still there long after they collapsed. China, the Chinese teach themselves, is eternal. It was, and always will be, the center ✳ of the civilized world.

We should remember this notion when we study China's regional geography because 4000 years of Chinese culture and perception will not change overnight—not even in a generation. Time and again, China overcame the invasions and depredations of foreign intruders, and afterward the Chinese would close off their vast country against the outside world. Just 30 years ago, in the early 1970s, there were just a few *dozen* foreigners in the entire country with its (then) nearly 1 billion inhabitants. The institutionalization of communism required this insularity, and even the Soviet advisors had been thrown out. But by the early 1970s, China's rulers decided that an opening to the Western world would be advantageous, and so U.S. President Richard Nixon was invited to visit Beijing. That historic occasion, in 1972, ended this latest period of isolation—as always, on China's terms. Since then, China has been open to tourists and business people, teachers and investors. Tens of thousands of Chinese students have been sent to study at American and other Western institutions. Long-suppressed ideas flowed into China, and the pro-democracy movement arose and climaxed in 1989. China's rulers knew that their violent repression of this movement would anger the world, but that did not matter because they deemed foreign condemnation irrelevant. Foreigners in China had done much worse. Moreover, Westerners had no business interfering in China's domestic affairs.

RELATIVE LOCATION

Throughout their nation's history, the Chinese have at times decided to close their country to any and all foreign influences; the most recent episode of exclusion occurred just a few decades ago. Exclusion is one of China's recurrent traditions, made possible by China's relative location and Asia's physiography. In other words, China's "splendid isolation" was made possible by geography.

Earlier we noted the role of relief and desert in encircling the culture hearth of East Asia, but equally telling is the factor of distance. Until recently, China lay far from the modern source areas of innovation and change. True, China—as the Chinese emphasize—was itself such a hearth, but China's contributions to the outside world remained limited, essentially, to finely made arts and crafts. China did interact with Korea, Japan, Taiwan, and parts of Southeast Asia, and eventually millions of Chinese emigrated to neighboring countries. But compare these regional links to those of the Arabs, who ranged worldwide and who brought their knowledge, religion, and political influence to areas from Mediterranean Europe to Bangladesh and from West Africa to Indonesia. Later, when Europe became the center of intellectual and material innovation, China found itself farther removed, by land or sea, than almost any other part of the world.

Today, modern communications notwithstanding, China still is distant from almost anywhere else on Earth. Going by rail from Beijing to Moscow, the capital of China's Eurasian neighbor, involves a tedious journey that takes the better part of a week. Direct surface connections with India are practically nonexistent. Overland linkages with Southeast Asian countries, though improving, remain tenuous.

But for the first time in its history, China now lies near a world-class hearth of technological innovation and financial power: Japan. This proximity to an industrial and financial giant is critical for the momentous economic developments taking place in China's coastal provinces. Japanese investments and business partnerships have transformed the economic landscape of Pacific-coast China. American and European trade links also are important, but Japan's role was crucial. Japan's economic success set the Pacific Rim engine in motion, and Japan's best financial years happened to coincide with China's reopening to foreigners in the 1970s. Geographic and economic circumstances combined to transform the map and made "Pacific Rim" a household word around the world.

EXTENT AND ENVIRONMENT

China's total area is slightly smaller than that of the United States including Alaska: each country has about 3.7 million square miles (9.6 million sq km). As Figure 9-5 reveals, the longitudinal extent of China and the 48 contiguous U.S. States also is similar. Latitudinally, however, China is considerably wider. Miami, near the southern limit of the United States, lies halfway between Shanghai and Guangzhou. Thus China's lower-latitude southern region takes on characteristics of tropical Asia. In the Northeast, too, China incorporates much of what in North America would be Quebec and Ontario. Westward, China's land area becomes narrower and physiographic similarities increase. But, of course, China has no west coast.

Now compare the climate maps of China and the United States in Figure 9-6 (which are enlargements of the appropriate portions of the world climate map

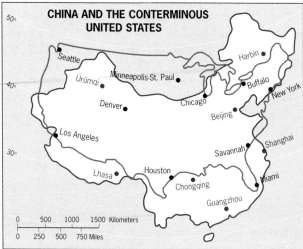

FIGURE 9-5

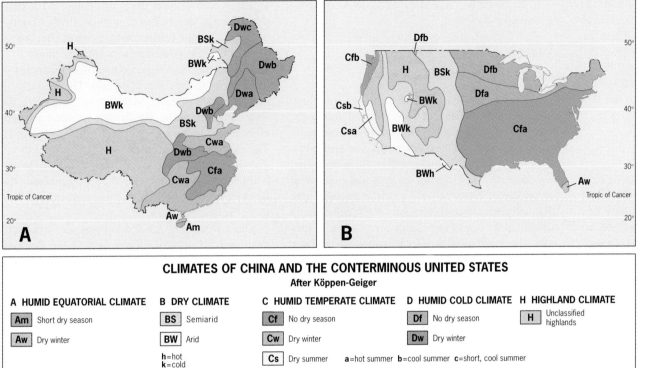

FIGURE 9-6

CLIMATES OF CHINA AND THE CONTERMINOUS UNITED STATES
After Köppen-Geiger

A HUMID EQUATORIAL CLIMATE	B DRY CLIMATE	C HUMID TEMPERATE CLIMATE	D HUMID COLD CLIMATE	H HIGHLAND CLIMATE
Am Short dry season	**BS** Semiarid	**Cf** No dry season	**Df** No dry season	**H** Unclassified highlands
Aw Dry winter	**BW** Arid	**Cw** Dry winter	**Dw** Dry winter	
	h=hot k=cold	**Cs** Dry summer a=hot summer b=cool summer c=short, cool summer		

in Figure I-8). Note that both have a large southeastern climatic region marked *Cfa* (that is, humid, temperate, warm-summer), flanked in China by a zone of *Cwa* (where winters become drier). Westward in both countries, the *C* climates yield to colder, drier climes. In the United States, moderate *C* climates develop again along the Pacific coast. China, however, stays dry and cold as well as high in elevation at equivalent longitudes.

Note especially the comparative location of the U.S. and Chinese *Cfa* areas in Figure 9-6. China's lies much farther to the south. In the United States, the *Cfa* climate extends beyond 40° North latitude, but in China, cold and generally winter-dry *D* climates take over at the latitude of Virginia. Beijing has a warm summer but a bitterly cold and long winter. Northeast China, in the general latitudinal range of Canada's lower Quebec and Newfoundland, is

much more severe than its North American equivalent. Harsh environments prevail over vast regions of China, but, as we will see, nature compensates in spectacular fashion. From the climatic zone marked *H* (for highlands) in the west come the great life-giving rivers whose wide basins contain enormous expanses of fertile soils. Without these waters, China would not have a population more than four times that of the United States.

EVOLVING CHINA

Even if China is not the world's longest continuous civilization (Egypt may claim this distinction), no other state on Earth can trace its cultural heritage as far back as China can. China's fortunes rose and fell, but

over more than 40 centuries its people created a society with strong traditions, values, and philosophies. The teachings of Kongfuzi (551–479 BC), **2** still known as **Confucius**, dominated Chinese life and thought for ✵ over 20 centuries. Kongfuzi not only taught and championed the poor and indigent: his revolutionary ideas extended to the rulers as well as the ruled. He abhorred supernatural mysticism and dismissed notions of divine ancestries of dynastic rulers. Competence and merit, not some godly legacy, he insisted, should determine a person's place in society. He wrote lengthy treatises from which emerged the so-called Confucian Classics, 13 texts that became the basis for education in China until Western influences during the Qing Dynasty began to erode their relevance. Ranging from government to morality and from law to religion, the Classics, especially the revised *Analects*, were Chinese civilization's guide. But the communists, who took power in 1949, attacked Kongfuzi thought on all fronts, substituting indoctrination for education and relegating even the family, one of Kongfuzi's predilections, to the "dustbin of history." Here the communists miscalculated. They were unable to eradicate two millennia of cultural conditioning in a few decades, and even today the spirit of Kongfuzi still haunts China's physical and mental landscapes.

As we try to gain a better understanding of China's present political, economic, and social geography, we should keep its continuities in mind. Yes, China is changing radically today, but it has undergone convulsive change in the past, and has endured and regained its coherence. Larger and more populous than Europe, China, too, had its divisive feudal periods, but unlike Europe, China always came together again under a single flag. In the process, often-dictatorial China incorporated minorities ranging from Koreans and Mongols to Uyghurs and Ti-betans; in the far south, it now includes several peoples with Southeast Asian affinities. Like the Chinese citizenry itself, these minorities have experienced both benevolent government and brutal subjugation. But it has always been China's wish **3** to **Sinicize** them, to endow them with the elements of Chinese culture.

A Century of Convulsion

When the European colonialists appeared in East Asia, China long withstood them with a self-assured superiority based on the strength of its culture and the reassuring continuity of the state. There was no market for the British East India Company's rough textiles in a country long used to finely fabricated silks and cottons. There was little interest in the toys and trinkets the Europeans produced in the hope of barter for Chinese tea and porcelain. Even key European inventions, such as the mechanical clock, though considered amusing and entertaining, were ignored and even deprecated as irrelevant to Chinese culture.

The Ming emperors were particularly dismissive of European manufactures, but the Ming rulers' confidence in their home-made products was sometimes misplaced. When the Manchu forces invaded in 1644, their bows and arrows proved superior to Chinese-manufactured muskets, which were so heavy and difficult to load that they were almost useless. As we know, the Manchu emperors kept Ming administrators in office, and the prevailing attitude continued. Even when Europe's sailing ships made way for steam-driven vessels and newer and better European products (including weapons) were offered in trade for China's tea and silk, China continued to reject European imports and resisted commerce in general. The Chinese kept the Europeans confined to small peninsular outposts, such as Macau, and minimized interaction with them. Long after India had succumbed to mercantilism and economic imperialism, China maintained its established order. This was no surprise to the Chinese. After all, they had held a position of undisputed superiority in their Celestial Kingdom as long as

could be remembered, and they had dealt with foreign invaders before.

A (Lost) War on Drugs and Its Aftermath

The Manchus, however, had the misfortune of ruling China when the balance of power shifted decisively in favor of the colonialists. On two fronts in particular, the economic and the political, the European powers destroyed China's invincibility. Economically, they succeeded in lowering the cost and improving the quality of manufactured goods, especially textiles, and the handicraft industries of China began to collapse in the face of unbeatable competition. Politically, the demands of the British merchants and the growing English presence in China led to conflicts. In the early part of the nineteenth century, the central issue was the importation into China from British India of opium, a dangerous and addictive intoxicant. Opium was destroying the very fabric of Chinese culture, weakening the society, and rendering China easy prey for colonial profiteers. As the Manchu government moved to stamp out the opium trade in 1839, armed hostilities broke out, and soon the Chinese found themselves losing a war on their own territory. The First Opium War (1839–1842) ended in disaster: China's rulers were forced to yield to British demands, and the breakdown of Chinese sovereignty was under way.

British forces penetrated up the Chang Jiang and controlled several areas south of it (Fig. 9-7); Beijing hurriedly sought a peace treaty by which it granted leases and concessions to foreign merchants. In addition, China ceded Hong Kong Island to the British and opened five ports, including Guangzhou (Canton) and Shanghai, to foreign commerce. No longer did the British have to accept a status that was inferior to the Chinese in order to do business; henceforth, negotiations would be pursued on equal terms. Opium now flooded into China, and its impact on Chinese society became even more devastating. Fifteen years after the First Opium War, the Chinese again tried to stem the disastrous narcotic tide, and the foreigners who had attached themselves to their country again defeated them. Now the government

legalized cultivation of the opium poppy in China itself. Chinese society was disintegrating; the scourge of this drug abuse was not defeated until after the revival of Chinese power in the twentieth century.

But before China regained control of its destiny, it lost heavily to the colonial powers. The Germans in 1898 obtained a "lease" on the city of Qingdao on the Shandong Peninsula. The French acquired a sphere of influence in the far south at Zhanjiang (Fig. 9-7). The Portuguese confirmed their hold over Macau. The Russians took control over Liaodong in the Northeast. Even Japan got in the act by annexing the Ryukyu Islands (1879) and, more importantly, Formosa (Taiwan) in 1895.

One of the most humiliating practices the Europeans forced upon the Chinese was the doctrine **4** known as **extraterritoriality**. Under this doctrine, foreign states and their representatives are immune from the jurisdiction of the country in which they are based, for example, in the case of embassies and diplomatic personnel. But in Qing (Manchu) China, it went far beyond that. The European and Japanese invaders established as many as 90 so-called *treaty ports* which were, in effect, extraterritorial enclaves where traders as well as diplomats were exempt from Chinese law. The best residential areas in major cities were declared "extraterritorial" and made inaccessible to Chinese citizens, including Sha Mian Island on the Pearl River waterfront in the city of Canton (now Guangzhou). Christian missionaries fanned out into China, their residences and churches fortified with extraterritorial security. In many places, Chinese citizens found themselves unable to enter parks and buildings without permission from foreigners. This involved a loss of face and built a resentment that exploded in the Boxer Rebellion of 1900.

Figure 9-7 shows the fate of the territory originally acquired by the Manchu rulers. Much of the Northeast was taken by Russia; the Japanese colonized Korea; Mongolia became independent under foreign auspices; large parts of Turkestan were lost, also to Russia; the French took over in Indochina and the British in Burma. But in the early twentieth century, the Chinese had had enough. Bands of revolutionaries roamed city and countryside, attacking the hated foreigners as well as Chinese who had adopted Western

**CHINA:
COLONIAL SPHERES,
TERRITORIAL LOSSES**

⎯⎯⎯ Manchu Dynasty at its greatest extent

⟵ The Long March, 1934–1935

⎯⎯⎯ Administered by Japan, 1937–1945

▨ Territorial losses 19th and 20th Century

–·–·– People's Republic of China (1949–)

19TH CENTURY COLONIAL INFLUENCE

Russian   British   French   German

0   300   600   900 Kilometers
0 100 200 300 400 500 Miles

FIGURE 9-7

ways. The Boxer Rebellion (after a loose translation of the Chinese name for these revolutionaries) was put down with much bloodshed by what today would be called a "multinational force": British, French, German, Italian, Russian, Japanese, and American soldiers participated. But at the same time another revolutionary movement was gaining support, aimed against the Qing Dynasty itself. In 1911, the emperor's garrisons were attacked all over China, and in a few months the 267-year-old dynasty was overthrown.

The so-called Nationalist forces that ousted the last emperor proclaimed a republican China and negotiated an end to the extraterritorial treaties, but China remained a badly divided country. Even while the famous Chinese revolutionary leader Sun Yat-sen tried to unify the country from his base in Canton (Guangzhou) in the south, another government tried to rule from Beijing, the old imperial headquarters. Meanwhile, a group of intellectuals in Shanghai founded the Chinese Communist Party. A prominent member of this group was a man named Mao Zedong.

Nationalists and Communists

During the chaotic 1920s the Nationalists and the Communist Party at first cooperated, with the remaining foreign presence their joint target. After Sun Yat-sen's death in 1925, Chiang Kai-shek became the Nationalists' leader, and by 1927 the foreigners were on the run, escaping by boat and train or falling victim to rampaging Nationalist forces. But soon the Nationalists began purging communists even as they pursued foreigners, and in 1928, when Chiang established his Nationalist capital in the city of Nanjing, it appeared that the Nationalists would emerge victorious from their campaigns. They had driven the communists ever deeper into the interior, and by 1933 the Nationalist armies were on the verge of encircling the last communist stronghold in the area of Ruijin in Jiangxi Province.

This led to a momentous event in Chinese history: the *Long March*. Nearly 100,000 people—soldiers, peasants, leaders—marched westward from Ruijin in 1934, a communist column that included Mao Zedong and Zhou Enlai. The Nationalist forces rained attack after attack on the marchers, and of the original 100,000, about three-quarters were killed. But new sympathizers joined along the way (see the route marked on Fig. 9-7), and the 20,000 survivors found a refuge in the mountainous interior of Shaanxi Province, 6000 miles (10,000 km) away. There, they prepared for a renewed campaign that would bring them to power.

Japan in China

Although many foreigners fled China during the 1920s and 1930s, others seized the opportunity presented by the contest between the Nationalists and communists. The Japanese took control over the Northeast, and when the Nationalists proved unable to dislodge them, they set up a puppet state there, appointed a Manchu ruler to represent them, and called their possession Manchukuo.

The inevitable full-scale war between the Chinese and the Japanese broke out in 1937, with the Nationalists bearing the brunt of it (which gave the communists an opportunity to regroup). The gray boundary on Figure 9-7 shows how much of China the Japanese conquered. The Nationalists moved their capital to Chongqing, and the communists controlled the area centered on Yanan. China had been broken into three pieces.

The Japanese committed unspeakable atrocities in their campaign in China. Millions of Chinese citizens were shot, burned, drowned, subjected to gruesome chemical and biological experiments, and otherwise wantonly victimized. Years later, when China's economic reforms of the 1980s and 1990s led to a renewed Japanese presence in China, the Chinese public and its leaders called for Japan to acknowledge and apologize for these wartime abuses. In Japan, this pitted apologists against strident nationalists, causing a political crisis. In 1992, Emperor Akihito visited China and referred to the war but stopped short of a formal apology. The book is not yet closed on this most sensitive issue.

Communist China Arises

After the U.S.-led Western powers defeated Japan in 1945, the civil war in China quickly resumed. The United States, hoping for a stable and friendly government in China, sought to mediate the conflict but at the same time recognized the Nationalists as the legitimate government. The United States also aided the Nationalists militarily, destroying any chance of genuine and impartial mediation. By 1948, it was clear that Mao Zedong's well-organized militias would defeat Chiang Kai-shek. Chiang kept moving his capital—back to Guangzhou, seat of Sun Yat-sen's first Nationalist government, then back to Chongqing. Late in 1949, after a series of disastrous defeats in which hundreds of thousands of Nationalist forces were killed, the remnants of Chiang's faction gathered Chinese treasures and valuables and fled to the island of Taiwan. There, they took control of the government and proclaimed their own Republic of China.

Meanwhile, on October 1, 1949, standing in front of the assembled masses at the Gate of Heavenly Peace on Beijing's Tiananmen Square, Mao Zedong proclaimed the birth of the People's Republic of China.

CHINA'S HUMAN GEOGRAPHY

After more than a half-century of communist rule, China is a society transformed. It has been said that the year 1949 actually marked the beginning of a new dynasty not so different from the old, an autocratic system that dictated from the top. In that view, Mao Zedong simply bore the mantle of his dynastic predecessors. Only the family lineage had fallen away; now communist "comrades" would succeed each other.

And certainly some of China's old traditions continued during the communist era, but in many other ways Chinese society was totally overhauled. Benevolent or otherwise, the dynastic rulers of old China headed a country in which—for all its splendor, strength, and cultural richness—the fate of landless people and of serfs often was undescribably miserable; in which floods, famines, and diseases could decimate the populations of entire regions without any help from the state; in which local lords could (and often did) repress the people with impunity; in which children were sold and brides were bought. The European intrusion made things even worse, bringing slums, starvation, and deprivation to millions who had moved to the cities.

The communist regime, dictatorial though it was, attacked China's weaknesses on many fronts, mobilizing virtually every able-bodied citizen in the process. Land was taken from the wealthy; farms were collectivized; dams and levees were built with the hands of thousands; the threat of hunger for millions receded; health conditions improved; child labor was reduced. But China's communist planners also made terrible mistakes. The *Great Leap Forward*, requiring the reorganization of the peasantry into communal brigade teams to speed industrialization and make farming more productive, had the opposite effect and so disrupted agriculture that between 20 and 30 million people died of starvation between 1958, when the program was implemented, and 1962, when it was abandoned.

Mao ruled China from 1949 to 1976, long enough to leave lasting marks on the state. Another of his communist dictums had to do with population. Like the Soviets (and influenced by a horde of Soviet advisors and planners), Mao refused to impose or even recommend any population policy, arguing that such a policy would represent a capitalist plot to constrain

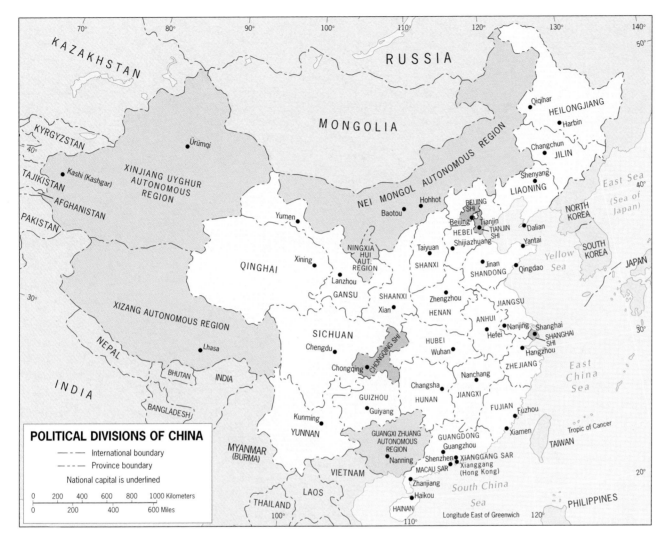

POLITICAL DIVISIONS OF CHINA

— · — · — International boundary
- - - - - Province boundary
National capital is underlined

0 200 400 600 800 1000 Kilometers
0 200 400 600 Miles

FIGURE 9-8

disastrous: Red Guard factions took to fighting among themselves, and anarchy, terror, and economic paralysis followed. Thousands of China's leading intellectuals died, moderate leaders were purged, and teachers, elderly citizens, and older revolutionaries were tortured to make them confess to "crimes" they did not commit. As the economy suffered, food and industrial production declined. Violence and famine killed as many as 30 million people as the Cultural Revolution spun out of control. One of those who survived was a Communist Party leader who had himself been purged and then been reinstated—Deng Xiaoping. Deng was destined to lead the country in the post-Mao period of economic transformation.

Political and Administrative Divisions

Before we investigate the emerging human geography of contemporary China, we should acquaint ourselves with the country's political and administrative framework (Fig. 9-8). For administrative purposes, China is divided into the following units:

4 Central-government-controlled Municipalities (*Shi's*)
5 Autonomous Regions
22 Provinces
2 Special Administrative Regions

The four central-government-controlled municipalities are the capital, Beijing; its nearby port city, Tianjin; China's largest metropolis, Shanghai; and the Chang River port of Chongqing, in the interior. These *Shi's* form the cores of China's most populous and important subregions, and direct control over them from the capital entrenches the central government's power.

China's human resources. As a result, China's population grew explosively during his rule.

Yet another costly episode of Mao's rule was the so-called *Great Proletarian Cultural Revolution*, launched by Mao Zedong during his last decade in power (1966–1976). Fearful that Maoist communism was being contaminated by Soviet "deviationism" and worried about his own stature as its revolutionary architect, Mao unleashed a campaign against what he

viewed as emerging elitism in society. He mobilized young people living in cities and towns into cadres known as Red Guards and ordered them to attack "bourgeois" elements throughout China, criticize Communist Party officials, and root out "opponents" of the system. He shut down all of China's schools, persecuted untrustworthy intellectuals, and encouraged the Red Guards to engage in what he called a renewed "revolutionary experience." The results were

We should note that the administrative map of China continues to change—and to pose problems for geographers. The city of Chongqing was made a *shi* in 1996, and its "municipal" area was enlarged to incorporate not only the central urban area but a huge hinterland covering all of eastern Sichuan Province. As a result, the "urban" population of Chongqing is officially 30 million, making this the world's largest metropolis—but in truth, the central urban area has no more than about 6 million inhabitants. And because Chongqing's population is officially *not* part of the province that borders it to the west (Fig. 9-8), the official population of Sichuan declined by 30 million when the *Chongqing Shi* was created.

The five Autonomous Regions were established to recognize the non-Han minorities living there. Some laws that apply to Han Chinese do not apply to certain minorities. As we saw in the case of the former Soviet Union, however, demographic changes and population movements affect such regions, and the policies of the 1940s may not work in the twenty-first century. Han Chinese immigrants now outnumber several minorities in their own regions. The five Autonomous Regions (A.R.'s) are: (1) Nei Mongol A.R. (Inner Mongolia); (2) Ningxia Hui A.R. (adjacent to Inner Mongolia); (3) Xinjiang Uyghur A.R. (China's northwest corner); (4) Guangxi Zhuang A.R. (far south, bordering Vietnam); and (5) Xizang A.R. (Tibet).

China's 22 Provinces, like U.S. States, tend to be smallest in the east and largest toward the west. The territorially smallest are the three easternmost provinces on China's coastal bulge: Zhejiang, Jiangsu, and Fujian. The two largest are Qinghai, flanked by Tibet, and Sichuan, China's Midwest.

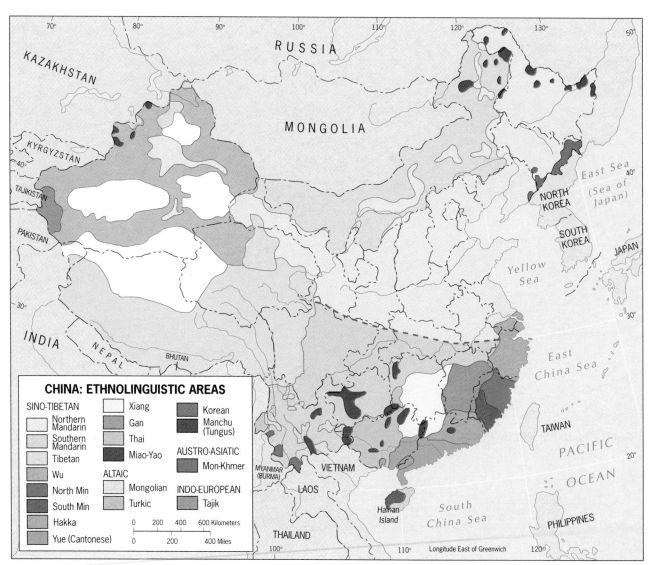

FIGURE 9-9

As with all large countries, some provinces are more important than others. The Province of Hebei nearly surrounds Beijing and occupies much of the core of the country. The Province of Shaanxi is centered on the great ancient city of Xian. In the southeast, momentous economic developments are occurring in the Province of Guangdong, whose urban focus is Guangzhou. When, in the pages that follow, we refer to a particular province or region, Figure 9-8 is a useful locational guide.

In 1997, the British dependency of Hong Kong (Xianggang) was taken over by China and became **5** the country's first **Special Administrative Region (SAR)**. In 1999, Portugal similarly transferred

Macau, opposite Hong Kong on the Pearl River estuary, to Chinese control, creating the second SAR under Beijing's administration.

Population Issues

Many Chinese provinces, like many of India's States, have populations larger than most of the world's countries. With Chongqing Shi, Sichuan Province has more than 120 million inhabitants. Approaching 100 million are Henan and Shandong Provinces (see Fig. 9-8). There are almost as many people in Guangdong Province as in Germany. Jiangsu Province has 20 million more inhabitants than France.

With a population of over 1.3 billion in 2002, China is the largest nation on Earth, and it inherited from the communist period a high rate of natural increase. Aware of the economic costs of rapid population growth, China after Mao's death embarked on a vigorous population-control program. In the early 1970s, the annual rate of natural increase was about 3 percent; by the mid-1980s, it was down to 1.2 percent. Families were ordered to have one child only, and those who violated the policy were penalized by losing tax advantages, educational opportunities, and even housing privileges. Today, China's census bureau officially reports a growth rate of 0.9 percent.

The Minorities

When the Soviet Union collapsed, some observers described that event as the "end of empire"—the breakdown of the last empire on Earth. But those observers forgot about China. Diminished from its Qing Dynasty dimensions as China may be, the Beijing government controls territories and peoples that are in effect colonized. The ethnolinguistic map (Fig. 9-9) apparently confirms this proposition. This map should be seen in context, however. It does not show local-area majority populations but instead reveals where minorities are concentrated. For example, the Mongolian population in China is shown to be clustered along the southeastern border of Mongolia in the Autonomous Region called Nei Mongol (see Fig.

9-8). But even in that A.R., the Han Chinese, not the Mongols, are now in the majority.

Nevertheless, the map gives definition to the term *China Proper* as the home of the People of Han, the ethnic (Mandarin-speaking) Chinese depicted in tan and light orange in Figure 9-9. From the upper Northeast to the border with Vietnam and from the Pacific coast to the margins of Xinjiang, this Chinese majority dominates. When you compare this map to that of population distribution (to be discussed shortly), it will be clear that the minorities constitute only a small percentage of the country's total. The Han Chinese form the largest and densest clusters.

In any case, China controls non-Chinese areas that are vast, if not populous. The Tibetan group numbers under 3 million, but it extends over all of settled Xizang. Turkic peoples inhabit large areas of Xinjiang. Thai, Vietnamese, and Korean minorities also occupy areas on the margins of Han China. As we will see later, the Southeast Asian minorities in China have participated strongly in the Pacific Rim developments on their doorsteps. Hundreds of thousands have migrated from their Autonomous Regions to the economic opportunities along the coast.

Numerically, Chinese dominate in China to a far greater degree than Russians dominated their Soviet Empire. But territorially, China's minorities extend over a proportionately larger area. The Ming and Manchu rulers bequeathed the People of Han an empire.

PEOPLE AND PLACES OF CHINA PROPER

A map of China's population distribution (Fig. 9-10) reveals the continuing relationship between the physical stage and its human occupants. In technologically advanced countries, we have noted, people shake off their dependence on what the land can provide; they cluster in cities and in other areas of economic opportunity. This depopulates rural areas that may once have been densely inhabited. In China, that stage has not yet been reached. While China has large cities, as in India the great majority of the peo-

ple (69 percent) still live on—and from—the land. Thus the map of population distribution reflects the livability and productivity of China's basins, lowlands, and plains. Compare Figures 9-6A and 9-10, and China's continuing dependence on soil, water, and warmth will be evident.

The population map also suggests that in certain areas environmental limitations are being overcome. Industrialization in the Northeast, irrigation in the Inner Mongolia Autonomous Region, and oilwell drilling in Xinjiang enabled millions of Chinese to migrate from China Proper into these frontier zones, where they now outnumber the indigenous minorities.

Nevertheless, physiography and demography remain closely related in China. To grasp this relationship, it is useful to compare Figures 9-2 (physiography) and 9-10 (population) as our discussion proceeds. On the population map, the darker the color, the denser the population: in places we can follow the courses of major rivers through this scheme. Look, for example, at China's Northeast. A ribbon of population follows the Liao and the Songhua rivers. Also note the huge, nearly circular population concentration on the western edge of the red zone; this is the Sichuan Basin, in the upper course of the Chang Jiang. Four major river basins contain more than three-quarters of China's nearly 1.3 billion people:

1. The Liao-Songhua Basin or the Northeast China Plain
2. The Lower Huang He (Yellow River) Basin, known as the North China Plain
3. The Upper and Lower Basins of the Chang Jiang (Yangzi River)
4. The Basins of the Xi (West) and Pearl River

Not all the people living in these river plains are farmers, of course. China's great cities also have developed in these populous areas, from the industrial centers of the Northeast to the Pacific Rim upstarts of the South. Both Beijing and Shanghai lie in major river basins. And, as the map shows, the hilly areas between the river basins are not exactly sparsely

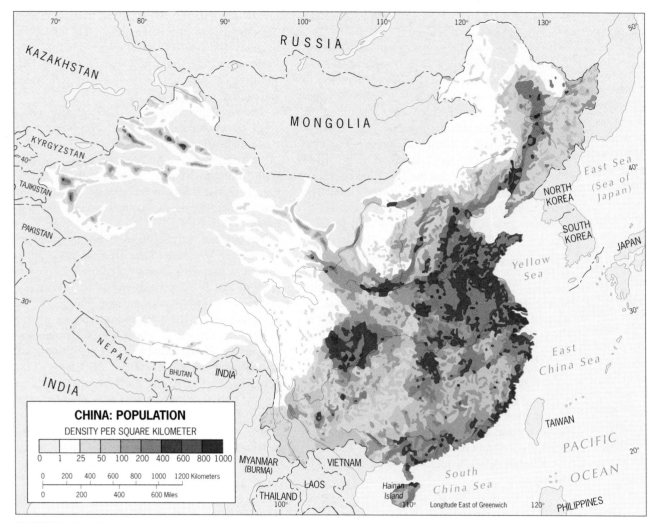

FIGURE 9-10

CHINA: POPULATION

DENSITY PER SQUARE KILOMETER

0   1   25   50   100   200   400   600   800   1000

gional infrastructure. (At one time, half of the entire railroad mileage in China was in the Northeast.) After the Japanese were ousted, the Soviets looted the area of machinery, equipment, and other goods. During the late 1940s, the Northeast was a ravaged frontier. But then the communists took power and made the industrial development of the Northeast a priority. From the 1950s until the 1970s, the Northeast led the nation in manufacturing growth. Its population, just a few million in the 1940s, mushroomed to more than 100 million. Towns and cities grew exponentially.

All this growth was based on the Northeast's considerable mineral wealth (see the symbols in Fig. 9-11), and for a time, it worked. During the 1970s, the Northeast contributed fully one-quarter of the entire country's industrial output.

6 But then came the **economic restructuring** of the post-Mao era, resulting from a new, market-driven economic order. The huge, inefficient, state-supported industries could not adapt, and in short order the Northeast became a Rustbelt, its contribution to China's industrial output below 10 percent. Still, as the map shows, the Northeast is not only a storehouse of resources but has a relative location (to a future-unified Korea, to a future-thriving Russian Far East) that may yet produce still another era of economic prosperity.

populated either. Note that the hill country south of the Chang Basin (opposite Taiwan) still has a density of 130 to 260 people per square mile (50 to 100 per sq km).

Northeast China Plain

The Northeast China Plain is the heartland of China's Northeast, ancestral home of the Manchus who founded China's last dynasty, battleground between Japanese and Russian invaders, Japanese colony,

heartland of communist industrial development, and now losing ground to the industries of the Pacific Rim. This used to be called Manchuria. As the administrative map (Fig. 9-8) shows, there are three provinces here: Liaoning in the south, facing the Yellow Sea; Jilin in the center; and Heilongjiang, by far the largest, in the north.

The Northeast has experienced many ups and downs. Although Japanese colonialism was ruthless and exploitive, Japan did build railroads, roads, bridges, factories, and other components of the re-

North China Plain

The North China Plain is one of the world's most heavily populated agricultural areas. Figure 9-10 shows that most of it has a density of more than 1000 people per square mile (400 per sq km), and in some parts the density is twice as high. Here, the ultimate hope of the Beijing government lay less in land redis-

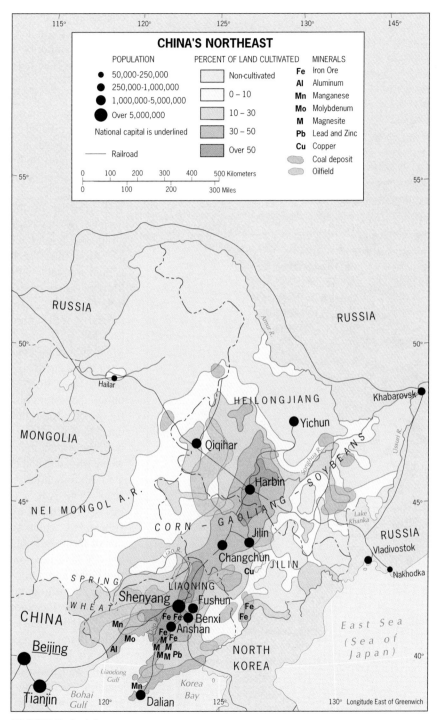

FIGURE 9-11

tribution than in raising yields through improved fertilization, expanded irrigation facilities, and the more intensive use of labor. A series of dams on the Huang River, including the Xiaolangdi Dam upstream in Henan Province, now reduce the flood danger, but outside the irrigated areas the ever-present problem of rainfall variability and drought persists. The North China Plain has not produced any substantial food surplus even under normal circumstances; thus when the weather turns unfavorable the situation soon becomes precarious. The specter of famine may have receded, but the food situation is still uncertain in this very critical part of China Proper.

The North China Plain is an excellent example of **7** the concept of a national **core area**. Not only is it a densely populated, highly productive agricultural zone, but it also is the site of the capital and other major cities, a substantial industrial complex, and several ports, among which Tianjin ranks as one of China's largest (Fig. 9-12). Tianjin, on the Bohai Gulf, is linked by rail and highway (less than a two-hour drive) to Beijing. Like many of China's harbors, that of Tianjin's river port is not particularly good; but Tianjin is well situated to serve not only the northern sector of the North China Plain and the capital, but also the Upper Huang Basin and Inner Mongolia beyond (see Fig. 9-1). Tianjin, again like several other Chinese ports, had its modern start as a treaty port, but the city's major growth awaited communist rule. For decades it was a center for light industry and a flood-prone harbor, but after 1949 the communists constructed a new artificial port and flood canals. They also chose Tianjin as a site for major industrial development and made large investments in the chemical industry (in which Tianjin still leads China), iron and steel production, heavy machine manufacturing, and textiles. Today, with a population of 9.3 million, Tianjin is China's fourth largest metropolis and the center of one of its leading industrial complexes.

✳ Beijing, unlike Tianjin, is China's political, cultural, and educational center. Its industrial development has not matched Tianjin's. The communist administration did, however, greatly expand the municipal area of Beijing, which (as we noted), is not

FIGURE 9-12

controlled by the Province of Hebei but is directly under the central government's authority. In one direction, Beijing was enlarged all the way to the Great Wall—30 miles (50 km) to the north—so that the "urban" area includes hundreds of thousands of farmers. Not surprisingly, this has circumscribed an enormous total population, enough to rank Beijing, with its 10.9 million inhabitants, among the world's larger megacities.

Northwest of the core area as defined by the North China Plain, along the border with the state of Mongolia, lies Inner Mongolia, administratively defined as the Nei Mongol Autonomous Region (see Fig. 9-8). Originally established to protect the rights of the approximately 5 million Mongols who live outside the Mongolian state, Inner Mongolia has been the scene of massive immigration by Han Chinese. Today Chinese outnumber Mongols here by nearly four to one. Near the Mongolian border, Mongols still traverse the steppes with their tents and herds, but elsewhere irrigation and industry have created an essentially Chinese landscape. The A.R.'s capital, Hohhot, has been eclipsed by Baotou on the Huang He, which supports a corridor of farm settlements as it crosses the dry land here on the margins of the Gobi and Ordos deserts. With about 29 million inhabitants, Inner Mongolia still cannot compare to the huge numbers that crowd the North China Plain, but recent mineral discoveries have boosted industry in Baotou and livestock herding is expanding. Nei Mongol may retain its special administrative status, but it functions as part of Han China in all but name.

Basins of the Chang/Yangzi

In contrast to the contiguous, flat agricultural-urban-industrial North China Plain defined by the sediment-laden Huang River, the basins and valleys of the Chang Jiang (Long River) display variation in elevation and relief. The Chang River, whose lower-course name becomes the Yangzi, is an artery like no other in China. Near its mouth lies the  country's largest city, Shanghai. Part of its middle course is now being transformed by a gigantic engineering project. Farther upstream, the Chang crosses the populous, productive Sichuan Basin. And unlike the Huang, the Chang Jiang is navigable to oceangoing ships for over 600 miles (1000 km) from the coast all the way to Wuhan (Fig. 9-12). Smaller ships can reach Chongqing, even after dam construction. Several of the Chang River's tributaries also are navigable, so that 18,500 miles (30,000 km) of water transport routes serve its drainage basin. Thus the Chang Jiang constitutes one of China's leading transit corridors. With its tributaries it handles the trade of a vast area, including nearly all of middle China and sizeable parts of the north and south. The North China Plain may be the core area of China, but in many ways the Lower Chang Basin is its heart.

As Figure 9-13 shows, the Lower Chang Basin is an area of both rice and wheat farming, offering further proof of its pivotal situation between south and north in the heart of China Proper. ✦ Shanghai lies at the coastal gateway to this productive region on a small tributary of the Chang (now Yangzi) River, the Huangpu. The city has an immediate hinterland of some 20,000 square miles (50,000 sq km)—half the size of Ohio—containing more than 50 million people. About two-thirds of this population are farmers who

produce food, silk filaments, and cotton for the city's industries.

Travel upriver along the Yangzi/Chang Jiang, and you meet an unending stream of vessels large and small, including numerous "barge trains"—as many as six or more barges pulled by a single tug in an effort to save fuel (they are slow, but time is not the primary concern). The traffic between Shanghai and Wuhan—Wuhan is short for Wuchang, Hanyang, and Hankou, a coalescing conurbation of three cities—makes the Yangzi one of the world's busiest waterways. Smoke-belching boat engines create a more or less permanent plume of pollution here, worsening

the regional smog created by the factories of Nanjing and others perched on the waterfront. Here China's Industrial Revolution is in full gear, with all its environmental consequences.

Above Wuhan the river traffic dwindles because the depth of the Chang reduces the size of vessels that can reach Yichang. River boats carry coal, rice, building materials, barrels of fuel, and many other items of trade. But the middle course of the Chang River is becoming more than a trade route. In the northward bend of the great river between Yichang and Chongqing, where the Chang has cut deep troughs on its way to the coast, one of the world's

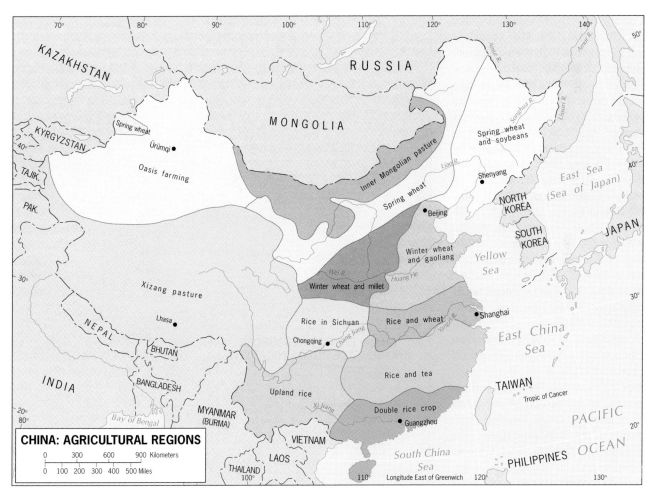

FIGURE 9-13

largest engineering projects is in progress (with a completion date of 2009). It has several names: the Sanxia Project, the Chang Jiang Water Transfer Project, the Three Gorges Dam, and the New China Dam. It is most commonly called the Three Gorges Dam because it is here that the great river flows through a 150-mile (240-km) series of steep-walled valleys less than 360 feet (110 m) wide. Near the lower end of this natural trough, the dam will rise to a height of over 600 feet (180 m) above the valley floor to a width of 1.3 miles (2.1 km), creating a reservoir that will inundate the Three Gorges and extend more than 380 miles (over 600 km) upstream. China's project engineers say that the dam will end the river's rampaging flood cycle, enhance navigation, stimulate development along the new lake perimeter, and provide at least one-tenth, and perhaps as much as one-eighth, of China's electrical power supply. As such it will transform the heart of China, which is why many Chinese like to compare this gigantic project to the Great Wall and the Grand Canal, and call it the New China Dam.

The Three Gorges Dam project will strongly affect the fortunes of the city of Chongqing, upstream from the reservoir. A diversion channel around the dam will enhance Chongqing's river-port functions, allowing bigger vessels to reach it. Although Chengdu is Sichuan's capital, Chongqing, whose status was recently elevated to *shi*, is the province's key outlet. With 120 million people, energy resources, and farms that produce crops ranging from grains to tea, sugarcane, fruits, and vegetables, Sichuan is a Chinese breadbasket, and Chongqing may become the country's number-one growth pole.

Basins of the Xi (West) and Pearl River

As Figure 9-12 shows, the Xi River and its basins are no match for the Chang or Huang, not even for the Liao. This southernmost river even seems to have the wrong name: Xi means West! It reaches the coast in a complex area where it forms a delta immediately adjacent to the estuary of the Pearl River. In this subtropical part of China, local relief is higher than in the lowlands and basins of the center and north, so that farmlands are more confined. Especially in the interior areas, water supply is a recurrent problem. On the other hand, its warm climate permits the double-cropping of rice. Regional food production, however, has never approached that of the North

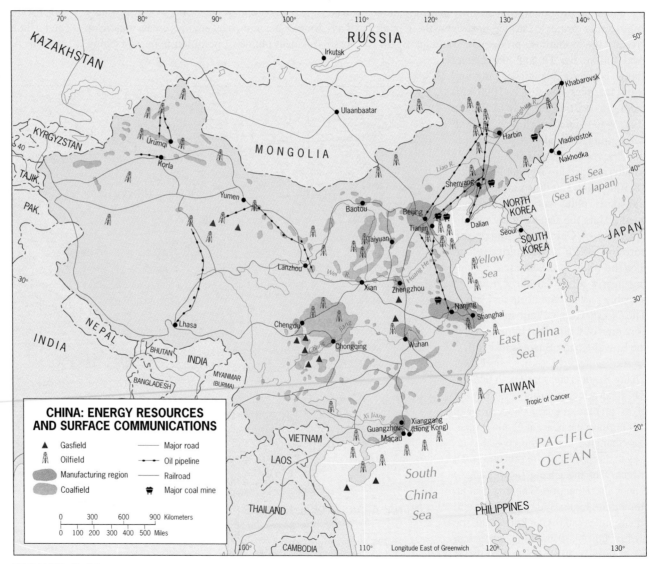

FIGURE 9-14

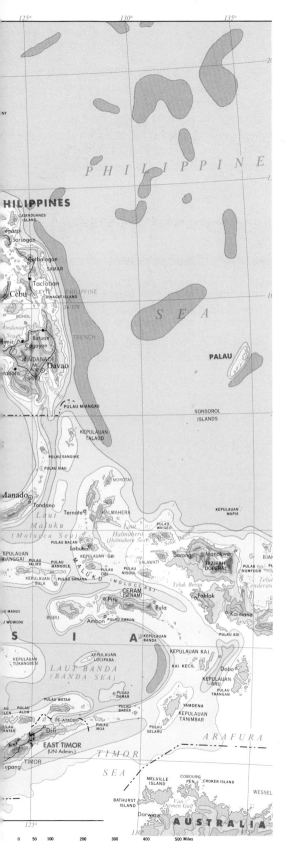

chapter 10 / Southeast Asia

CONCEPTS, IDEAS, AND TERMS

1 Buffer zone

2 Shatter belt

3 Overseas Chinese

4 Organic theory

5 State boundaries

6 Antecedent boundary

7 Subsequent boundary

8 Superimposed boundary

9 Relict boundary

10 State territorial morphology

11 Compact state

12 Protruded state

13 Elongated state

14 Fragmented state

15 Perforated state

16 Domino theory

17 *Entrepôt*

18 Archipelago

19 Transmigration

REGIONS

▷ MAINLAND SOUTHEAST ASIA

▷ INSULAR SOUTHEAST ASIA

FIGURE 10-1

Southeast Asia is a realm of peninsulas and islands, a corner of Asia bounded by India on the northwest and China on the northeast (Fig. 10-1). Its western coasts are washed by the Indian Ocean, and to the east stretches the vast Pacific. From all these directions, Southeast Asia has been penetrated by outside forces. From India came traders; from China, settlers; from across the Indian Ocean, Arabs to engage in commerce and Europeans to build empires; and from across the Pacific, the Americans. Southeast Asia has been the scene of countless contests for power and primacy—the competitors have come from near and far.

Southeast Asia's geography in some ways resembles that of Eastern Europe. It is a mosaic of smaller countries on the periphery of one of the world's **1** largest states. It has been a **buffer zone** between **2** powerful adversaries. It is a **shatter belt** in which stresses and pressures from without and within have fractured the political geography. Like Eastern Europe, Southeast Asia exhibits great cultural diversity. It is a realm of hundreds of cultures, numerous languages and dialects, and several major religions.

DEFINING THE REALM

Figure 10-2 shows the dimensions of the Southeast Asian geographic realm, but note the disconformity between the eastern boundary of the realm and the eastern limits of its most populous state, Indonesia. The easternmost part of Indonesia is the western half of the island of New Guinea, where indigenous cultures are not Southeast Asian but Pacific. Today Indonesia rules what is in effect a Pacific island colony, although Irian Jaya (West Irian) is officially one of its provinces. While we refer in this chapter to Irian Jaya because of its association with Indonesia, we discuss all of New Guinea under the Pacific Realm in Chapter 12.

Because the politico-geographical map (Fig. 10-2) is so complicated, it should be studied atten-

---

◆ Major Geographic Qualities of Southeast Asia

1. Southeast Asia extends from the peninsular mainland to the archipelagos offshore. Because Indonesia controls part of New Guinea, its functional region reaches into the neighboring Pacific geographic realm.

2. Southeast Asia, like Eastern Europe, has been a shatter belt between powerful adversaries and has a fractured cultural and political geography shaped by foreign intervention.

3. Southeast Asia's physiography is dominated by high relief, crustal instability marked by volcanic activity and earthquakes, and tropical climates.

4. A majority of Southeast Asia's more than half-billion people live on the islands of just two countries: Indonesia, with the world's fourth-largest population, and the Philippines. The rate of population increase in the insular region of Southeast Asia exceeds that of the mainland.

5. Although the overwhelming majority of Southeast Asians have the same ancestry, cultural divisions and local traditions abound, which the realm's divisive physiography sustains.

6. The legacies of powerful foreign influences, Asian as well as non-Asian, continue to affect the cultural landscapes of Southeast Asia.

7. Southeast Asia's political geography exhibits a variety of boundary types and several categories of state territorial morphology.

8. The Mekong River, Southeast Asia's Danube, has its source in China and borders or crosses five Southeast Asian countries, sustaining tens of millions of farmers, fishing people, and boat owners.

9. The realm's giant in terms of territory as well as population, Indonesia, has not asserted itself as the dominant state because of mismanagement and corruption; but Indonesia has enormous potential.

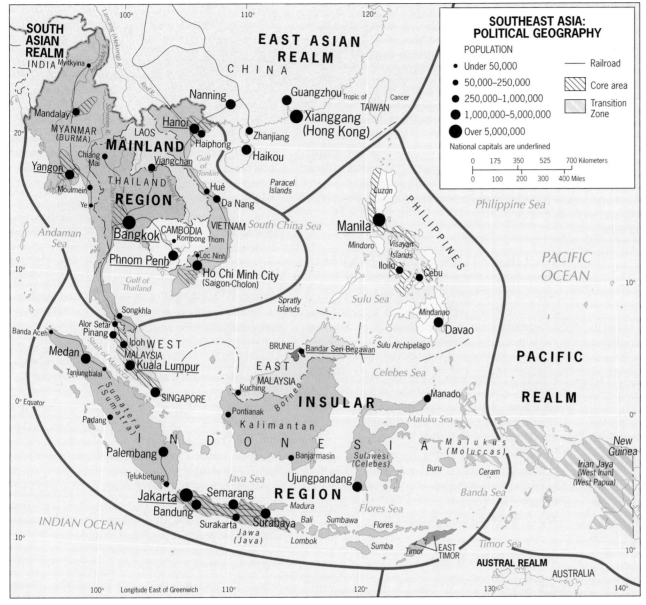

FIGURE 10-2

Map content:

SOUTH ASIAN REALM
INDIA

EAST ASIAN REALM
CHINA

SOUTHEAST ASIA: POLITICAL GEOGRAPHY

POPULATION
● Under 50,000
● 50,000–250,000
● 250,000–1,000,000
● 1,000,000–5,000,000
● Over 5,000,000
National capitals are underlined

— Railroad
▨ Core area
▨ Transition Zone

0   175   350   525   700 Kilometers
0   100   200   300   400 Miles

Myitkyina
Mandalay
MYANMAR (BURMA)
Yangon
Moulmein
Ye
Chiang Mai
THAILAND
REGION
MAINLAND
LAOS
Hanoi
Haiphong
Viangchan
Gulf of Tonkin
Hué
Da Nang
Nanning
Zhanjiang
Haikou
Guangzhou
Xianggang (Hong Kong)
Tropic of Cancer
TAIWAN
Paracel Islands
VIETNAM
South China Sea
Luzon
PHILIPPINES
Manila
Mindoro
Visayan Islands
Iloilo
Cebu
Philippine Sea
PACIFIC OCEAN
Bangkok
CAMBODIA
Kompong Thom
Phnom Penh
Loc Ninh
Ho Chi Minh City (Saigon-Cholon)
Gulf of Thailand
Andaman Sea
Songkhla
Spratly Islands
Sulu Sea
Mindanao
Davao
Sulu Archipelago
PACIFIC REALM
Alor Setar
Pinang
Ipoh
WEST MALAYSIA
Kuala Lumpur
Medan
Tanjungbalai
Banda Aceh
Strait of Malacca
Sumatera (Sumatra)
Padang
Palembang
SINGAPORE
BRUNEI
Bandar Seri Begawan
EAST MALAYSIA
Kuching
Celebes Sea
Pontianak
Kalimantan
Borneo
INSULAR
Banjarmasin
Manado
Maluku Sea
Sulawesi (Celebes)
Buru
Malukus (Moluccas)
Ceram
New Guinea
Irian Jaya [West Irian] (West Papua)
Telukbetung
Jakarta
Bandung
Semarang
Surakarta
Surabaya
Jawa (Java)
Madura
Bali
Lombok
Ujungpandang
REGION
Java Sea
Flores Sea
Sumbawa
Flores
Sumba
Banda Sea
Timor
EAST TIMOR
Timor Sea
AUSTRAL REALM
AUSTRALIA
INDIAN OCEAN
Equator

Cambodia. Still moving generally eastward, we reach Vietnam, a strip of land that extends all the way to the Chinese border. And surrounded by its neighbors is landlocked Laos, remote and isolated. This leaves the islands that constitute *insular* Southeast Asia: the Philippines in the north and Indonesia in the south, and between them the offshore portion of Malaysia, situated on the largely Indonesian island of Borneo. Also on Borneo lies the ministate of Brunei, small but, as we will see, important in the regional picture. And finally, a new state has just appeared on the map in the realm's southeastern corner: East Timor, a former Portuguese colony annexed by Indonesia in 1976 and released to United Nations supervision in 1999.

These are countries of a geographic realm that has no dominant state—no China, no India, no Brazil—although one country, Indonesia, contains 40 percent of its total population and has the potential to emerge as a commanding force. Neither did any single, dominant core of indigenous culture develop here as it did in East Asia. In the river basins and on the plains of the mainland, as well as on the islands offshore, a flowering of cultures produced a diversity of societies whose languages, religions, arts, music, foods, and other achievements formed an almost infinitely varied mosaic—but none of those cultures rose to imperial power. The European colonizers forged empires here, often by playing one state off against another; the Europeans divided and ruled. Out of this foreign intervention came the modern map of Southeast Asia, as only Thailand (formerly Siam) survived the colonial era as an independent entity. Thailand was useful to two competing powers, the French to the east and the British to the west.

tively. One good way to strengthen your mental map of this realm is to follow the mainland coastline from west to east. The westernmost state in the realm is Myanmar (called Burma before 1989 and still referred to by that name), the only country in Southeast Asia that borders both India and China. Myanmar shares the "neck" of the Malay Peninsula with Thailand, heart of the *mainland* region. The south of the peninsula is part of Malaysia—except for Singapore, at the very tip of it. Facing the Gulf of Thailand is

📷 Photo Gallery   ✴ More to Explore   **20** Concepts, Ideas, and Terms   𝐆𝐄𝐎𝐃𝐈𝐒𝐂𝐎𝐕𝐄𝐑𝐈𝐄𝐒 Geodiscoveries Module

It was a convenient buffer, and while the colonists carved pieces off Thailand's domain, the kingdom endured.

Indeed, the Europeans accomplished what local powers could not: the formation of comparatively large, multicultural states that encompassed diverse peoples and societies and welded them together. Were it not for the colonial intervention, it is unlikely that the 17,000 islands of far-flung Indonesia would today constitute the world's fourth largest country in terms of population. Nor would the nine sultanates of Malaysia have been united, let alone with the peoples of northern Borneo across the South China Sea. For good or ill, the colonial intrusion consolidated a realm of few culture cores and numerous ministates into less than a dozen countries.

PHYSICAL GEOGRAPHY

As Figure 10-1 shows, Southeast Asia is a realm in which high relief dominates the physiography. From the Arakan Mountains in western Myanmar (Burma) to the glaciers (yes, glaciers!) of the Indonesian part of New Guinea, elevations rise above 10,000 feet (3300 m) in many locales.

The relief map reminds us that this is not only the Pacific Rim but also the Pacific Ring of Fire, where the crust is unstable, earthquakes are common, and volcanoes are active. Among the islands, Borneo is the sole exception. Borneo is a slab of ancient crust, pushed high above sea level by tectonic forces and eroded into its present mountainous topography.

As Figure 10-1 underscores, rivers rise in the highland backbones of the islands and peninsulas, and deposit their sediments as they wind their way toward the coast; the physiography of Sumatera demonstrates this unmistakably. The volcanic hills, plateaus, and better-drained lowlands are fertile and, in the warmth of tropical climates, can yield multiple crops of rice.

On the peninsular mainland we see a pattern that is already familiar: rivers rising in the Asian interior that create alluvial plains and deltas. The Mekong River is the Chang/Yangzi of Southeast Asia: you can trace it all the way from China via Laos, Thai-

land, and Cambodia into southern Vietnam, where it forms a massive and populous delta. In the west, Myanmar's key river is the Irrawaddy; Thailand's is the Chao Phraya. In the north, the Red River Basin is the breadbasket of northern Vietnam.

No survey of the physical geography of Southeast Asia would be complete without reference to the realm's seas, gulfs, straits, and bays. Irregular and indented coastlines such as these, with thousands of islands near and far, create difficult problems when it comes to drawing boundaries in the waters offshore (*maritime boundaries*). Southeast Asia has one of the most complex maritime boundary frameworks in the world.

POPULATION GEOGRAPHY

Compared to the huge population numbers and densities in the habitable regions of South Asia and China, demographic totals for the countries of Southeast Asia, with the exception of Indonesia, seem modest. Again, comparisons with Europe come to mind. Three countries—Thailand, the Philippines, and Vietnam—have populations between 60 and 85 million. Laos, quite a large country territorially (comparable to the United Kingdom), had just 5.5 million inhabitants in 2002. Cambodia, substantially larger than Greece, had 12.7 million. It is noteworthy that of Southeast Asia's 546 million inhabitants, well over half (303 million, 55 percent) live on the islands of Indonesia and the Philippines, leaving the realm's mainland countries with just 45 percent of the population.

The Ethnic Mosaic

Southeast Asia's peoples come from a common stock just as (Caucasian) Europeans do, but this has not prevented the emergence of regionally or locally discrete ethnic or cultural groups. Figure 10-3 displays the broad distribution of ethnolinguistic groups in the realm, but be aware that this is a generalization. At the scale of this map, numerous small groups cannot be depicted.

The map shows the rough spatial coincidence, on the mainland, between major ethnic group and

modern political state. The Burman dominate in the country once known as Burma (Myanmar); the Thai occupy the state once known as Siam (now Thailand); the Khmer form the nation of Cambodia and extend northward into Laos; and the Vietnamese inhabit the long strip of territory facing the South China Sea.

Territorially, by far the largest population is classified in Figure 10-3 as Indonesian, the inhabitants of the great archipelago that extends from Sumatera* west of the Malay Peninsula to the Malukus (Moluccas) in the east and from the lesser Sunda Islands in the south to the Philippines in the north. Collectively, all these peoples—the Filipinos, Malays, and Indonesians—shown in Figure 10-3 are known as Indonesians, but they have been divided by history and politics. Note, on the map, that the Indonesians in Indonesia itself include Javanese, Madurese, Sundanese, Balinese, and other large groups; hundreds of smaller ones are not shown. In the Philippines, too, island isolation and contrasting ways of life are reflected in the cultural mosaic. Also part of this Indonesian ethnic-cultural complex are the Malays, whose heartland lies on the Malay Peninsula but who form minorities in other areas as well. Like most Indonesians, the Malays are Muslims, but Islam is a more powerful force in Malay society than, in general, in Indonesian culture.

In the northern part of the mainland region, numerous minorities inhabit remote parts of the countries in which the Burman (Burmese), Thai, and Vietnamese dominate. Those minorities, as a comparison of Figures 10-2 and 10-3 indicates, tend to occupy areas on the peripheries of their countries, away from the core areas, where the terrain is mountainous and the forest is dense, and where the governments of the national states do not have complete control. This remoteness and sense of detachment give rise to notions of secession, or at least resistance to governmental ef-

---

*As in Africa, names and spellings have changed with independence. In this chapter, we will use the contemporary spellings, except when we refer to the colonial period. Thus Indonesia's four major islands are Jawa, Sumatera, Kalimantan (the Indonesian part of Borneo), and Sulawesi. The Dutch called them Java, Sumatra, Dutch Borneo, and Celebes, respectively.

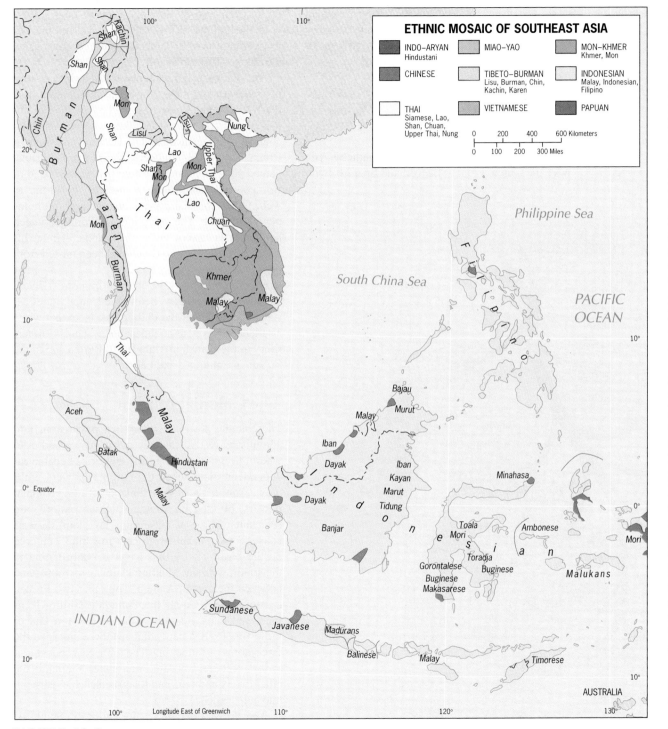

FIGURE 10-3

ETHNIC MOSAIC OF SOUTHEAST ASIA

INDO–ARYAN
Hindustani

CHINESE

THAI
Siamese, Lao,
Shan, Chuan,
Upper Thai, Nung

MIAO–YAO

TIBETO–BURMAN
Lisu, Burman, Chin,
Kachin, Karen

VIETNAMESE

MON–KHMER
Khmer, Mon

INDONESIAN
Malay, Indonesian,
Filipino

PAPUAN

0   200   400   600 Kilometers

0   100   200   300 Miles

forts to establish authority, often resulting in bitter ethnic conflict.

Immigrants

Figure 10-3 also reminds us that, again
like Eastern Europe, Southeast Asia is
home to major ethnic minorities from
outside the realm. On the Malay Peninsula, note the South Asian (Hindustani) cluster. Hindu communities with
Indian ancestries exist in many parts of
the peninsula, but in the southwest they
form the majority in a small area. In
Singapore, too, South Asians form a
significant minority. These communities arose during the colonial period,
but South Asians arrived in this realm
many centuries earlier, propagating
Buddhism and leaving architectural
and cultural imprints on places as far
away as Jawa and Bali.

By far the largest immigrant minority in Southeast Asia, however, is
Chinese. The Chinese began arriving
here during the Ming and early Qing
(Manchu) Dynasties, but the largest
exodus occurred during the late colonial period (1870–1940), when as
many as 20 million immigrated. The
European powers at first encouraged this influx, using the Chinese
in administration and trade. But soon
3 these **Overseas Chinese** began
to move into the major cities, where
they established "Chinatowns"
and gained control over much of the
commerce. By the time the Europeans
tried to reduce Chinese immigration,
World War II was about to start and
the colonial era would soon end.

Today, Southeast Asia may accommodate as many as 30 million Overseas Chinese, more than half the world
total. Their lives have often been
difficult. The Japanese relentlessly

persecuted those Chinese who lived in Malaya during World War II; during the 1960s Chinese in Indonesia were accused of communist sympathies, and hundreds of thousands were killed. In the late 1990s Indonesian mobs again attacked Chinese and their property, this time because of their relative wealth and because many Chinese became Christians during the colonial period, now targeted by Islamic throngs.

Figure 10-4 shows the migration routes and current concentrations of Chinese in Southeast Asia. Most originated in China's Fujian and Guangdong prov-

inces, and many invested much of their wealth in China when it opened up to foreign businesses. The Overseas Chinese of Southeast Asia played a major role in the economic miracle of the Pacific Rim.

HOW THE POLITICAL MAP EVOLVED

The leading colonial competitors in Southeast Asia were the Dutch, French, British, and Spanish (with

the last replaced by the Americans in their stronghold, the Philippines). The Japanese had colonial objectives here as well, but these came and went during the course of World War II.

The Dutch acquired the greatest prize: control over the vast archipelago now called Indonesia (formerly the Netherlands East Indies). France established itself on the eastern flank of the mainland, controlling all territory east of Thailand and south of China. The British conquered the Malay Peninsula, gained power over the northern part of the island of Borneo, and established themselves in Burma as well. Other colonial powers also gained footholds but not for long. The exception was Portugal, which held on to its eastern half of the island of Timor (Indonesia) until after the Dutch had been ousted from their East Indies.

Figure 10-5 shows the colonial framework in the late nineteenth century, before the United States assumed control over the Philippines. Note that while Thailand survived as an independent state, it lost territory to the British in Malaya and Burma and to the French in Cambodia and Laos.

The Colonial Imprint

The colonial powers divided their possessions into administrative units as they did in Africa and elsewhere. Some of these political entities became independent states when the colonial powers withdrew. France, one of the mainland's leading colonial powers, divided its Southeast Asian empire into five units. Three of these units lay along the east coast: Tonkin in the north next to China, centered on the basin of the Red River; Cochin China in the south, with the Mekong Delta as its focus; and between these two, Annam. The other two French territories were Cambodia, facing the Gulf of Thailand, and Laos, landlocked in the interior. Out of these five French dependencies there emerged the three states of Indochina. The three east coast territories ultimately became one state, Vietnam; the other two (Cambodia and Laos) each achieved separate independence.

The British ruled two major entities in Southeast Asia (Burma and Malaya) in addition to a large part

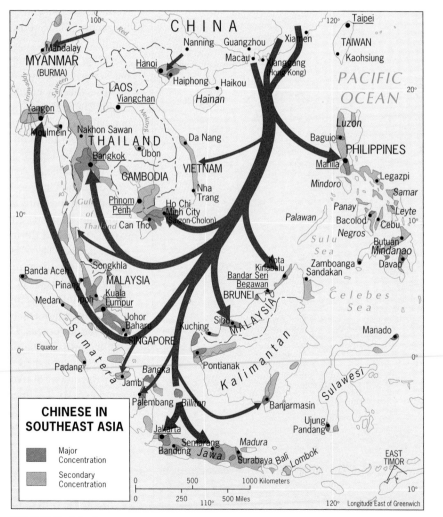

FIGURE 10-4

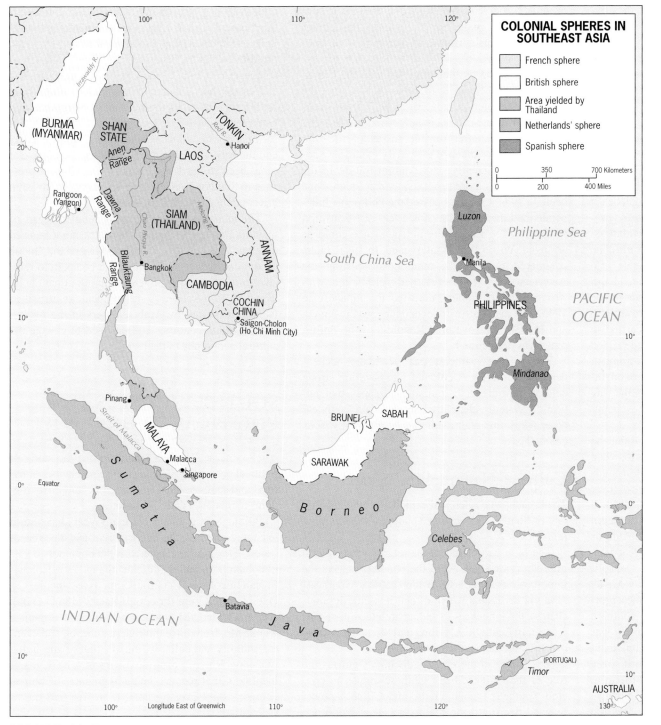

COLONIAL SPHERES IN SOUTHEAST ASIA

French sphere

British sphere

Area yielded by Thailand

Netherlands' sphere

Spanish sphere

FIGURE 10-5

of northern Borneo and many small islands in the South China Sea. Burma was attached to Britain's Indian empire; from 1886 until 1937 it was governed from distant New Delhi. But when British India became independent in 1947 and split into several countries, Burma was not part of the grand design that created West and East Pakistan (the latter now Bangladesh), Ceylon (now Sri Lanka), and India. Instead, in 1948 Burma (now Myanmar) was given the status of a sovereign republic.

In Malaya, the British developed a complicated system of colonies and protectorates that eventually gave rise to the equally complex, far-flung Malaysian Federation. Included were the former Straits Settlements (Singapore was one of these colonies), the nine protectorates on the Malay Peninsula (former sultanates of the Muslim era), the British dependencies of Sarawak and Sabah on the island of Borneo, and numerous islands in the Strait of Malacca and the South China Sea. The original Federation of Malaysia was created in 1963 by the political unification of recently independent mainland Malaya, Singapore, and the former British dependencies on the largely Indonesian island of Borneo. Singapore, however, left the Federation in 1965 to become a sovereign city-state, and the remaining units were later restructured into peninsular Malaysia and, on Borneo, Sarawak and Sabah. Thus the term *Malaya* properly refers to the geographic area of the Malay Peninsula, including Singapore and other nearby islands; the term *Malaysia* identifies the politico-geographical entity of which Kuala Lumpur is the capital city.

The Hollanders took control of the "spice islands" through their Dutch

East India Company, and the wealth that they extracted from what is today Indonesia brought the Netherlands its Golden Age. From the mid-seventeenth to the late-eighteenth century, the Dutch could develop their East Indies sphere of influence almost without challenge, for the British and French were preoccupied with the Indian subcontinent. By playing the princes of Indonesia's states against one another in the search for economic concessions and political influence, by placing the Chinese in positions of responsibility, and by imposing systems of forced labor in areas directly under its control, the Company had a ruinous effect on the Indonesian societies it subjugated. Java (Jawa), the most populous and productive island, became the focus of Dutch administration; from its capital at Batavia (now Jakarta), the Company extended its sphere of influence into Sumatra (Sumatera), Dutch Borneo (Kalimantan), Celebes (Sulawesi), and the smaller islands of the East Indies. This was not accomplished overnight, and the struggle for territorial control was carried on long after the Dutch East India Company had yielded its administration to the Netherlands government. Dutch colonialism thus threw a girdle around Indonesia's more than 17,000 islands, paving the way for the creation of the realm's largest and most populous nation-state (nearly 220 million today).

In the colonial tutelage of Southeast Asia, the Philippines, long under Spanish domination, had a unique experience. As early as 1571, the islands north of Indonesia were under Spain's control (they were named for Spain's King Philip II). Spanish rule began when Islam was reaching the southern Philippines via northern Borneo. The Spaniards spread their Roman Catholic faith with great zeal, and between them the soldiers and priests consolidated Hispanic dominance over the mostly Malay population. Manila, founded in 1571, became a profitable waystation on the route between southern China and western Mexico (Acapulco usually was the trans-Pacific destination for the galleons leaving Manila's port). There was much profit to be made, but the indigenous people shared little in it. Great landholdings were awarded to loyal Spanish civil servants and to men of the church. Oppression

eventually yielded revolution, and Spain was confronted with a major uprising in the Philippines when the Spanish-American War broke out elsewhere in 1898.

As part of the settlement of that war, the United States replaced Spain in Manila. That was not the end of the revolution, however. The Filipinos now took up arms against their new foreign ruler, and not until 1905, after terrible losses of life, did American forces manage to "pacify" their new dominion. Subsequently, U.S. administration in the Philippines was more progressive than Spain's had been. In 1934, Congress passed the Philippine Independence Law, providing for a ten-year transition to sovereignty. But before independence could be arranged, World War II intervened. In 1941, Japan conquered the islands, temporarily ousting the Americans; U.S. forces returned in 1944 and, with strong Filipino support, defeated the Japanese in 1945. The agenda for independence was resumed, and in 1946 the sovereign Republic of the Philippines was proclaimed.

Today, all of Southeast Asia's states are independent, but centuries of colonial rule have left strong cultural imprints. In their urban landscapes, their education systems, their civil service, and countless other ways, this realm still carries the marks of its colonial past.

Cultural-Geographic Legacies

The French, who ruled and exploited a crucial quadrant of Southeast Asia, had a name for their empire: *Indochina*.

The *Indo* part of Indochina refers to the cultural imprints from South Asia: the Hindu presence, the importance of Buddhism (which came to Southeast Asia via Sri Lanka [Ceylon] and its seafaring merchants), the influences of Indian architecture and art (especially sculpture), writing and literature, and social structures and patterns.

The *China* in the name Indochina signifies the role of the Chinese here. Chinese emperors coveted Southeast Asian lands, and China's power reached deep into the realm. Social and political

upheavals in China, combined with the opportunities created by the European colonists, sent millions of Sinicized people southward. Chinese traders, pilgrims, seafarers, fishermen, and others sailed from southeastern China to the coasts of Southeast Asia and established settlements there. Over time, those settlements attracted more Chinese emigrants, and Chinese influence in the realm grew (Fig. 10-4). Not surprisingly, relations between the Chinese settlers and the earlier inhabitants of Southeast Asia have at times been strained, even violent. The Chinese presence in Southeast Asia is long-term, but the invasion has continued into modern times.

The name *Indochina* can only refer to a part of mainland Southeast Asia, though. It cannot be used to refer to the realm as a whole. Although the "Indo" segment of this regional name can be taken to refer also to the Buddhist influences that dominate here, it makes no reference to the momentous arrival of Islam, introduced by Arab seafarers in the twelfth and thirteenth centuries and destined to change the cultural geography of the realm.

SOUTHEAST ASIA'S POLITICAL GEOGRAPHY

✵ Southeast Asia is a laboratory for the study of political geography. This realm is a patchwork of nations and states ranging from ancient Buddhist kingdoms to modern Islamic sultanates.

Political geographers for many years have studied the causes of the cyclical rise and decline of states. They have attributed this phenomenon to various factors including climate change (see Chapter 6) and ideological contradiction (see Chapter 2). One historic figure, Friedrich Ratzel (1844–1904), conceptualized the state as a biological organism whose life, from birth through maturation and eventual senility and collapse, mirrors that of any living **4** thing. Ratzel's **organic theory** of state development held that nations, being aggregates of human beings, would over the long term live and die as their citizens did.

To understand the state better, it is useful to learn more about its components. Southeast Asia's states display these in great variety. We focus first on national boundaries on land (deferring maritime boundaries until later), and then we concentrate on the territorial morphology, or shape, of this realm's states.

The Boundaries

GEODISCOVERIES Boundaries are sensitive parts of a state's anatomy: just as people are territorial about their individual properties, so nations and states are sensitive about their territories and borders. The saying that "good fences make good neighbors" certainly applies to states, but, as we know, the boundaries between states are not always "good fences."

5 **Boundaries**, in effect, are contracts between states. That contract takes the form of a treaty that contains the *definition* of the boundary in the form of elaborate description. Next, cartographers perform the *delimitation* of the treaty language, drawing the boundary on official, large-scale maps. And throughout human history, states have used those maps to build fences, walls, or other barriers in a process called *demarcation*.

Once established, we can classify boundaries geographically. Some are sinuous, conforming to rivers or mountain crests (*physiographic*) or coinciding with breaks or transitions in the cultural landscape (*anthropogeographic*). As any world political map shows, many boundaries are simply straight lines, delimited without reference to physical or cultural features. These *geometric* boundaries can produce problems when the cultural landscape changes where they exist.

In general, the boundaries of Southeast Asia were better defined than those of several other postcolonial areas of the world, notably Africa, the Arabian Peninsula, and Turkestan. The colonial powers that established the original treaties tried to define boundaries to lie in remote and/or sparsely peopled areas: for example, across interior Borneo. Nevertheless, certain Southeast Asian boundaries have produced problems, among them the geometric boundary between Irian Jaya, the portion of New

FROM THE FIELD NOTES

"I stood on the Laotian side of the great Mekong River which, during the dry season, did not look so great! On the opposite side was Thailand, and it was rather easy for people to cross here at this time of the year. But, the locals told me, it is quite another story in the wet season. Then the river inundates the rocks and banks you see here, it rushes past, and makes crossing difficult and even dangerous. The buildings where the canoes are docked are built on floats, and rise and fall with the seasons. The physiographic-political boundary between Thailand and Laos lies in the middle of the valley we see here."

Guinea ruled by Indonesia, and Papua New Guinea, the eastern component of the island.

Even on a small-scale map of the kind we use in this chapter, we can categorize the boundaries of this realm. A comparison between Figures 10-2 and 10-3 reveals that the boundary between Thailand and Myanmar over long segments is anthropogeographic, notably where the name *Karen*, the Myanmar minority, appears on Figure 10-3. Figure 10-1 shows that a large segment of the Vietnam-Laos boundary is physiographic-political, coinciding with the Annamese Cordillera (Mountains).

Boundaries also can be classified genetically, that is, as their evolution relates to the cultural landscapes they traverse. A leading political geographer, Richard

Hartshorne (1899–1992), proposed a four-level *genetic boundary classification*. All four of these boundary types can be observed in Southeast Asia.

Certain boundaries, Hartshorne reasoned, were defined and delimited before the present-day human landscape developed. In Figure 10-6 (upper-left map), the boundary between Malaysia and Indonesia on the island of Borneo is an example of the first

6 boundary type, the **antecedent boundary**. Most of this border passes through sparsely inhabited tropical rainforest, and the break in settlement can even be detected on the small-scale world population map (Fig. I-9).

A second category of boundaries evolved as the cultural landscape of an area took shape, part of the

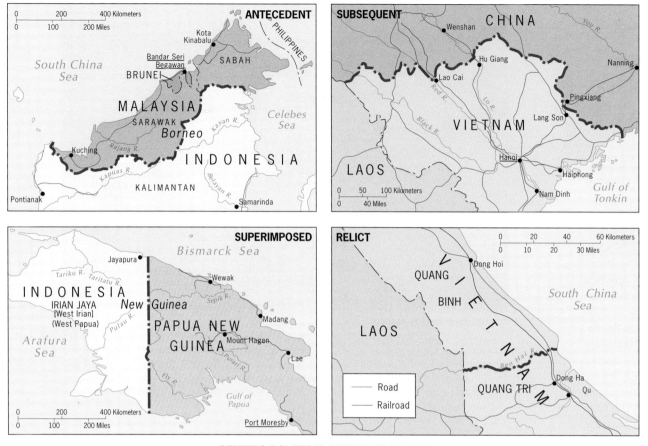

GENETIC POLITICAL BOUNDARY TYPES

FIGURE 10-6

from Southeast Asia into the Pacific Realm. Geographically, all of New Guinea forms part of the Pacific Realm.

The fourth genetic boundary type is **9** the so-called **relict boundary**—a border that has ceased to function but whose imprints (and sometimes influence) are still evident in the cultural landscape. The boundary between the former North and South Vietnam (Fig. 10-6, lower-right map) is a classic example: once demarcated militarily, it has had relict status since 1976 following the reunification of Vietnam in the aftermath of the Indochina War (1964–1975).

Southeast Asia's boundaries have colonial origins, but they have continued to influence the course of events in postcolonial times. Take one instance: the physiographic boundary that separates the main island of Singapore from the rest of the Malay Peninsula, the Johor Strait (see Fig. 10-11). That physiographic-political boundary facilitated, perhaps crucially, Singapore's secession from the state of Malaysia. Without it, Malaysia might have been persuaded to stop the separation process; at the very least, territorial issues would have arisen to slow the sequence of events. As it was, no land boundary needed to be defined. The Johor Strait demarcated Singapore and left no question as to its limits.

State Territorial Morphology

Boundaries define and delimit states; they also create the mosaic of often interlocking territories that give individual countries their shape, also known as their **10** *morphology*. The **territorial morphology** of a state affects its condition, even its survival. Vietnam's extreme elongation has influenced its existence since time immemorial. As we will see, Indonesia has tried

7 ongoing process of accommodation. These **subsequent boundaries** are represented in Southeast Asia by the map in the upper right of Figure 10-6, which shows in some detail the border between Vietnam and China. This border is the result of a long process of adjustment and modification, the end of which may not yet have come.

The third category involves boundaries drawn forcibly across a unified or at least homogeneous cultural landscape. The colonial powers did this when they divided the island of New Guinea by delimiting a boundary in a nearly straight line (curved in only one place to accommodate a bend in the Fly River), as shown in the lower-left map of Figure **8** 10-6. The **superimposed boundary** they delimited gave the Netherlands the western half of New Guinea. When Indonesia became independent in 1949, the Dutch did not yield their part of New Guinea, which is peopled mostly by ethnic Papuans, not Indonesians. In 1962 the Indonesians invaded the territory by force of arms, and in 1969 the United Nations recognized its authority there. This made the colonial, superimposed boundary the eastern border of Indonesia and had the effect of extending Indonesia

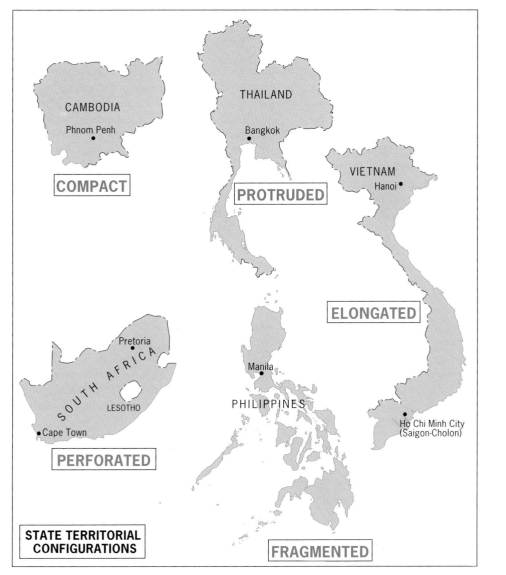

CAMBODIA

Phnom Penh

COMPACT

THAILAND

Bangkok

PROTRUDED

VIETNAM

Hanoi

ELONGATED

Pretoria

SOUTH AFRICA

LESOTHO

Cape Town

PERFORATED

Manila

PHILIPPINES

Ho Chi Minh City
(Saigon-Cholon)

STATE TERRITORIAL
CONFIGURATIONS

FRAGMENTED

FIGURE 10-7

to redress its fragmentation into thousands of islands by promoting unity through the "transmigration" of Jawanese from the most populous island to many of the others.

Political geographers identify five dominant state territorial configurations, all of which we have encountered in our world regional survey but which we have not categorized until now. All but one of these shapes are represented in Southeast Asia, and Figure 10-7 provides the terminology and examples:

- **11 Compact states** have territories shaped somewhere between round and rectangular, without major indentations. This encloses a maximum amount of territory within a minimum length of boundary. Southeast Asian example: Cambodia.

- **12 Protruded states** (sometimes called *extended*) have a substantial, usually compact territory from which extends a peninsular corridor that may be landlocked or coastal. Southeast Asian examples: Thailand, Myanmar.

- **13 Elongated states** (also called *attenuated*) have territorial dimensions in which the length is at least six times the average width, creating a state that lies astride environmental or cultural transitions. Southeast Asian example: Vietnam.

- **14 Fragmented states** consist of two or more territorial units separated by foreign territory or by water. Subtypes are mainland-mainland, mainland-island, and island-island. Southeast Asian examples: Malaysia, Indonesia, Philippines, and East Timor.

- **15 Perforated states** completely surround the territory of other states, so that they have a "hole" in them. No Southeast Asian example; the most illustrative current case is South Africa, perforated by Belgium-sized Lesotho.

In the discussion that follows, we will have frequent occasion to refer to this geographic property of Southeast Asia's states. For so comparatively small a realm with so few states, Southeast Asia displays a considerable variety of state morphologies. When we link these features to other geographic aspects (such as relative location), we obtain useful insights into the regional framework.

One point of caution: states' territorial morphologies do not determine their viability, cohesion, unity, or lack thereof; they can, however, influence these qualities. Cambodia's compactness has not ameliorated its divisive political geography, for example. But as we will find in the pages that follow, shape plays a key role in the still-unfolding political and economic geography of Southeast Asia.

REGIONS OF THE REALM

Southeast Asia's first-order regionalization must be based on its mainland-island fragmentation. But as we have noted there are physiographic, historical, and cultural reasons to include the Malaysian (southern) part of the Malay Peninsula in the insular region, as shown in Figure 10-2. Using the political framework as our grid, we see that the regions of Southeast Asia are constituted as follows:

Mainland Region Vietnam, Cambodia, Laos, Thailand, Myanmar (Burma)

Insular Region Malaysia, Singapore, Indonesia, East Timor, Brunei, Philippines

Note, however, that the realm boundary excludes the Indonesian zone of New Guinea (Irian Jaya), which is part of the Pacific geographic realm.

MAINLAND SOUTHEAST ASIA

Five countries form the mainland region of Southeast Asia: two of them protruded, one compact, one elongated, and one landlocked. Two colonial powers, buffered by Thailand, shaped its modern historical geography. One religion, Buddhism, dominates cultural landscapes, but this is a multicultural, multiethnic region. Although one of the least urbanized regions in the world, it contains several major cities. And as Figure 10-2 shows, two countries (Vietnam and Myanmar) possess more than one core area each. We approach the region from the east.

COUNTRIES OF INDOCHINA

Former French Indochina gave rise to three modern states: Vietnam, Cambodia and Laos. Here the United States fought and lost a disastrous war that ended in 1975 but whose impact on America continues to be felt today.

After the Indochina War started formally in 1964 (U.S. involvement actually began earlier), some scholars warned that the conflict might spill over from Vietnam into Laos and Cambodia, and hence into Thailand, Malaysia, and even Myanmar (Burma). **16** This view was based on the **domino theory**, which holds that destabilization and conflict from any cause in one country can result in the collapse of order in one or more neighboring countries, triggering a chain of events that can affect a series of contiguous states in a region.

History proved these scholars wrong; Cambodia and Laos were affected but not the other states. This seemed to invalidate the domino theory, which was grounded in the capitalist-communist struggle of the twentieth century. But communist insurgency was (and is) only one way a country may be destabilized. Ethnic conflict (Equatorial Africa) and cultural strife (Yugoslavia)—even environmental and economic causes—can set the domino effect in motion.

Elongated Vietnam still carries the scars of war, but consider this: about 60 percent of Vietnam's population of some 80 million is under 21 years of age, and the vast majority of Vietnamese have no personal memory of that terrible conflict. The more immediate concerns in Vietnam today are to reconnect the country with the outside world and to integrate its 1200-mile (2000-km) strip of attenuated territory through better infrastructure (Fig. 10-8).

The French colonizers recognized that Vietnam, whose average width is under 150 miles (240 km), was not a homogeneous colony, so they divided it into three units: (1) Tonkin, land of the Red River Delta and centered on Hanoi in the north; (2) Cochin China, region of the Mekong Delta and centered on Saigon in the south; and (3) Annam, focused on the ancient city of Hué, in the middle (Fig. 10-5). Today, the Vietnamese prefer to use *Bac Bo*, *Nam Bo*, and *Trung Bo* to designate these areas.

The Vietnamese (or Annamese, also Annamites, after their cultural heartland) speak the same language, although the northerners can easily be distinguished from southerners by their accent. As else-

where in their colonial empire, the French made their language the *lingua franca* of Indochina, but their tenure was cut short by the Japanese, who invaded Vietnam in 1940. During the Japanese occupation, Vietnamese nationalism became a powerful force, and after the Japanese defeat in 1945, the French could not regain control. In 1954, the French suffered a disastrous final trouncing on the battlefield at Dien Bien Phu in the northwest and were ousted from the country.

But even after its forces routed the colonizers, Vietnam did not become a unified state. Separate regimes took control: a communist one in Hanoi and a noncommunist counterpart in Saigon. Vietnam's pronounced elongation had made things difficult for the French; now it played its role during the postcolonial period. Note, in Figure 10-8, that Vietnam is widest in the north and south, with a narrow "waist" in its middle zone. North and South Vietnam were worlds apart, and those worlds were represented in Hanoi by communism and in Saigon by anticommunism. For more than a decade the United States tried to prop up the Saigon regime that controlled the south, but the communists prevailed, and like China, Vietnam has a communist government today. As many as 2 million Vietnamese refugees set out on often-flimsy boats onto the South China Sea; of those who survived, a majority were settled in the United States.

Today, contrasts between north and south continue. The capital, Hanoi, carries the imprints of its Soviet-era tutelage in the form of Ho Chi Minh's mausoleum and rows of faceless apartment buildings representative of the Soviet "socialist city." With 4 million residents, Hanoi anchors the northern (Tonkin) core area of Vietnam, the basin of the Red River (its agricultural hinterland). In the paddies, irrigation water is still raised by bucket. On the roads, goods move by human- or animal-drawn cart. As yet little has changed here—except people's expectations.

Elongated Vietnam's southern headquarters, Saigon-Cholon (officially named Ho Chi Minh City),

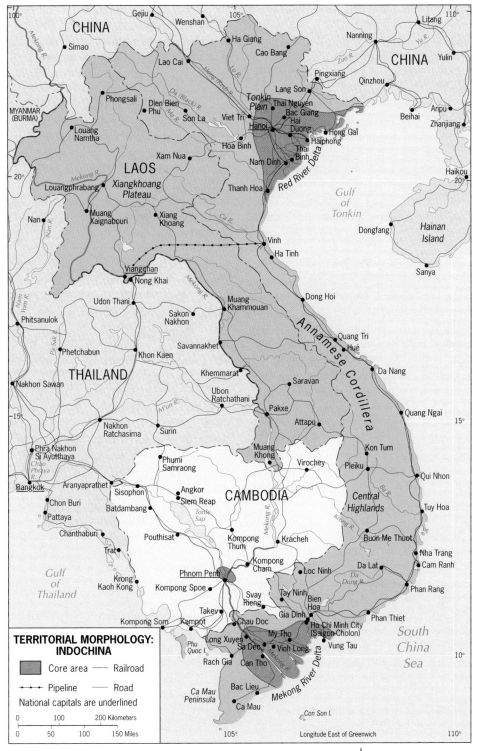

FIGURE 10-8

is far ahead of the capital. More than 8 million people inhabit this urban agglomeration on the Saigon River, which contains about 10 percent of the country's entire population—and a far larger share of its best-educated and most capable people. Saigon-Cholon (Cholon is the city's Chinatown, much diminished after the war but now reviving) is changing rapidly as high-rises built by foreign investors tower over the still-colonial townscape, modern hotels are opening, and a large Special Economic Zone has been laid out just downstream from the port. Unlike Hanoi, Saigon can be reached by oceangoing vessels, and it lies north of, rather than within, the Mekong Delta that is part of its hinterland. The city's streets are choked with bicycles, mopeds, handcarts, buses, taxis, and even a few private cars; consumer goods are everywhere, and the city throbs with commercial activity. Saigon, too, has serious infrastructural problems, but it is recovering from decades of neglect. One great need is a bridge to link the right bank of the Saigon River, on which downtown and most of Saigon lie, to the left bank, which remains merely a village.

Unlike China's rulers, Vietnam's communist leaders have not pursued market reforms with much enthusiasm, and Vietnam is no economic tiger. But Vietnam is self-sufficient in staples and is a major exporter of rice; it has recently become the world's second-ranking exporter of coffee (grown in the mountains of the country's "waist," where the lands of minorities are being cleared over their protests). Hanoi's state planners, well aware of the historic contrasts between north and south, do not want regional disparities resulting from Pacific Rim development to threaten the unity of their country, so prone to centrifugal forces.

Compact Cambodia is heir to the ancient Khmer Empire whose capital was Angkor and whose legacy is a vast landscape of imposing monuments including Buddhist-inspired Angkor Wat. Today, 90 percent of Cambodians are Khmers, with small minorities of Vietnamese and Chinese. The present capital, Phnom Penh, lies on the Mekong River (Fig. 10-8), which crosses Cambodia before it enters and forms its great delta in Vietnam.

Geographically, Cambodia enjoys several advantages; compact states enclose a maximum of territory within a minimum of boundary, and cultural homogeneity tends to diminish centrifugal forces. But neither spatial morphology nor homogeneous ethnicity could withstand the impact of the Indochina War, which led to communist revolution and the systematic murder of as many as 2 million Cambodians by the Maoist terror group, the *Khmer Rouge*. Once self-sufficient and a food exporter, Cambodia today must import food. Chronic instability and rural dislocation make this Southeast Asia's poorest country. Its postwar trauma continues.

Landlocked Laos has no fewer than five neighbors, one of which is East Asia's giant, China. The Mekong River forms a long stretch of its western boundary, and the important sensitive border with Vietnam to the east lies in mountainous terrain. With 5.5 million people (about half of them ethnic Lao, related to the Thai of Thailand), Laos lies surrounded by comparatively powerful states. The country has no railroads, just a few miles of paved roads, and very little industry; it is only 17 percent urbanized (the capital, Viangchan, lies on the Mekong and has an oil pipeline to Vietnam's coast). Like Vietnam, Laos has a communist government, and like Cambodia, the country continues to be plagued by social unrest as political opposition takes the form of anarchy.

PROTRUDED THAILAND AND MYANMAR

In virtually every way, Thailand is the leading state of the mainland region. In contrast to its neighbors, Thailand has been a strong participant in the Pacific Rim's economic development. Thailand's capital, Bangkok, is one of the two largest urban centers in the region and one of the world's most prominent primate cities. The country's population, 63.1 million in 2002, is growing at just about the slowest rate in the entire realm, almost equivalent to that of fully urbanized Singapore. Over the past few decades, only political instability and uncertainty have inhibited economic progress. Thailand is a constitutional monarchy; its moves toward stable democracy have

been thwarted by graft, corruption, and, sometimes, violent confrontation.

Thailand is the textbook example of a protruded state. From a relatively compact heartland, in which lie the core area, capital, and major areas of productive capacity, a 600-mile (1000-km) corridor of land, in places less than 20 miles (32 km) wide, extends southward to the border with Malaysia (Fig. 10-9). The boundary that defines this protrusion runs down the length of the Malay Peninsula to the Kra Isthmus, where neighboring Myanmar peters out and Thailand fronts the Andaman Sea (an arm of the Indian Ocean) as well as the Gulf of Thailand. In the entire country, no place lies farther from the capital than the southern end of this tenuous protrusion.

As Figure 10-2 shows, Thailand occupies the heart of the mainland region of Southeast Asia. While Thailand has no Red, Mekong, or Irrawaddy Delta, its central lowland is watered by a set of streams that flow off the northern highlands and the Khorat Plateau in the east. One of these streams, the Chao Phraya, is the Rhine of Thailand. From the head of the Gulf of Thailand to Nakhon Sawan, this river is a highway of traffic. Barge trains loaded with rice head for the coast, ferry boats head upstream, freighters transport tin and tungsten (of which Thailand is among the world's largest producers). Bangkok sprawls on both sides of the lower Chao Phraya, here flanked by skyscrapers, pagodas, factories, boatsheds, ferry landings, luxury hotels, and modest dwellings in crowded confusion.

As Figure 10-9 shows, Thailand's core area, anchored by Bangkok, opens to the fishing grounds and oil reserves of the Gulf of Thailand and thence to the Pacific Ocean. It is also well connected to the northeast, centered on Nakhon Ratchasima, and the northwest, where the urban focus is sprawling, temple-studded Chiang Mai (Fig. 10-9). The magnificent architecture of Buddhism, on display in these cities, attracts millions of visitors annually, and the warm coasts and beaches of the south have long made Thailand a favorite destination for tourists from countries around the world, notably Germany and Japan. Thailand's relaxed attitude toward sex has contributed to this traffic, but by the mid-1990s

the country was suffering from one of the world's worst outbreaks of AIDS.

Thailand's territorial morphology creates both problems and opportunities. Problems include the effective integration of the national territory through surface communications, which must cover long distances to reach remote locales; the influx of refugees from neighboring countries dislocated by internal conflict; and control over contraband (opium) drug production and trade in the interior "Golden Triangle" where Myanmar, Thailand, and Laos meet. The opportunities relate in part to the still-remote Kra Isthmus, where proximity to Malaysia and a potentially stable Indonesia could lead to significant coordinated regional development.

Thailand's also-protruded neighbor, Myanmar (still referred to as Burma in some anti-regime quarters), is one of the world's poorest countries where, it seems, time has stood still for centuries. Long languishing under one of the world's most corrupt military dictatorships, Myanmar shares with Thailand a long stretch of the Kra Isthmus—but look again at Figure 10-9 and note the contrast in surface communications.

Myanmar's territorial morphology is complicated by a shift in the Burmese core area that took place during colonial times. Prior to the colonial period, the focus of embryonic Burma lay in the so-called dry zone between the Arakan Mountains and the Shan Plateau. The urban focus of the state was Mandalay, which had a central situation and relative proximity to the non-Burmese highlands all around (Fig. 10-1). Then the British developed the agricultural potential of the Irrawaddy Delta, and Rangoon (now called Yangon) became the hub of the colony. The Irrawaddy waterway links the old and the new core areas, but the center of gravity has shifted to the south.

The political geography of Myanmar constitutes a particularly good example of the role and effect of territorial morphology on internal state structure. Not only is Myanmar a protruded state: its core area is also surrounded on the west, north, and east by a horseshoe of great mountains—where many of the country's 11 minority peoples had their homelands before the British occupation. The colonial boundaries had the effect of incorporating these peoples into the domain of the Burman people, who constitute about two-thirds of the population (50.9 million today). When the British departed and Burma became independent in 1948, the less numerous peoples traded one master for another. In 1976, nine of these indigenous peoples formed a union to demand the right to self-determination in their homelands.

As Figure 10-3 shows, the peripheral peoples of Myanmar occupy a significant part of the state. The Shan of the northeast and far north, who are related to the neighboring Thai, account for about 9 percent of the population, or 4.6 million. The Karen (7 percent, 3.6 million) live in the neck of Myanmar's protrusion and have proclaimed that they wish to create an autonomous territory within a federal Myanmar. Although the powerful military have dealt these aspirations a series of setbacks, centrifugal forces continue to bedevil the central regime. Its response has been to exert power by all available means rather than to accommodate these forces, even to the point of stifling political discourse (let alone opposition) among the Burman themselves.

▷ INSULAR SOUTHEAST ASIA

On the peninsulas and islands of Southeast Asia's southern and eastern periphery lie six of the realm's 11 states (Fig. 10-2). Few regions in the world contain so diverse a set of countries. Malaysia, the former British colony, consists of two major areas separated by hundreds of miles of South China Sea. The realm's southernmost state, Indonesia, sprawls across thousands of islands from Sumatera in the west to New Guinea in the east. North of the Indonesian archipelago lies the Philippines, a nation that once was a U.S. colony. These are three of the most severely fragmented states on Earth, and each has faced the challenges that such politico-spatial division brings. This insular region of Southeast Asia also contains two small but important sovereign entities: a city-state and a sultanate. The city-state is Singapore, once a part of Malaysia (and one instance in

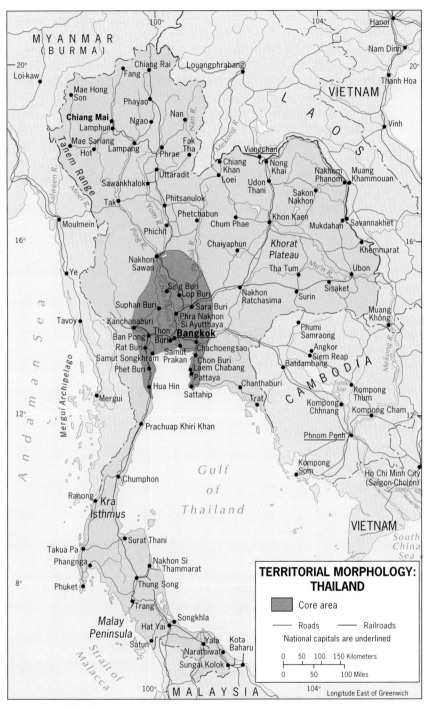

FIGURE 10-9

TERRITORIAL MORPHOLOGY: THAILAND

■ Core area

Roads ——— Railroads ———

National capitals are underlined

0   50   100   150 Kilometers

0   50   100 Miles

Longitude East of Greenwich

which internal centrifugal forces were too great to be overcome). The sultanate is Brunei, an oil-rich Muslim territory on the island of Borneo that seems transplanted from the Persian Gulf. In addition, a third small entity, East Timor, has just achieved statehood. Few parts of the world are more varied or interesting geographically.

MALAYSIA

The state of Malaysia represents one of the three types of fragmented states discussed earlier: the mainland-island type, in which one part of the national territory lies on a continent and the other on an island. Malaysia is a colonial political artifice that combines two quite disparate components into a single state: the southern end of the Malay Peninsula and the northern part of the island of Borneo. These are known, respectively, as West Malaysia and East Malaysia (Fig. 10-2). The name *Malaysia* came into use in 1963, when the original Federation of Malaya, on the Malay Peninsula, was expanded to incorporate the areas of Sarawak and Sabah in Borneo. When the name Malaya is used, it refers to the peninsular part of the Federation, whereas Malaysia refers to the total entity.

The Malays of the peninsula, traditionally a rural people, displaced older aboriginal communities there and today make up about 58 percent of the country's population of 24.2 million. They possess a strong cultural identity expressed in adherence to the Muslim faith, a common language, and a sense of territoriality that arises from their perceived Malayan origins and their collective view of Chinese, Indian, European, and other foreign intruders.

The Chinese came to the Malay Peninsula and to northern Borneo in substantial numbers during the colonial period, and today they constitute about one-fourth of Malaysia's population (they are the largest single group in Sarawak).

Hindu South Asians were in this area long before the Europeans, and for that matter before the Arabs and Islam arrived on these shores. Today they still form a substantial minority of over 7 percent of the

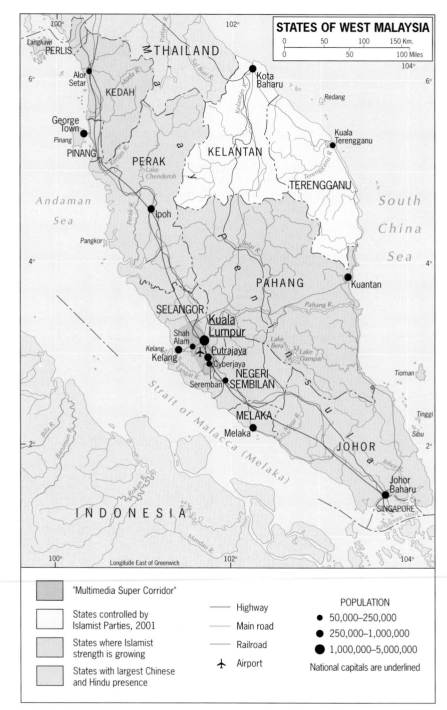

FIGURE 10-10

population, clustered, like the Chinese, on the western side of the peninsula (Fig. 10-3).

The populous peninsular part of Malaysia remains the country's dominant sector with 11 of its 13 States and nearly 80 percent of its population. Here the Malay-dominated government has strictly controlled economic and social policies while pushing the country's modernization. During the Asian economic boom of the 1990s, Malaysia's planners embraced the notion of symbols: the (then) capital, Kuala Lumpur, was endowed with the world's tallest building; a space-age airport outpaced Malaysia's needs; a high-tech administrative capital was built at Putrajaya, and a nearby development was called Cyberjaya—all part of a so-called *Multimedia Super Corridor* to anchor Malaysia's core area (Fig. 10-10).

The chief architect of this program was Malaysia's autocratic leader, Dr. Mahathir Mohamad. Not only did he have the support of a majority of the peninsula's Malays, but also his economic policies garnered the endorsement of most of Malaysia's ethnic Chinese. But Malaysia's headlong rush to modernize, coupled with some personal issues involving Mahathir and his rivals, caused a backlash among more conservative Muslims in the north, where an Islamic fundamentalist party made headway. By 2001, two northern States, tin-producing Kelantan and energy-rich but socially poor Terengganu, had Islamist State governments, and the fundamentalists' appeal in neighboring States was growing (Fig. 10-10).

The Malay Peninsula's primacy began long ago, during colonial times, when the British created a substantial economy based on rubber plantations, palm-oil extraction, and mining (tin, bauxite, copper, iron). The Strait of Malacca (Melaka) became one of the world's busiest and most strategic waterways, and Singapore, at the southern end of it, a prized possession (Singapore seceded from the Malaysian Federation in 1965).

Malaysia, despite the loss of Singapore and notwithstanding its recurrent ethnic troubles, became a major player on the burgeoning Pacific Rim. The strong skills and modest wages of the local workforce attracted many companies, and the government capitalized on its opportunities, for example, by encourag-

ing the creation of a high-technology manufacturing complex on the island of Pinang, where Chinese outnumber Malays by two to one and a future Singapore may be in the making.

The decision to combine Malaya with the States of Sabah and Sarawak on Borneo, creating the country now called Malaysia, had far-reaching consequences. These two States make up 60 percent of Malaysia's territory (although they represent only 20 percent of the population). They endowed Malaysia with major energy resources and huge stands of timber. They also complicated Malaysia's ethnic makeup because each State is home to more than two dozen indigenous groups (in fact, the immigrant Chinese form the largest single group in Sarawak). These locals complain that the federal government in Kuala Lumpur treats East Malaysia as a colony, and politics here are contentious and fractious. It is likely that Malaysia will confront devolutionary forces in East Malaysia as time goes on.

Also located on Borneo—where Sarawak and Sabah meet—is Brunei, a rich, oil-exporting Islamic sultanate far from the Persian Gulf. Brunei, the remnant of a former Islamic kingdom that once controlled all of Borneo and areas beyond, came under British control and was granted independence in 1984. Just slightly larger than Delaware and with a population of about 350,000, the sultanate is a mere ministate—except for the discovery of oil in 1929 and natural gas in 1965, which made this Southeast Asia's richest country except Singapore. And there are indications that further discoveries will be made in the offshore zone owned by Brunei. The Sultan of Brunei rules as an absolute monarch; his palace in the capital Bandar Seri Begawan, is reputed to be the world's largest. He will have no difficulty finding customers for his oil in energy-poor eastern Asia.

SINGAPORE

As the map shows, Singapore lies where the Strait of Malacca (Melaka), leading westward to India, opens into the South China Sea and the waters of Indonesia (Fig. 10-11 inset). The port developed as part of the

British Southeast Asian empire, and when independence came in 1963, it was made a part of the Malaysian Federation. However, two years later Singapore seceded from Malaysia and became a genuine city-state on the Pacific Rim.

Benefiting from its relative location, the old port of Singapore had become one of the world's busiest (by numbers of ships served) even before independence. **17** It thrived as an *entrepôt* between the Malay Peninsula, Southeast Asia, Japan, and other emerging economic powers on the Pacific Rim and beyond. Crude oil from Southeast Asia still is unloaded and refined at Singapore, then shipped to Asian destinations. Raw rubber from the adjacent peninsula and from Indonesia's island of Sumatera is shipped to Japan, the United States, China, and other countries. Timber from Malaysia, rice, spices, and other foodstuffs are processed and forwarded via Singapore. In return, automobiles, machinery, and equipment are imported into Southeast Asia through Singapore.

But that is the old pattern. Singapore's leaders want to redirect the city-state's economy and move toward high-tech industries for the future. In Singapore the government tightly controls business as well as other aspects of life. (Some newspapers and magazines have been banned for criticizing the regime, and there are even fines for such things as eating on the subway and failing to flush a public toilet.) Its overall success after secession has tended to keep the critics quiet: while GNP per capita from 1965 to 2000 multiplied by a factor of more than 15—to over U.S. $30,000—that of neighboring Malaysia reached just U.S. $3670. Among other things, Singapore became (and for many years remained) the world's largest producer of disk drives for small computers.

To accomplish this revival, Singapore has moved in several directions. First, it will focus on three growth areas: information technology, automation, and biotechnology. Second, there are notions of a "Growth Triangle" involving Singapore's developing neighbors, Malaysia and Indonesia; those two countries would supply the raw materials and cheap labor, and Singapore the capital and technical know-how. Third, Singapore opened its doors to capitalists of Chinese ancestry who left Hong Kong when

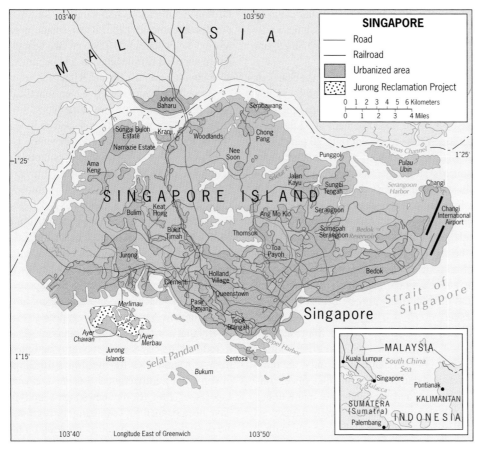

FIGURE 10-11

China took it over and who wanted to relocate their enterprises here. Singapore's population is 76 percent Chinese, 15 percent Malay, and 7 percent South Asian. The government is Chinese-dominated, and its policies have served to sustain Chinese control. Indeed, Singapore's combination of authoritarianism and economic success often is cited in China itself as proving that communism and market economics can coexist.

INDONESIA

The fourth-largest country in the world in terms of human numbers is also the globe's most expansive

 archipelago. Spread across more than 17,000 islands, Indonesia's 220 million people live both separated and clustered—separated by water and clustered on islands large and small.

The complicated map of Indonesia requires close attention (Fig. 10-12). Five large islands dominate the archipelago territorially, but one of these, New Guinea in the east, is not part of the Indonesian culture sphere, although its western half is under Indonesian control. The other four major islands are collectively known as the Greater Sunda islands: *Jawa* (Java), smallest but by far the most populous and important; *Sumatera* (Sumatra) in the west, directly across the Strait of Malacca from Malaysia; *Kalimantan,* the Indonesian sector of large, compact, minicontinent Borneo; and

wishbone-shaped, distended *Sulawesi* (Celebes) to the east. Extending eastward from Jawa are the Lesser Sunda Islands, including Bali and, near the eastern end, Timor. Another important island chain within Indonesia is the Maluku (Molucca) Islands, between Sulawesi and New Guinea. The central water body of Indonesia is the Java Sea.

Indonesia is a Dutch colonial creation, and the Dutch chose Jawa as their colonial headquarters, making Batavia (now Jakarta) their capital. Today, **Jawa** remains the core of Indonesia. With about 130 million inhabitants, Jawa is one of the world's most densely peopled places and one of the most agriculturally productive. Jawa also is the most highly urbanized part of a country in which more than 60 percent of the people still live on the land, and the Pacific Rim boom of the 1990s had strong impact here. The city of Jakarta, on the northwestern coast, became the heart of a larger conurbation now known as *Jabotabek*, consisting of the capital as well as Bogor, Tangerang, and Bekasi. During the 1990s the population of this megalopolis grew from 15 to 20 million, and it is predicted to reach 30 million by 2010. Already, Jabotabek is home to over 10 percent of Indonesia's entire population and 25 percent of its urban population. Thousands of factories, their owners taking advantage of low prevailing wages, were built in this area, straining its infrastructure and overburdening the port of Jakarta. On an average day, hundreds of ships lie at anchor, awaiting docking space to offload raw materials and take on finished products.

Always in Indonesia, Jawa is where the power lies. As a cultural group (though itself heterogeneous), the Jawanese constitute about 60 percent of the country's population. In the center of the island, a politically powerful sultanate centers on Yogyakarta; Jawa also is the main base for the two national Islamic movements, one comparatively moderate and the other increasingly revivalist.

Sumatera, Indonesia's westernmost island, forms the western shore of the busy Strait of Malacca; Singapore lies across the Strait from approximately the middle of the island. Although much larger than Jawa, Sumatera has only about one-third as many people (45 million). In colonial times the island be-

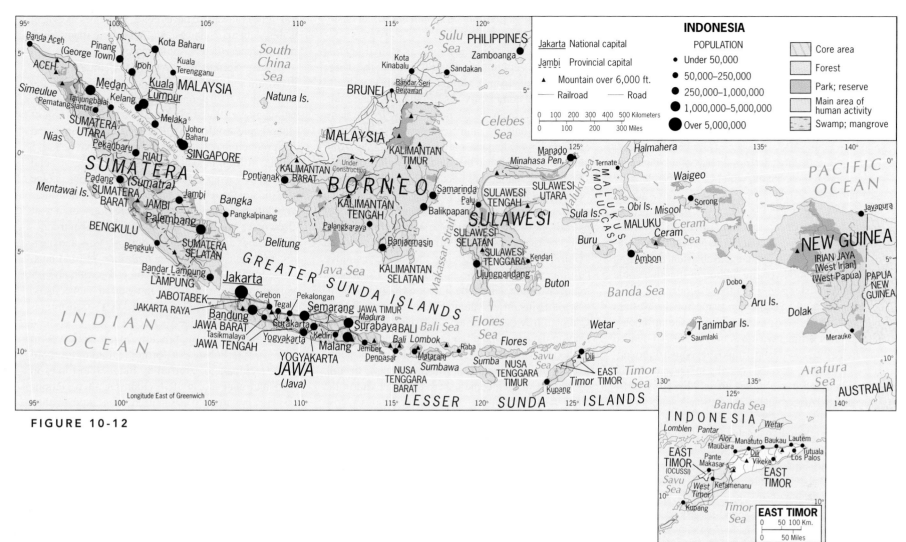

FIGURE 10-12

came a base for rubber and palm-oil plantations; its high relief makes possible the cultivation of a wide range of crops, and neighboring Bangka and Belitung yield petroleum and natural gas. Palembang is the key urban center in the south, but current attention focuses on the north. There, the Batak people accommodated colonialism and Westernization and made Medan one of Indonesia's Pacific Rim boom towns. Farther north the Aceh fought the Dutch into

the twentieth century and now demand greater autonomy, if not outright independence, for their province from the Indonesian government. The Province of Riau, too, is politically restive, and Sumatera presents Jakarta with economic assets and political liabilities.

Kalimantan is the Indonesian part of the island of Borneo, a slab of the Earth's crystalline crust whose backbone of tall mountains is of erosional, not vol-

canic, origin. Larger than Texas, Borneo has a deep, densely forested interior that is a last refuge for some 30,000 orangutans. Elephants, rhinoceroses, and tigers survive even as loggers attack their habitat. Borneo's Pleistocene heritage survives because its human population still is small (14 million on the Indonesian side, 5 million on the Malaysian) and because indigenous peoples, principally the Dayak clans, had less impact on the natural environment than the Indonesian and

Chinese immigrants who log the forests and clear land for farms. As Figure 10-12 shows, the only towns of any size in Kalimantan lie on or near the coast; routes into the interior still are few and far between.

Sulawesi consists of a set of intersecting, volcanic mountain ranges rising above sea level; the 500-mile (800-km) Minahasa Peninsula is still growing by volcanic action into the Philippine Sea. This northern peninsula, a favorite of the Dutch colonizers, remains the most developed part of an otherwise rugged and remote island, with Manado its relatively prosperous focus. Seven major ethnic groups inhabit the valleys and basins between the mountains, but the population of 17 million also includes immigrants from Jawa, especially in and around the southern center of Ujungpandang. Subsistence farming is the leading mode of life, although logging, some mining, and fishing augment the economy.

West Irian, the Indonesian name for the western part of New Guinea, has become an issue in Indonesian politics. **Irian Jaya** (*Jaya* means West), bounded to the east by a classic superimposed geometric boundary (Fig. 10-12), was taken over by Indonesia from the Dutch in 1969. It constitutes about 22 percent of Indonesia's territory, but its population is barely more than 2 million—just 1 percent of the nation. The indigenous inhabitants of this province, which is in effect a colony, are Papuan, most living in remote reaches of this mountainous and densely forested island. Irian Jaya is economically important to Indonesia, for it contains what is reputed to be the world's richest gold mine and its second-largest open-pit copper mine, but political consciousness has reached the Papuans. The Free Papua Movement has held small rallies in the capital, Jayapura, displaying a Papuan flag and demanding recognition.

Diversity in Unity

Indonesia's survival as a unified state is as remarkable as India's and Nigeria's. With more than 300 discrete ethnic clusters, over 250 languages, and just about every religion practiced on Earth (although Islam dominates), actual and potential centrifugal forces are powerful here. Wide waters and high mountains perpetuate cultural distinctions and differences. Indone-

sia's national motto is *bhinneka tunggal ika*: diversity in unity.

What Indonesia has achieved is etched against the country's continuing cultural complexity. There are dozens of distinct aboriginal cultures; virtually every coastal community has its own roots and traditions. And the majority, the rice-growing Indonesians, include not only the numerous Jawanese—who are Muslims largely in name only and have their own cultural identity—but also the Sundanese (who constitute 14 percent of Indonesia's population), the Madurese (8 percent), and others. Perhaps the best impression of the cultural mosaic comes from the string of islands that extends eastward from Jawa to Timor (Fig. 10-12). The rice-growers of Bali adhere to a modified version of Hinduism, giving the island a unique cultural atmosphere; the population of Lombok is mainly Muslim, with some Balinese Hinduism; Sumbawa is a Muslim community; Flores is mostly Roman Catholic. In western Timor, Protestant groups dominate; in the east, where the Portuguese ruled, Roman Catholicism prevails. Nevertheless, Indonesia nominally is the world's largest Muslim country: overall, 88 percent of the people adhere to Islam, and in the cities the silver domes of neighborhood mosques rise above the townscape. But Islam is not (perhaps not yet) the issue it is in Malaysia, where observance generally is stricter and where minorities fear Islamization as Malay power grows.

Transmigration and the Outer Islands

As noted earlier, Indonesia's population of 219 million makes it the world's fourth most populous country, but Jawa, we also noted, contains approximately 60 percent of it. With about 130 million people on an island the size of Louisiana, the population pressure is enormous here. Moreover, Indonesia's annual rate of population growth remains at 1.6 percent, resulting in a doubling time of 44 years. To deal with this problem, and at the same time to strengthen the core's power over outlying areas, the Indonesian government has long **19** pursued a policy known as **transmigration**, inducing Jawanese (and Madurese from the adjacent island of Madura) to relocate to other islands. Sev-

eral million Jawanese have moved to locales as distant as the Malukus, Sulawesi, and Sumatera; many Madurese were resettled in Kalimantan.

In the government's view, the transmigration policy would serve to counter centrifugal forces arising from Indonesia's geography, would spread the official language (Bahasa Indonesia, a modified form of Malay), and would bring otherwise unexploited land into production. It has done so, but it also has negative effects. In Kalimantan, the Muslim Madurese found themselves facing armed Dayaks, indigenous people who did not want to give up their land, and thousands were killed in what amounted to a regional civil war. Elsewhere, Jawanese often are blamed for introducing methods and practices of administration and business that run counter to local traditions. In Irian Jaya, where Indonesia is in effect a colonial power, aboriginal people have attacked Indonesian newcomers. Only the future will tell whether Jakarta's transmigration policy will ultimately succeed.

Certainly Indonesian policy did not work well in East Timor, Portugal's former colonial foothold in the archipelago, overrun by Indonesian forces in 1975 and annexed in 1976 against United Nations wishes. The largely Christianized, long-impoverished population of about 800,000 suffered as a bitter struggle for independence led to destruction and retaliation. Finally, in 1999 the people were allowed to vote on the issue, and most favored sovereignty. To this the Indonesians reacted with fury, and engaged in a campaign of devastation and killing that had to be countered by foreign intervention.

As the inset map in Figure 10-12 shows, East Timor's move toward independence under UN auspices creates some politico-geographical complications. East Timor consists of a main territory, where the capital named Dili is located, and a small exclave on the north coast of (still-Indonesian) West Timor. Relations between Indonesia and the East Timorese are such that overland links may not be feasible. And in the Timor Sea, the maritime boundary between East Timor and Australia crosses the so-called Timor Gap, where valuable oil reserves lie. That boundary was adjusted in 2002, giving East Timor 90 percent of the revenues derived from these reserves—a first ray of hope for the evolving state's economic future.

FROM THE FIELD NOTES

"Getting to Ambon was no easy matter, and my boat trip was enlivened by an undersea earthquake that churned up the waters and shook up all aboard. The next morning our first view of the town of Ambon, provincial capital of Maluku, showed a center dominated by a large mosque with a modern minaret and a large white dome to the right of the main street leading to the port. My host from the local university told me that, as on many of the islands of the Malukus, about half of the people were Christians, not Muslims, and that some large churches were situated in the outskirts. 'But the relationships between Muslims and Christians are worsening,' he said. It had to do with the arrival of Jawanese, some of whom were 'agitators' and Islamic fundamentalists, stirring up religious passions. 'Have you heard of the Taliban?' he asked. 'Well, we have similar so-called religious students here, educated in Islamic schools that teach extremism.' . . . We drove from the town into the countryside to the west, toward Wakisihu and past the old Dutch Fort Rotterdam. 'Let me show you something,' he said as we turned up a dirt road. At its end, in the middle of a field, sat a large single structure called University of Islam, distinguished by a pair of enormous stairways. 'Can you call a single building like this where all they teach is Islam a university?' he asked. 'What they teach here is Islamic fundamentalism and intolerance.' He was prophetic. Weeks later, religious conflict broke out, the mosque in town was burned along with Christian churches; his own university lay in ruins."

THE PHILIPPINES

North of Indonesia, across the South China Sea from Vietnam, and south of Taiwan lies an archipelago of more than 7000 islands (only about 460 of them larger than one square mile in area) inhabited by 83.9 million people. The inhabited islands of the Philippines can be viewed as three groups: (1) Luzon, largest of all, and Mindoro in the north, (2) the Visayan group in the center, and (3) Mindanao, second largest, in the south (Fig. 10-13). Southwest of Mindanao lies a small group of islands, the Sulu Archipelago, nearest to Indonesia,

where Muslim-based insurgencies have kept the area in turmoil.

Few of the generalizations we have been able to make for Southeast Asia could apply in the Philippines without qualification. The country's location relative to the mainstream of change in this part of the world has had much to do with this situation. The islands, inhabited by peoples of Malay ancestry with Indonesian strains, shared with much of the rest of Southeast Asia an early period of Hindu cultural influence, which was strongest in the south and southwest and diminished northward. Next came a Chinese invasion, felt more strongly on the largest island of Luzon in the northern

part of the Philippine archipelago. Islam's arrival was delayed somewhat by the position of the Philippines well to the east of the mainland and to the north of the Indonesian islands. The few southern Muslim beachheads were soon overwhelmed by the Spanish invasion during the sixteenth century. Today the Philippines, adjacent to the world's largest Muslim state (Indonesia), is 83 percent Roman Catholic, 9 percent Protestant, and only 5 percent Muslim.

Out of the Philippines melting pot, where Mongoloid-Malay, Arab, Chinese, Japanese, Spanish, and American elements have met and mixed, has emerged the distinctive Filipino culture. It is not a

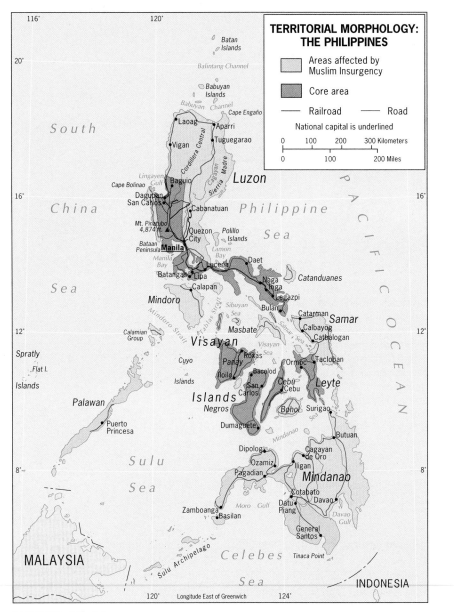

FIGURE 10-13

areas (Fig. I-9): (1) the northwestern and south-central part of Luzon, (2) the southeastern extension of Luzon, and (3) the islands of the Visayan Sea between Luzon and Mindanao. Luzon is the site of the capital, ✳ Manila-Quezon City (11.7 million, nearly one-seventh of the entire national population), a major metropolis facing the South China Sea. Alluvial as well as volcanic soils, together with ample moisture in this tropical environment, produce self-sufficiency in rice and other staples and make the Philippines a net exporter of farm products despite a high population growth rate of 2.2 percent.

In recent years, the population issue has divided this dominantly Roman Catholic society, with the government promoting family planning and the clergy opposing it. But behind this debate lies another of the Philippines' assets: in a realm of mostly undemocratic regimes, the Philippines has come out of its period of authoritarian rule a rejuvenated, if not yet robust, democracy.

The Philippines seems to get little mention in discussions of developments on the Pacific Rim, and yet it would seem to be well positioned to share in the Pacific Rim's economic growth. Governmental mismanagement and political instability have slowed the country's participation, but during the 1990s the situation improved. Despite a series of jarring events—the ouster of U.S. military bases, the damaging eruption of a volcano near the capital, the violence of Muslim insurgents, and a dispute over the nearby Spratly Islands in the South China Sea—the Philippines made substantial economic progress during the decade. Its electronics and textile industries (mostly in the Manila hinterland) expanded continuously, and more foreign investment arrived. But agriculture continues to dominate the Philippines' economy, unemployment remains high, further land reform is badly needed, and social restructuring (reducing the controlling influence over national affairs by a comparatively small group of families) must occur. However, progress is being made. The country now is a lower-middle-income economy, and given a longer period of stability and success in reducing the population growth rate, it will rise to the next level and finally take its place among Pacific Rim growth poles.

homogeneous or unified culture, as reflected by the nearly 90 Malay languages in use in the islands, but it is in many ways unique. At independence in 1946, the largest of the Malay languages, Tagalog (also called Pilipino), became the country's official language. But English is widely learned as a second language, and a Tagalog-English hybrid, "Taglish," is increasingly heard today.

The Philippines' population, concentrated where the good farmlands lie, is densest in three general

11 / The Austral Realm

Sydney, Australia: recently-emergent primate city, its famed Opera House an image of sea and sail at the waterfront gateway of an island continent.

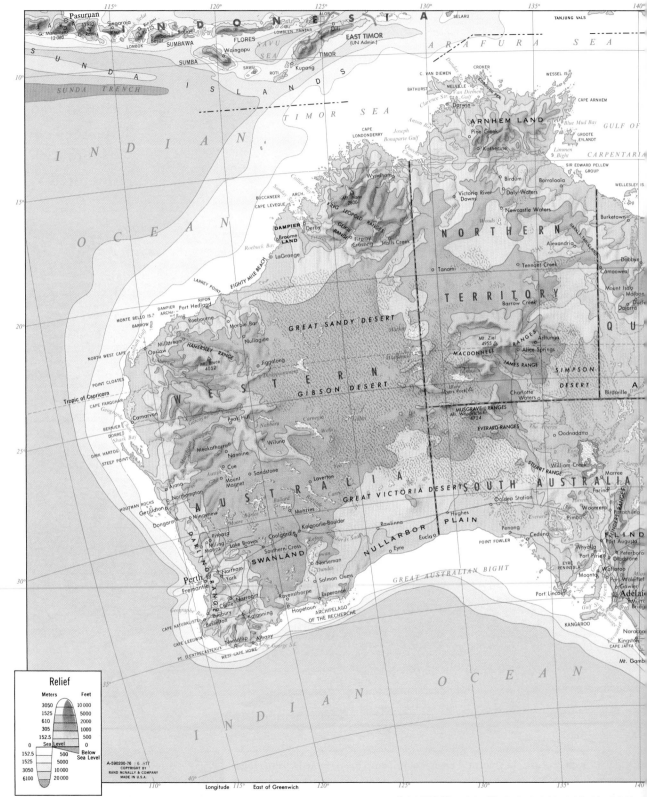

FIGURE 11-1

Scale 1:16 000 000; one inch to 250 miles. Lambert's Azimuthal, Equal Area Projection

Elevations and depressions are given in feet

chapter 11 / The Austral Realm

CONCEPTS, IDEAS, AND TERMS

1 Austral

2 Southern Ocean

3 Subtropical Convergence

4 West Wind Drift

5 Biogeography

6 Wallace's Line

7 Aboriginal population

8 Outback

9 Federation

10 Unitary state

11 Import-substitution industries

12 Aboriginal land issue

13 Immigration policies

14 Environmental degradation

15 Peripheral development

REGIONS

AUSTRALIA

  CORE AREA

  OUTBACK

NEW ZEALAND

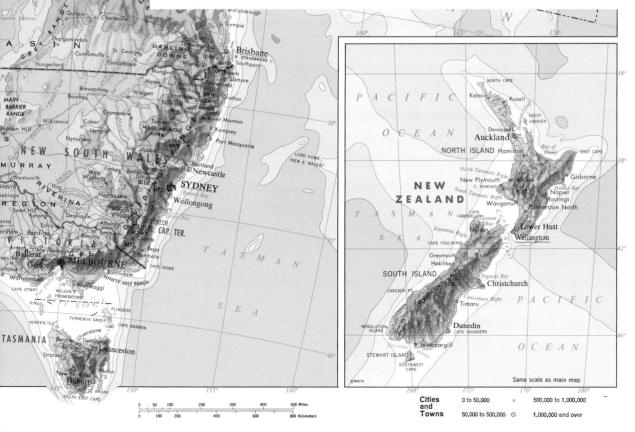

The Austral Realm is geographically unique. It is the only geographic realm that lies entirely in the Southern Hemisphere. It is also the only realm that has no land link of any kind to a neighboring realm, and is thus completely surrounded by ocean and sea. It is second only to the Pacific as the world's least populous realm. Appropriately, its name refers **1** to its location (**Austral** means south)—a location far from the sources of its dominant cultural heritage but close to its newfound economic partners on the western Pacific Rim.

DEFINING THE REALM

Two countries constitute this Austral Realm: Australia, in every way the dominant one, and New Zealand, physiographically more varied than its giant partner. Between them lies the Tasman Sea. To the west lies the Indian Ocean, to the east the Pacific, and to the south the frigid Southern Ocean.

This southern realm is at a crossroads. On the doorstep of populous Asia, its Anglo-European legacies are now infused by other cultural strains. Polynesian Maori in New Zealand and Aboriginal communities in Australia are demanding better terms of life. Pacific Rim markets are buying huge quantities of raw materials. Japanese and other Asian tourists fill hotels and resorts. Queensland's tropical Gold Coast resembles Honolulu's Waikiki. The streets of Sydney and Melbourne display a multicultural panorama unimagined just two generations ago. All these changes have stirred political debate. Issues ranging from immigration quotas to indigenous land rights dominate, exposing social fault lines (city versus Outback in Australia, North and South in New Zealand). Aborigines and Maori were here first, and the Europeans came next. Now Asia looms in Australia's doorway.

LAND AND ENVIRONMENT

Physiographic contrasts between massive, compact Australia and elongated, fragmented New Zealand are related to their locations with respect to the Earth's tectonic plates (consult Fig. I-4). Australia, with some of the geologically most ancient rocks on the planet, lies at the center of its own plate, the Australian Plate. New Zealand, younger and less stable, lies at the convulsive convergence of the Australian and Pacific Plates. Earthquakes are rare in Australia and volcanic eruptions are unknown; New Zealand has plenty of both. This locational contrast is also reflected by differences in relief (Fig. 11-1). Australia's highest relief occurs in what Australians call the Great Dividing Range, the mountains that line the east coast from the Cape York Peninsula to southern Victoria, with an outlier in Tasmania. The highest point along these old, now eroding moun-

◆ Major Geographic Qualities of the Austral Realm

1. Australia and New Zealand constitute a geographic realm by virtue of territorial dimension, relative location, and dominant cultural landscape.

2. Despite their inclusion in a single geographic realm, Australia and New Zealand differ physiographically. Australia has a vast, dry, low-relief interior; New Zealand is mountainous.

3. Australia and New Zealand are marked by peripheral development—Australia because of its aridity, New Zealand because of its topography.

4. The populations of Australia and New Zealand are not only peripherally distributed but also highly clustered in urban centers.

5. The realm's human geography is changing—in Australia because of Aboriginal activism and Asian immigration, and in New Zealand because of Maori activism and Pacific-islander immigration.

6. The economic geography of Australia and New Zealand is dominated by the export of livestock products (and in Australia also by wheat production and mining).

7. Australia and New Zealand are being integrated into the economic framework of the western Pacific Rim, principally as suppliers of raw materials.

tains is Mount Kosciusko, 7310 feet (2228 m) tall. In New Zealand, entire ranges are higher than this, and Mount Cook reaches 12,315 feet (3764 m).

West of Australia's Great Dividing Range, the physical landscape generally has low relief with some local exceptions such as the Macdonnell Ranges near the center; plateaus and plains dominate (Fig. 11-2). The Great Artesian Basin is a key physiographic region, providing underground water sources in what would otherwise be desert country, and crossed in the south by the continent's major river system, the Murray-Darling. The area mapped as *Shield* in Figure 11-2 contains Australia's oldest rocks and much of its mineral wealth.

Figure I-8 reveals the effects of latitudinal location and interior isolation on Australia's climatology. In this respect, Australia is far more varied than New Zealand, its climates ranging from tropical in the far north, where rainforests flourish, to Mediterranean in parts of the south. The interior is dominated by desert and steppe conditions, the steppes providing the grasslands that sustain tens of millions of livestock. Only in the east does Australia have an area of humid temperate climate, and here lies most of the country's economic core area. New Zealand, by contrast, is totally under the influence of the Southern and Pacific Oceans, creating moderate, moist conditions, temperate in the north and colder in the south.

2 Twice now we have referred to the **Southern Ocean**, but try to find this ocean on maps and globes published by famous cartographic organizations such as the National Geographic Society and Rand McNally. From their maps you would conclude that the Atlantic, Pacific, and Indian Oceans reach all the way to the shores of Antarctica.

Australians and New Zealanders know better. They experience the frigid waters and persistent winds of this great weathermaker on a daily basis.

For us geographers, it is a good exercise to turn the globe upside down now and then. After all, the usual orientation is quite arbitrary. Modern mapmaking started in the Northern Hemisphere, and the cartographers put their hemisphere on top and the other at the bottom. That is now the norm, and it can distort our view of the world. In bookstores in the Southern Hemisphere, you sometimes see tongue-in-cheek maps showing Australia and Argentina at the top, and Europe and Canada at the bottom. But this matter has a serious side. A reverse view of the globe shows us how vast the ocean encircling Antarctica is. The Southern Ocean may be remote, but its existence is real.

Where do the northward limits of the Southern Ocean lie? This ocean is bounded not by land but by a marine transition called **3** the **Subtropical Convergence**. Here the cold, extremely dense waters of the Southern Ocean meet the warmer waters of the Atlantic, Pacific, and Indian Oceans. It is quite sharply defined by changes in temperature, chemistry, salinity, and marine fauna. Flying over it, you can actually observe it in the changing colors of the water: the Antarctic side is a deep gray, the northern side a greenish blue.

Although the Subtropical Convergence moves seasonally, its position does not vary far from latitude 40° South, which also is the approximate northern limit of Antarctic icebergs. Defined this way, the great Southern Ocean is a huge body of water that moves clockwise (from west to east) around Antarctica, which is why we also call it the **4** **West Wind Drift**.

Biogeography

One of this realm's defining characteristics is its wildlife. Australia is the land of kangaroos and koalas, wallabies and wombats, possums and platypuses. These and numerous other *marsupials* (animals whose young are born very early in their development and then carried in an abdominal pouch) owe their survival to Australia's early isolation during the breakup of Gondwana (see Fig. 7-3). Before more advanced mammals could enter Australia and replace the marsupials, as happened in other parts of the world, the landmass was separated from Antarctica and India, and today it contains the world's largest assemblage of marsupial fauna.

Australia's vegetation also has distinctive qualities, notably the hundreds of species of eucalyptus trees native to this geographic realm. Many other plants form part of Australia's unique flora, some with unusual adaptation to the high temperatures and low humidity that characterize much of the continent.

The study of fauna and flora in spatial perspective combines the disciplines of biology and geography in **5** a field known as **biogeography**, and Australia is a giant laboratory for biogeographers. In the Introduction (pp. 11-13), we noted that several of the world's climatic zones are named after the vegetation that marks them: tropical savanna, steppe, tundra. When climate, soil, vegetation, and animal life reach a long-term, stable adjustment, vegetation forms the most visible element of this ecosystem.

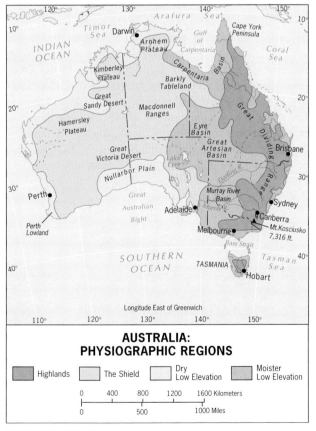

AUSTRALIA: PHYSIOGRAPHIC REGIONS

| Highlands | The Shield | Dry Low Elevation | Moister Low Elevation |

0 400 800 1200 1600 Kilometers
0   500   1000 Miles

FIGURE 11-2

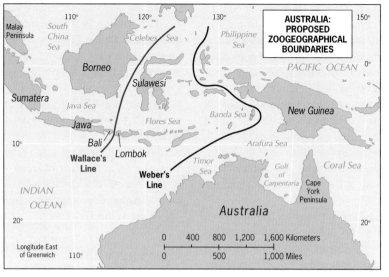

FIGURE 11-3

Biogeographers are especially interested in the distribution of plant and animal species, and in the relationships between plant and animal communities and their natural environments. (The study of plant life is called *phytogeography*; the study of animal life is called *zoogeography*.) These scholars seek to explain the distributions the map reveals. In 1876 one of the founders of biogeography, Alfred Russel Wallace, published a book entitled *The Geographical Distribution of Animals* in which he fired the first shot in a long debate: where does the zoogeographic boundary of Australia's fauna lie? Wallace's fieldwork in the area revealed that Australian forms exist not only in Australia itself but also in New Guinea and in some islands to the west. So Wallace proposed that the faunal boundary should lie between Borneo and Sulawesi, and just east of Bali (Fig. 11-3).

6 **Wallace's Line** soon was challenged by other researchers, who found species Wallace had missed and who visited islands Wallace had not. There was no question that Australia's zoogeographic realm ended somewhere in the Indonesian archipelago, but where? Western Indonesia was the habitat of non-marsupial animals such as tigers, rhinoceroses, and elephants, as well as primates; New Guinea clearly was part of the realm of the marsupials. How far had the more advanced mammals progressed eastward along the island stepping stones toward New Guinea? The zoogeographer Max Weber found evidence that led him to postulate his own *Weber's Line*, which, as Figure 11-3 shows, lay very close to New Guinea.

Not all research in zoogeography or phytogeography deals with such large questions. Much of it focuses on the relationships between particular species and their habitats, that is, the environment they normally occupy and of which they form a part. Such environments change, and the changes can spell disaster for the species. In Australia, the arrival of the **7** **Aboriginal population** (between 50,000 and 60,000 years ago) had limited effect on the habitats of the extant fauna. But the invasion of the European colonizers and the introduction of their livestock led to the destruction of habitats and the extinction of many native species. Whether in East Africa or in western Australia, the key to the conservation of what remains of natural flora and fauna lies in the knowledge embodied by the field of biogeography.

REGIONS OF THE REALM

Australia is the dominant component of the Austral Realm, a continent-scale country in a size category that also includes China, Canada, the United States, and Brazil. For two reasons, however, Australia has fewer regional divisions than do the aforementioned countries: Australia's relatively uncomplicated physiography and its diminutive human numbers. Our discussion, therefore, uses the core-periphery concept as a basis for investigating Australia and focuses on New Zealand as a region by itself.

AUSTRALIA

On January 1, 2001, Australia celebrated its 100th birthday as a state, the Commonwealth of Australia, recognizing (still) the British monarch as the head of state and entering its second century with a strong economy, stable political framework, high standard of living for most of its people, and favorable prospects ahead. Positioned on the Pacific Rim, ten times the size of Texas, well endowed with farmlands and vast pastures, major rivers, ample underground water, minerals, and energy resources, served by good natural

Regional Geography

▶ de Blij/Muller — *Geography: Realms, Region, and Concepts 10e* (0-471-40775-5)

▶ de Blij/Muller — *Concepts and Regions in Geography* (0-471-09303-3)

▶ Blouet/Blouet — *Latin America and the Caribbean 4e* (0-471-85103-5)

▶ Weightman — *Dragons and Tigers: Geography of South, East and Southeast Asia* (0-471-25358-8)

Physical Geography

▶ Strahler/Strahler — *Physical Geography: Science and Systems of the Human Environment 2e* (0-471-23800-7)

▶ Strahler/Strahler — *Introducing Physical Geography 3e* (0-471-41741-6)

▶ MacDonald — *Biogeography: Introduction to Space, Time and Life* (0-471-24193-8)

▶ Cutter/Renwick — *Exploration, Conservation, Preservation: A Geographic Perspective on Natural Resource Use* (0-471-15225-0)

Human Geography

▶ Kuby/Harner/Gober — *Human Geography in Action 2e* (0-471-40093-9)

▶ de Blij/Murphy — *Human Geography 7e* (0-471-44107-4)

Water Resources

▶ Cech — *Water Resources* (0-471-43861-8)

GIS

▶ Chrisman — *Exploring GIS 2e* (0-471-31425-0)

▶ DeMers — *GIS Modeling in Raster* (0-471-31965-1)

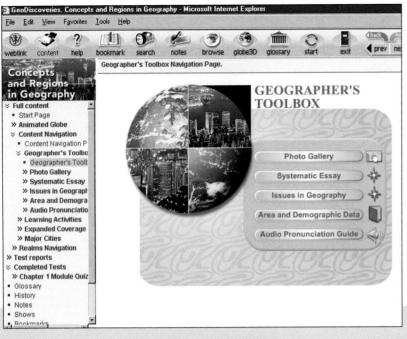

Geographer's Tool Box on CD

Additional visual, audio, and text resources, including:

 Photo Gallery Visual tour of the realm, highlighted by photos taken by Harm de Blij. From the Field Notes and Locator Maps provide valuable insights into how a geographer observes and interprets information in the field.

Interactive Globe Explore and understand the world by changing the face of this 3-D globe by using 5 distinct textures:
- World Realms Map that launch into specific realm-based learning modules
- Earth From Space
- Political Map with country borders
- Plate Tectonics
- Population Density

Systematic Essays Thirteen in-depth explorations of systematic specialties.

Issues in Geography Topical and sometimes controversial issues are highlighted for each world realm.

Audio Pronunciation Guide Index of over 2000 audio files with key words and place-names from the text.

Area & Demographic Data Interactive gazetteer providing key data for every country and world realm.

More to Explore

✳ **Systematic Essays**

✳ **Issues in Geography**

Chapter by Chapter Table of Contents

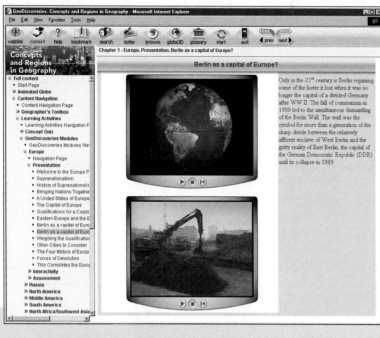

Berlin as a capital of Europe?

Only in the 21st century is Berlin regaining some of the luster it lost when it was no longer the capital of a divided Germany after WW II. The fall of communism in 1989 led to the simultaneous dismantling of the Berlin Wall. The wall was the symbol for more than a generation of the sharp divide between the relatively affluent enclave of West Berlin and the gritty reality of East Berlin, the capital of the German Democratic Republic (DDR) until its collapse in 1989.

Learning Activities on CD

In-depth interactive exercises designed to complement and further your exploration of each world realm.

1 Concepts, Ideas, and Terms Quiz Flashcard exercises and multiple tests help students review and quiz themselves on concepts, ideas, and terms highlighted in each chapter.

GeoDiscoveries Modules These learning activities allow you to explore important concepts in more depth. Each module has a three-part structure:
  • Presentations that use videos, animations, and other resources to focus on key concepts from the chapter.
  • Interactivities that engage students in concept-based exercises.
  • Assessment self-tests that allow students to measure their comprehension of the concept being explored.

Map Quizzes Mastery of place names is crucial to your success in this course. For each world realm, GeoDiscoveries offers three game-formatted pace name activities that quiz
  • Countries, States and Provinces
  • Cities
  • Physical Features

On the Web site www.wiley.com/college/deblij

Chapter Quizze's Link to our Web site for our on-line graded quizzes.

Virtual Field Trips Link to our Web site for these interesting and informative resources.

GeoResources Link to our Web site for on-line resources including Lonely Planet country guides, live radio and Webcams, and annotated Web links.

CHAPTER-SPECIFIC MODULES

Chapter 1   GeoDiscoveries: Supranationalism and devolution

Chapter 2   GeoDiscoveries: Industrialization and transportation

Chapter 3   GeoDiscoveries: Urbanization and sprawl

Chapter 4   GeoDiscoveries: Maquiladoras and border relationships

Chapter 5   GeoDiscoveries: Land use patterns

Chapter 6   GeoDiscoveries: Oil resources of the Caspian Sea

Chapter 7   GeoDiscoveries: Diffusion and medical geography

Chapter 8   GeoDiscoveries: Agriculture and climate

Chapter 9   GeoDiscoveries: Enterprise zones and traditional economics

Chapter 10 GeoDiscoveries: The benefits and drawbacks of tourism

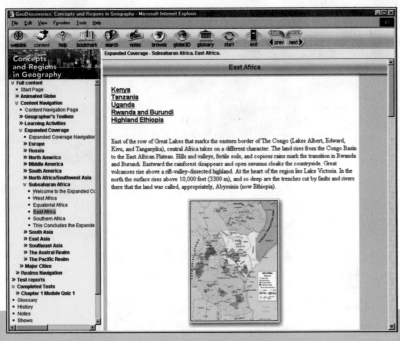

Expanded Coverage on CD

▷ **Regions of the Realm** Use the CD to expand your exploration of each realm's distinctive regions. Additional text, illustrations, and maps are available to help you prepare for tests and complete research papers.

Chapter 1 Western Europe, The British Isles, Northern (Nordic) Europe, Mediterranean Europe, Eastern Europe

Chapter 2 Russian Core and Peripheries, Eastern Frontier, Siberia, Russian Far East

Chapter 3 North American Core, Maritime Northeast, French Canada, Continental Interior, South, Southwest, Western Frontier, Northern Frontier, Pacific Hinge

Chapter 4 Caribbean Basin (Greater Antilles & Lesser Antilles), Mexico, Central America

Chapter 5 Brazil, Caribbean North, Andean West, Southern Cone

Chapter 6 Egypt and the Lower Nile Basin, Maghreb and its neighbors, African Transition Zone, Middle East, Arabian Peninsula, Empire States, Turkestan

Chapter 7 West Africa, Equatorial Africa, Southern Africa

Chapter 8 Pakistan, India, Bangladesh, Mountainous North, Southern Islands

Chapter 9 China Proper, Xizang (Tibet), Xinjiang, Mongolia, Jakota Triangle (Japan-Korea- Taiwan)

Chapter 10 Mainland Southeast Asia, Insular Southeast Asia

Chapter 11 Australia (Core Area, Outback), New Zealand

Chapter 12 Melanesia, Micronesia, Polynesia

Major Cities on CD

This feature reflects the growing process and influence of urbanization worldwide. More than thirty profiles of the world's leading cities are presented, each accompanied by specially drawn maps.

- **Paris, London, Rome, Athens**
- **Moscow, St. Petersburg**
- **Toronto, Chicago, New York, Montreal, Los Angeles**
- **Mexico City**
- **Rio de Janeiro, São Paulo, Lima, Buenos Aires**
- **Cairo, Istanbul**
- **Lagos, Nairobi, Johannesburg**
- **Mumbai (Bombay), Kolkata (Calcutta), Delhi New and Old**
- **Xian, Beijing, Shanghai, Tokyo, Seoul, Saigon, Bangkok**
- **Jakarta, Manila**
- **Sydney**

Web site www.wiley.com/college/deblij

The Concepts and Regions Web site offers additional resources that compliment the textbook in ways that help teachers teach and students learn. Enhance your understanding of world regional geography and improve your grade by using the following resources:

- *The Learning Styles Survey* identifies your strengths and creates a customized learning plan that strategically integrates text and media.

- *Annotated Learning Objectives* help you focus your efforts and maximize the time that you spend studying.

- *Web Quizzes* provide immediate feedback and help you master key concepts.

- *Annotated Web Links* put useful electronic resources into context.

- *Lonely Planet Web Links* keep you informed about travel conditions and put you into contact with student travelers.

- *Webcams and Live Radio Links* let you see and hear what is happening all over the world.

- *Virtual Field Trips* compliment the textbook by relating geographic themes and concepts to daily life and everyday people.

- *Demographic and Economic Data* is available for every country in the world.

- *Blank Outline Maps* help you study and master the locations of countries, states, provinces, cities, and physical features.

- *The Audio Pronunciation Guide* helps you learn to pronounce difficult words and place-names.

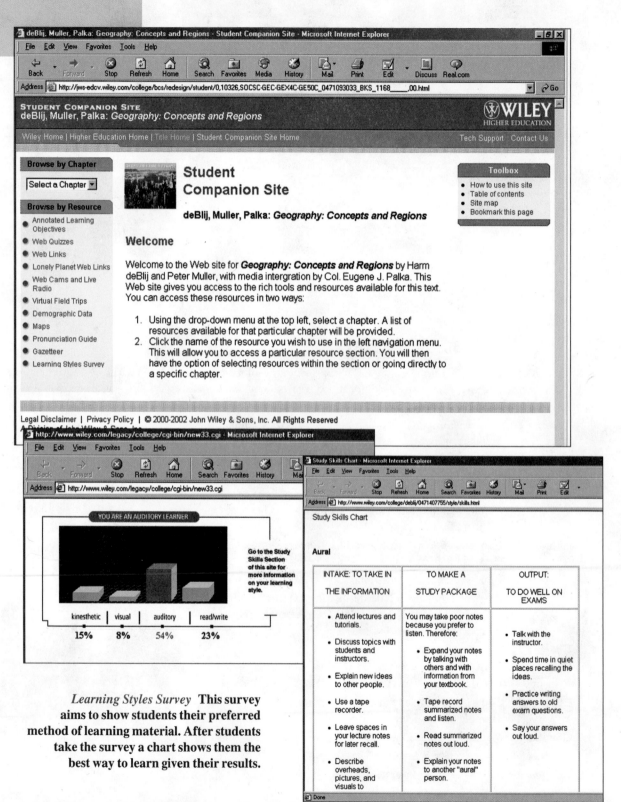

Learning Styles Survey **This survey aims to show students their preferred method of learning material. After students take the survey a chart shows them the best way to learn given their results.**